Advances in Intelligent Systems and Computing

Volume 295

Series editor

Janusz Kacprzyk, Polish Academy of Sciences, Warsaw, Poland
e-mail: kacprzyk@ibspan.waw.pl

For further volumes:
http://www.springer.com/series/11156

About this Series

The series "Advances in Intelligent Systems and Computing" contains publications on theory, applications, and design methods of Intelligent Systems and Intelligent Computing. Virtually all disciplines such as engineering, natural sciences, computer and information science, ICT, economics, business, e-commerce, environment, healthcare, life science are covered. The list of topics spans all the areas of modern intelligent systems and computing.

The publications within "Advances in Intelligent Systems and Computing" are primarily textbooks and proceedings of important conferences, symposia and congresses. They cover significant recent developments in the field, both of a foundational and applicable character. An important characteristic feature of the series is the short publication time and world-wide distribution. This permits a rapid and broad dissemination of research results.

Advisory Board

Thomas Villmann · Frank-Michael Schleif
Marika Kaden · Mandy Lange
Editors

Advances in Self-Organizing Maps and Learning Vector Quantization

Proceedings of the 10th International
Workshop, WSOM 2014,
Mittweida, Germany, July, 2–4, 2014

 Springer

Editors

Thomas Villmann
Department of Mathematics
University of Applied Sciences
Mittweida
Germany

Frank-Michael Schleif
University of Applied Sciences Mittweida
Mittweida
Germany

Marika Kaden
University of Applied Sciences Mittweida
Mittweida
Germany

Mandy Lange
University of Applied Sciences Mittweida
Mittweida
Germany

ISSN 2194-5357   ISSN 2194-5365   (electronic)
ISBN 978-3-319-07694-2   ISBN 978-3-319-07695-9   (eBook)
DOI 10.1007/978-3-319-07695-9
Springer Cham Heidelberg New York Dordrecht London

Library of Congress Control Number: 2014940407

Printed on acid-free paper

Springer is part of Springer Science+Business Media (www.springer.com)

Preface

This book contains all refereed contributions presented at the 10th Workshop on Self-Organizing Maps (WSOM 2014) held at the University of Applied Sciences Mittweida, Mittweida (Germany, Saxony), on July 24, 2014. Starting with the first WSOM-workshop 1997 in Helsinki this workshop series attract many researchers to present newest results in the field of supervised and unsupervised vector quantization and related topics.

This 10th WSOM brought together more than 50 researchers, experts and practitioners in the beautiful small town Mittweida in Saxony (Germany) nearby the mountains *Erzgebirge* to discuss new developments in the field of self-organizing vector quantization systems. The book collects the accepted papers of the workshop after a careful review process. Among the book chapters there are excellent examples of the use of self-organizing maps (SOMs) in agriculture, computer science, data visualization, health systems, economics, engineering, social sciences, text and image analysis, and time series analysis. Other chapters present the latest theoretical work on SOMs as well as Learning Vector Quantization (LVQ) methods.

Our deep appreciation is extended to Teuvo Kohonen, for serving as Honorary General Chair. We warmly thank the members of the Steering Committee and the Executive Committee. Our sincere thanks go to Michael Biehl (University Groningen), Erzsébet Merényi (Rice University Houston) and Fabrice Rossi (Université Paris 1, Pantheón-Sorbonne) for their plenary talks. We are grateful to the members of the Program Committee and other reviewers for their excellent and timely work, and above all to the authors whose contributions made this book possible.

We deeply acknowledge the support of the workshop by the University of Applied Sciences Mittweida under the guidance of the rector Prof. Dr. Ludwig Hilmer. Last but not least we cordially thank Dr. Ellen Weißmantel (University of Applied

Sciences Mittweida) and Dr. Sven Hellbach (University of Applied Sciences Dresden) as well as the *Computational Intelligence Group Mittweida* (K. Domaschke, M. Gay, Dr. T. Geweniger, M. Kaden, M. Lange, D. Nebel, M. Riedel) for local organization.

Mittweida, 2nd July 2014

Thomas Villmann
Frank-Michael Schleif
Marika Kaden
Mandy Lange

Organization

WSOM'14 was held during July 02–04, 2014 in Mittweida, Saxony (Germany). It was organized by the Computational Intelligence Group of the Faculty for Mathematics, Natural and Computer Sciences at the University of Applied Sciences Mittweida.

Executive Committee

Honorary Chair:	Teuvo Kohonen Academy of Finland, Finland
General Chair:	Thomas Villmann, University of Applied Sciences Mittweida, Germany
Program Chair:	Frank-Michael Schleif, University of Birmingham, Birmingham, UK
Local Chairs:	Marika Kaden, Ellen Weißmantel, University of Applied Sciences Mittweida, Germany

Steering Committee

Teuvo Kohonen	Academy of Finland, Finland
Marie Cottrell	Université Paris 1, Pantheón-Sorbonne, France
Pablo Estévez	University of Chile, Chile
Timo Honkela	Aalto University, Finland
Erkki Oja	Aalto University, Finland
José Príncipe	University of Florida, USA
Helge Ritter	Bielefeld University, Germany
Thomas Villmann	University of Applied Sciences Mittweida, Germany
Takeshi Yamakawa	Kyushu Institute of Technology, Japan
Hujun Yin	University of Manchester, UK

Program Committee

Michael Biehl	Marie Cottrell
Pablo Estévez	Baretto Guilherme
Barbara Hammer	Tom Heskes
Timo Honkela	Marika Kaden
Ryotaro Kamimura	Markus Koskela
John Aldo Lee	Paulo Lisboa
Thomas Martinetz	Erzsébet Merényi
Risto Miikulainen	Tim Nattkemper
Erkki Oja	Madalina Olteanu
Jaakko Peltonen	José Príncipe
Andreas Rauber	Helge Ritter
Fabrice Rossi	Udo Seiffert
Marc Strickert	Peter Tino
Alfred Ultsch	Marc van Hulle
Michel Verleysen	Hujun Yin

Additional Referees

Andreas Backhaus	Kerstin Bunte	Tina Geweniger
Andrej Gisbrecht	Sven Hellbach	Matthias Klingner
Mandy Lange	Amaury Lendasse	Bassam Mokbel
David Nebel	Martin Riedel	Sambu Seo
Kadim Taşdemir	Nathalie Villa-Vialaneix	Xibin Zhu
Dietlind Zühlke		

Sponsoring Institutions

- German Chapter of the European Neural Network Society (GNNS)
- Institut für intelligente Datenanalyse e.V. (CIID), Mittweida
- University of Applied Sciences Mittweida

Contents

Part I: SOM-Theory and Visualization Techniques

How Many Dissimilarity/Kernel Self Organizing Map Variants Do We Need? 3
Fabrice Rossi

Dynamic Formation of Self-Organizing Maps 25
Jérémy Fix

MS-SOM: Magnitude Sensitive Self-Organizing Maps 35
Enrique Pelayo, David Buldain

Bagged Kernel SOM 45
Jérôme Mariette, Madalina Olteanu, Julien Boelaert,
Nathalie Villa-Vialaneix

Probability Ridges and Distortion Flows: Visualizing Multivariate Time Series Using a Variational Bayesian Manifold Learning Method 55
Alessandra Tosi, Iván Olier, Alfredo Vellido

Short Review of Dimensionality Reduction Methods Based on Stochastic Neighbour Embedding 65
Diego H. Peluffo-Ordóñez, John A. Lee, Michel Verleysen

Part II: Prototype Based Classification

Attention Based Classification Learning in GLVQ and Asymmetric Misclassification Assessment 77
Marika Kaden, W. Hermann, Thomas Villmann

Visualization and Classification of DNA Sequences Using Pareto Learning
Self Organizing Maps Based on Frequency and Correlation Coefficient   89
Hiroshi Dozono

Probabilistic Prototype Classification Using t-norms   99
Tina Geweniger, Frank-Michael Schleif, Thomas Villmann

Rejection Strategies for Learning Vector
Quantization – A Comparison of Probabilistic and Deterministic
Approaches   109
*Lydia Fischer, David Nebel, Thomas Villmann, Barbara Hammer,
Heiko Wersing*

Part III: Classification and Non-Standard Metrics

Prototype-Based Classifiers and Their Application in the Life Sciences   121
Michael Biehl

Generative versus Discriminative Prototype Based Classification   123
*Barbara Hammer, David Nebel, Martin Riedel,
Thomas Villmann*

Some Room for GLVQ: Semantic Labeling of Occupancy Grid Maps   133
*Sven Hellbach, Marian Himstedt, Frank Bahrmann, Martin Riedel,
Thomas Villmann, Hans-Joachim Böhme*

Anomaly Detection Based on Confidence Intervals Using SOM with an
Application to Health Monitoring   145
Anastasios Bellas, Charles Bouveyron, Marie Cottrell, Jerome Lacaille

RFSOM – Extending Self-Organizing Feature Maps with Adaptive
Metrics to Combine Spatial and Textural Features for Body Pose
Estimation   157
*Mathias Klingner, Sven Hellbach, Martin Riedel, Marika Kaden,
Thomas Villmann, Hans-Joachim Böhme*

Beyond Standard Metrics – On the Selection and Combination of
Distance Metrics for an Improved Classification of Hyperspectral Data   167
Uwe Knauer, Andreas Backhaus, Udo Seiffert

Part IV: Advanced Applications of SOM and LVQ

The Sky Is Not the Limit   181
Erzsébet Merényi

Development of Target Reaching Gesture Map in the Cortex and Its
Relation to the Motor Map: A Simulation Study   187
Jaewook Yoo, Jinho Choi, Yoonsuck Choe

A Concurrent SOM-Based Chan-Vese Model for Image Segmentation 199
Mohammed M. Abdelsamea, Giorgio Gnecco, Mohamed Medhat Gaber

Five-Dimensional Sentiment Analysis of Corpora, Documents
and Words 209
Timo Honkela, Jaakko Korhonen, Krista Lagus, Esa Saarinen

SOMbrero: An R Package for Numeric and Non-numeric
Self-Organizing Maps 219
Julien Boelaert, Laura Bendhaiba, Madalina Olteanu,
Nathalie Villa-Vialaneix

K-Nearest Neighbor Nonnegative Matrix Factorization for Learning
a Mixture of Local SOM Models 229
David Nova, Pablo A. Estévez, Pablo Huijse

Comparison of Spectrum Cluster Analysis with PCA and Spherical SOM
and Related Issues Not Amenable to PCA 239
Masaaki Ohkita, Heizo Tokutaka, Kazuhiro Yoshihara,
Matashige Oyabu

Exploiting the Structures of the U-Matrix 249
Jörn Lötsch, Alfred Ultsch

Partial Mutual Information for Classification of Gene Expression Data
by Learning Vector Quantization 259
Mandy Lange, David Nebel, Thomas Villmann

Composition of Learning Patterns Using Spherical Self-Organizing Maps
in Image Analysis with Subspace Classifier 271
Nobuo Matsuda, Fumiaki Tajima, Hedeaki Sato

Self-Organizing Map for the Prize-Collecting Traveling Salesman
Problem 281
Jan Faigl, Geoffrey A. Hollinger

A Survey of SOM-Based Active Contour Models for Image
Segmentation 293
Mohammed M. Abdelsamea, Giorgio Gnecco, Mohamed Medhat Gaber

A Biologically Plausible SOM Representation of the Orthographic Form
of 50,000 French Words 303
Claude Touzet, Christopher Kermorvant, Hervé Glotin

Author Index 313

Part I

SOM-Theory and Visualization Techniques

How Many Dissimilarity/Kernel Self Organizing Map Variants Do We Need?

Fabrice Rossi

SAMM (EA 4543), Université Paris 1,
90, rue de Tolbiac, 75634 Paris Cedex 13, France
fabrice.rossi@univ-paris1.fr

Abstract. In numerous applicative contexts, data are too rich and too complex to be represented by numerical vectors. A general approach to extend machine learning and data mining techniques to such data is to really on a dissimilarity or on a kernel that measures how different or similar two objects are.

This approach has been used to define several variants of the Self Organizing Map (SOM). This paper reviews those variants in using a common set of notations in order to outline differences and similarities between them. It discuss the advantages and drawbacks of the variants, as well as the actual relevance of the dissimilarity/kernel SOM for practical applications.

Keywords: Self Organizing Map, Dissimilarity data, Pairwise data, Kernel, Deterministic annealing.

1 Introduction

Complex data are frequently too rich and too elaborate to be represented in a simple tabular form where each object is described via a fixed set of attributes/variables with numerical and/or nominal values. This is especially the case for relational data when objects of different categories are interconnected by relations of different types. For instance online retailers have interconnected customers and products databases, in which a customer can buy one or several copies of a product, and can also leave some score and/or review of said products.

Adapting data mining and machine learning methods to complex data is possible, but time consuming and complex, both at the theoretical level (e.g., consistency of the algorithms is generally proved only in the Euclidean case) and on a practical point of view (new implementations are needed). Therefore, it is tempting to build generic methods that use only properties that are shared by all types of data.

Two such generic approaches have been used successfully: the dissimilarity based approach and the kernel based approach [42]. Both are based on fairly generic assumptions: the analyst is given a data set on which either a dissimilarity or a kernel is defined. A dissimilarity measures how much two objects differs, while a kernel can be seen as a form a similarity measure, at least in

T. Villmann et al. (eds.), *Advances in Self-Organizing Maps and Learning
Vector Quantization,* Advances in Intelligent Systems and Computing 295,
DOI: 10.1007/978-3-319-07695-9_1, © Springer International Publishing Switzerland 2014

the correlation sense. Dozens of dissimilarities and kernels have been proposed over the years, covering many types of complex data (see e.g. [15]). Then one needs only to adapt a classical data mining or machine learning method to the dissimilarity/kernel setting in order to obtain a fully generic approach. As a dissimilarity can always be constructed from a kernel, dissimilarity algorithms are probably the more generic ones. A typical example is the k nearest neighbor method which is based only on dissimilarities.

We review in this paper variants of the Self Organizing Map (SOM) that have been proposed following this line of research, that is SOM variants that operate on dissimilarity/kernel data. We discuss whether those variants are really usable and helpful in practice. The paper is organized as follows. Section 2 describes our general setting: dissimilarity data, kernel data and the Self Organizing Map. Section 3 is dedicated to the oldest dissimilarity variant of the SOM, the Median SOM, while Section 4 focuses on the modern variant, the relational SOM. Section 5 presents a different approach to SOM extensions based on the deterministic annealing principle. Section 6 describes kernel based variants of the SOM. An unifying view is provided in Section 7 which shows that the differences between the SOM variants are mainly explained by the optimization strategy rather than by the data properties. Finally Section 8 gathers our personal remarks and insights on the dissimilarity/kernel SOM variants.

2   General Setting

The data set under study comprises N data points $x_1, \ldots, x_N$ from an abstract space $\mathcal{X}$. We specifies below the two options, namely dissimilarity data and kernel data. We also recall the classical SOM algorithms.

2.1   Dissimilarity Data

In the dissimilarity data setting (a.k.a. the pairwise data setting), it is assumed that the data are described indirectly by a square $N \times N$ symmetric matrix D that contains dissimilarities between the data points. The convention is that $D_{ij} = d(x_i, x_j)$, a non negative real number, is high when x_i and x_j are different and low when they are similar. Minimal assumptions on D are symmetry and non negativity of each element. It is also natural to assume some basic ordering, that is that $D_{ii} \leq D_{ij}$ for all i and j, but this is not use in SOM variants. Some theoretical results also need $D_{ii} = 0$ (e.g. [20]), but this again not a very strong constraint. Notice that one can be given either the dissimilarity function d from $\mathcal{X}^2$ to $\mathbb{R}^+$ or directly the matrix D.

2.2   Kernel Data

In the kernel data setting, one is given a *kernel* function k from $\mathcal{X}^2$ to $\mathbb{R}$ which satisfies the following properties:

1. k is symmetric: for all x and y in $\mathcal{X}$, $k(x, y) = k(y, x)$;

2. k is non negative: for all $m > 0$, all $(x_1, \ldots, x_m) \in \mathcal{X}^m$ and all $(\alpha_1, \ldots, \alpha_m) \in \mathbb{R}^m$, $\sum_{i=1}^m \sum_{j=1}^m \alpha_i \alpha_j k(x_i, x_j) \geq 0$.

The most important aspect of the kernel setting lays in the Moore-Aronszajn theorem [3]. It states that a Reproducing Kernel Hilbert Space (RKHS) $\mathcal{H}$ can be associated to $\mathcal{X}$ and k through a mapping function ϕ from $\mathcal{X}$ to $\mathcal{H}$ such that $\langle \phi(x), \phi(y) \rangle_{\mathcal{H}} = k(x, y)$ for all x and y in $\mathcal{X}$. The mapping ϕ is called the *feature map*. It enables one to leverage the Hilbert structure of $\mathcal{H}$ in order to build machine learning algorithms on $\mathcal{X}$ indirectly. This can be done in general without using ϕ but rather by relying on k only: this is known as the *kernel trick* (see e.g. [42]).

Notice that the kernel can be used to define a dissimilarity on $\mathcal{X}$ by transporting the Hilbert distance from $\mathcal{H}$. Indeed, it is natural to define d_k on $\mathcal{X}$ by

$$d_k(x, y) = \langle \phi(x) - \phi(y), \phi(x) - \phi(y) \rangle_{\mathcal{H}}. \tag{1}$$

Elementary algebraic manipulations show that

$$d_k(x, y) = k(x, x) + k(y, y) - 2k(x, y), \tag{2}$$

which is an example of the use of the kernel trick to avoid using explicitly ϕ.

The construction of d_k shows that the dissimilarity setting is more general than the kernel setting. It is always possible to use a kernel as the basis of a dissimilarity: all the dissimilarity variants of the SOM can used on kernel data. Therefore, we will focus mainly on dissimilarity algorithms, and then discuss how they relate to their kernel counterparts.

Notice finally that as in the case of the dissimilarity setting, the kernel can be given as a function from $\mathcal{X}$ to $\mathbb{R}$ or as a kernel matrix $K = (K_{ij}) = (k(x_i, x_j))$. In the latter case, K is symmetric and positive definite and is associated to a dissimilarity matrix D_K via equation (2).

2.3 SOM

To contrast its classical setting with the dissimilarity and kernel ones, and to introduce our notations, we briefly recall the SOM principle and algorithm [28]. A SOM is a low dimensional clustered representation of a data set.

One needs first to specify a low dimensional prior structure, in general a regular lattice of K units/neurons positioned in $\mathbb{R}^2$, the $(r_k)_{1 \leq k \leq K}$. The structure induces a time dependent neighborhood function $h_{kl}(t)$ which measures how much the prototype/model associated to unit r_k should be close to the one associated to unit r_l, at step t of the learning algorithm (from 0 for unrelated models to 1 for maximally related ones). We will not discuss here the numerous possible variants for this neighborhood function [28]: if the lattice is made of points r_k in $\mathbb{R}^2$ a classical choice is

$$h_{kl}(t) = exp\left(-\frac{\|r_k - r_l\|^2}{2\sigma^2(t)}\right),$$

where σ increases over time to reduce gradually the influences of the neighbors during learning.

The SOM attaches to each unit/neuron r_k in the prior structure a proto-type/model in the data space m_k. The objective of the SOM algorithm is to adapt the values of the models in such a way that each data point is as close as possible to its closest model in the data space (at standard goal in prototype based clustering). In addition if the closest model for the data point x is m_k, then m_l should also be close to x if r_k and r_l are close in the prior structure. In other words proximities in the prior structure should reflect proximities in the data space and vice versa. The unit/neuron associated to the closest model of a data point is called the *best matching unit* (BMU) for this point. The set of points for which r_k is the BMU defines a cluster in the data space, denoted C_k.

This is essentially achieved via two major algorithms (and dozens of variants). Let us assume that the data space is a classical normed vector space. Then both algorithms initialize the prototypes $(m_k)_{1 \leq k \leq K}$ in an "appropriate way" and proceed then iteratively. We will not discuss initialization strategies in this paper.

In the stochastic/online SOM (SSOM), a data point x is selected randomly[1] at each iteration t. Then $c \in \{1, \ldots, K\}$ is determined as the index of the best matching unit, that is

$$c = \arg \min_{k \in \{1,\ldots,K\}} \|x - m_k(t)\|^2, \tag{3}$$

and all prototypes are updated via

$$m_k(t+1) = m_k(t) + \epsilon(t) h_{kc}(t)(x - m_k(t)), \tag{4}$$

where $\epsilon(t)$ is a learning rate.

In the batch SOM (BSOM), each iteration is made of two steps. In the first step, the best matching unit for each data point x_i is determined as:

$$c_i(t) = \arg \min_{k \in \{1,\ldots,K\}} \|x_i - m_k(t)\|^2. \tag{5}$$

Then all prototypes are updated via a weighted average

$$m_k(t+1) = \frac{\sum_{i=1}^{N} h_{kc_i(t)}(t) x_i}{\sum_{i=1}^{N} h_{kc_i(t)}(t)}. \tag{6}$$

Obviously, neither algorithm can be applied *as is* on non vector data.

---

[1] or data points are looped through.

3   The Median SOM

3.1   General Principle

It is well known (and obvious) that the prototype update step of the Batch SOM can be considered as solving an optimization problem, namely

$$\forall \, k \in \{1,\ldots,K\}, \ m_k(t+1) = \arg\min_s \sum_{i=1}^{N} h_{kc_i(t)}(t)\|s - x_i\|^2. \qquad (7)$$

This turns the vector space operations involved in equation (6) into an optimization problem that uses only the squared Euclidean norm between prototypes and observations. In an arbitrary space $\mathcal{X}$ with a dissimilarity, $\|s_k - x_i\|^2$ can be replaced by the dissimilarity between s_k and x_i which turns problem (7) into

$$\forall \, k \in \{1,\ldots,K\}, \ m_k(t+1) = \arg\min_{s\in\mathcal{X}} \sum_{i=1}^{N} h_{kc_i(t)}(t)d(s, x_i), \qquad (8)$$

which is a typical generalized median problem.

However, the most general dissimilarity setting only assumes the availability of dissimilarities between *observations* not between arbitrary points in $\mathcal{X}$. In fact, generating new points in $\mathcal{X}$ might be difficult for complex data such as texts. Then the most general solution consists in looking for the optimal prototypes into the data set rather than in $\mathcal{X}$. The Median SOM [27,29,30] and its variants [12,13] are based on this principle. The Median SOM consists in iterating two steps. In the first step, the best matching unit for each data point x_i is determined as

$$c_i(t) = \arg\min_{k\in\{1,\ldots,K\}} d(x_i, m_k(t)). \qquad (9)$$

Then all prototypes are updated by solving the generalized median problem

$$\forall \, k \in \{1,\ldots,K\}, \ m_k(t+1) = \arg\min_{x_j} \sum_{i=1}^{N} h_{kc_i(t)}(t)D_{ij}. \qquad (10)$$

Notice that each prototype is a data point which means that in equation (9) $d(x_i, m_k(t))$ is in fact a D_{ik} for some k.

A variant of the Median SOM was proposed in [1]: rather than solving problem (10), it associates to each unit the generalized median of the corresponding cluster (in other words, it does not take into account the neighborhood structure at this point). Then the BMU of a data point is chosen randomly using the neighborhood structure and the dissimilarities. This means that a data point can be moved from its natural BMU to a nearby one. As far as we know, this variant has not been studied in details.

3.2   Limitations of the Median SOM

The Median SOM has numerous problems. As a batch SOM it is expected to request more iterations to converge than a potential stochastic version (which is not possible in the present context, unfortunately). In addition, it will also exhibit sensitivity to its initial configuration.

There are also problems more specific to the Median SOM. Each iteration of the algorithm has a rather high computational cost: a naive implementation leads to a cost of $O(N^2 K + NK^2)$ per iteration, while a more careful one still costs $O(N^2 + NK^2)$ [10]. Numerous tricks can be used to reduce the actual cost per iteration [7,8] but the N^2 factor cannot be avoided without introducing approximations.

Arguably the two main drawbacks of the Median SOM are of a more intrinsic nature. Firstly, restricting the prototypes to be chosen in the data set has some very adverse effects. A basic yet important problem comes from collisions in prototypes [36]: two different units can have the same optimal solution according to equation (10). This corresponds to massive folding of the two dimensional representation associated to the SOM and thus to a sub-optimal data summary. In addition, equation (9) needs a tie breaking rule which will in general increase the cost of BMU determination (see [30] for an example of such a rule). The solution proposed in [36] can be used to avoid those problems at a reasonable computational cost.

A more subtle consequence of the restriction of prototypes to data points is that no unit can remain empty, apart from collided prototypes. Indeed, the BMU of a data point that is used as a prototype should be the unit of which it is the prototype. This means that no interpolation effect can take place in the Median SOM [43,44] a fact that limits strongly the usefulness of visual representations such as the U-matrix [45,46]. For some specific data types such as strings, this can be avoided by introducing ways of generating new data points by some form of interpolations. This was studied in [43,44] together with a stochastic/online algorithm.

A generic solution to lift the prototype restriction is provided by the relational SOM described in Section 4.

3.3   Non Metric Dissimilarities

The second intrinsic problem of the Median SOM is its reliance on a prototype based representation of a cluster in the dissimilarity context, while this is only justified in the Euclidean context. Indeed let us consider that the N data points $(x_i)_{1 \leq i \leq N}$ belong to a Euclidean space. Then for any positive weights β_i, the well known König-Huygens identity states:

$$\sum_{i=1}^{N} \beta_i \left\| \frac{\sum_{j=1}^{N} \beta_j x_j}{\sum_{j=1}^{N} \beta_j} - x_i \right\|^2 = \frac{1}{2} \frac{1}{\sum_{i=1}^{N} \beta_i} \sum_{i=1}^{N} \sum_{j=1}^{N} \beta_i \beta_j \|x_i - x_j\|^2. \quad (11)$$

This means that

$$\min_m \sum_{i=1}^N \beta_i \|m - x_i\|^2 = \frac{1}{2} \frac{1}{\sum_{i=1}^N \beta_i} \sum_{i=1}^N \sum_{j=1}^N \beta_i \beta_j \|x_i - x_j\|^2. \tag{12}$$

Applied to the SOM, this means that solving[2]

$$(m(t), c(t)) = \arg\min_{m,c} \sum_{k=1}^K \sum_{i=1}^N h_{kc_i}(t) \|m_k - x_i\|^2, \tag{13}$$

where $m(t) = (m_1(t), \ldots, m_K(t))$ denotes the prototypes and $c = (c_1, \ldots, c_n)$ denotes the BMU mapping, is equivalent to solving

$$c(t) = \arg\min_c \frac{1}{2} \sum_{k=1}^K \frac{1}{\sum_{i=1}^N h_{kc_i}(t)} \sum_{i=1}^N \sum_{j=1}^N h_{kc_i}(t) h_{kc_j}(t) \|x_i - x_j\|^2. \tag{14}$$

This second problem makes clear that the classical SOM is not only based on *quantization* but is also optimizing the within pairwise distances in the clusters defined by the BMU mapping. Here h_{kc_i} is considered as a form of membership value of x_i to cluster k, which give the "size" $\sum_{i=1}^N h_{kc_i}$ to the cluster k. Then the sum of pairwise distances in each cluster measures the compactness of the cluster in terms of within variance. As the SOM minimizes the sum of those quantities, it can be seen as a *clustering* algorithm[3].

However, the König-Huygens identity does not apply to arbitrary dissimilarities. In other words, the natural dissimilarity version of problem (14) that is

$$c(t) = \arg\min_c \frac{1}{2} \sum_{k=1}^K \frac{1}{\sum_{i=1}^N h_{kc_i}(t)} \sum_{i=1}^N \sum_{j=1}^N h_{kc_i}(t) h_{kc_j}(t) d(x_i, x_j), \tag{15}$$

is not equivalent to the Median SOM problem given by

$$(m(t), c(t)) = \arg\min_{m \in \{x_1, \ldots, x_N\}^K, c} \sum_{k=1}^K \sum_{i=1}^N h_{kc_i}(t) d(x_i, m_k). \tag{16}$$

When the dissimilarity satisfies the triangular inequality this is not a major problem. In by virtue of the triangular inequality, we have for all m

$$d(x_i, x_j) \le d(x_i, m) + d(m, x_j), \tag{17}$$

and therefore for all m

$$\sum_{i=1}^N \sum_{j=1}^N h_{kc_i}(t) h_{kc_j}(t) d(x_i, x_j) \le 2 \left(\sum_{i=1}^N h_{kc_i}(t) \right) \sum_{j=1}^N h_{kc_j}(t) d(x_j, m), \tag{18}$$

---

[2] The quantity optimized in equation (13) is the energy defined in [25].

[3] This classical analysis mimics the one used to see the k-means algorithm both as a clustering algorithm and as a quantization algorithm.

which shows that

$$\frac{1}{2\sum_{i=1}^{N} h_{kc_i}(t)} \sum_{i=1}^{N} \sum_{j=1}^{N} h_{kc_i}(t) h_{kc_j}(t) d(x_i, x_j) \leq \min_{m} \sum_{j=1}^{N} h_{kc_j}(t) d(x_j, m). \quad (19)$$

Then the Median SOM is optimizing an upper bound of the cluster oriented quality criterion for dissimilarities. In practice, this means that a good quantization will give compact clusters.

However, when the dissimilarity does not satisfy the triangular inequality, the two criteria are not directly related any more. In fact, one prototype can be close to a set of data points while those points remain far apart from each other. Then doing of form of *quantization* by solving problem (16) is not the same thing as doing a form of *clustering* by solving problem (15). By choosing the prototype based solution, the Median SOM appears to be a quantization method rather than a clustering one. If the goal is to display *prototypes* in an organized way, then this choice make sense (but must be explicit). If the goal is to display *clusters* in an organized way, this choice is intrinsically suboptimal. As pointed out in Section 8, dissimilarity SOMs are not very adapted to prototype display, which puts in question the interest of the Median SOM in particular and of the quantization approach in general.

4   The Relational SOM

The quantification of the prototypes induced by restricting them to data points has quite negative effects described in Section 3.2. The relational approach is a way to address this problem. It is based on the simple following observation [23]. Let the $(x_i)_{1,...,N}$ be N points in a Hilbert space equipped with the inner product $\langle ., . \rangle$ and let $y = \sum_{i=1}^{N} \alpha_i x_i$ for arbitrary real valued coefficients $\alpha^T = (\alpha_i)_{1,...,N}$ with $\sum_{i=1}^{N} \alpha_i = 1$. Then

$$\langle x_i - y, x_i - y \rangle = (D\alpha)_i - \frac{1}{2}\alpha^T D\alpha, \quad (20)$$

where D is the squared distance matrix given by $D_{ij} = \langle x_i - x_j, x_i - x_j \rangle$. This means that computing the (squared) distance between a linear combination of some data points and any of those data points can be done using only the coefficients of the combination and the (squared) distance matrix between those points.

4.1   Principle

But as shown by equation (6), prototypes in the classical SOM are exactly linear combinations of data points whose coefficients sum to one. It is therefore possible to express the Batch SOM algorithm without using directly the values of the x_i, but rather by keeping the coefficients of the prototypes and using equation (20) and the squared distance matrix to perform all calculations.

Then one can simply apply the so called *relational* version of the algorithm to an arbitrary dissimilarity matrix as if it were a squared euclidean one. This is essentially what is done in [22,23] for the c-means (a fuzzy variant of the k-means) and in [21] for the Batch SOM (and the Batch Neural Gas [11]). Using the concept of pseudo-Euclidean spaces, it was shown in [20] that this general approach can be given a rigorous derivation: it amounts to using the original algorithm (SOM, k-means, etc.) on a pseudo-Euclidean embedding of the data points.

In practice, the Batch relational SOM proceeds by iterating two steps that are very similar to the classical batch SOM steps. The main difference is that each prototype $m_k(t)$ (at iteration t) is given by a vector of $\mathbb{R}^N$, $\alpha_k(t)$, which represents the coefficients of the linear combination of the x_i in the pseudo-Euclidean embedding. Then the best matching unit computation from equation (5) is replaced by

$$c_i(t) = \arg \min_{k \in \{1,...,K\}} \left((D\alpha_k(t))_i - \frac{1}{2}\alpha_k(t)^T D\alpha_k(t) \right), \qquad (21)$$

while the prototype update becomes

$$\alpha_k(t+1)_i = \frac{h_{kc_i}(t)}{\sum_{l=1}^{N} h_{kc_l}(t)}. \qquad (22)$$

A stochastic/online variant of this algorithms was proposed in [34]. As for the classical SOM, it consists in selecting randomly a data point x_i, computing its BMU c_i (using equation (21)) and updating all prototypes as follows:

$$\alpha_k(t+1)_j = \alpha_k(t)_j + \epsilon(t)h_{kc_i}(t)(\delta_{ij} - \alpha_k(t)_j), \qquad (23)$$

where δ_{ij} equals 1 when $i = j$ and 0 in other cases. Notice that is the α_k are initialized so as to sum to one, this is preserved by this update. As shown in [34], the stochastic variant tends to be less sensitive to the initial values of the prototypes. However [34] overlooks that both batch and online relational SOM algorithms share the same computational cost per iteration[4] which negates the traditional computational gain provided by online versions.

4.2 Limitations of the Relational SOM

The Relational SOM solves several problems of the Median SOM. In particular, it is not subject to the quantization effect induced by constraining the prototypes to be data points. As a consequence, it exhibits in practice the same interpolation effects as the classical SOM. The availability of a stochastic version provides also a simple way to reduce the adverse effects of a bad initialization.

However, the relational SOM is very computationally intensive. Indeed, the evaluation of all the $\alpha_k(t)^T D\alpha_k(t)$ costs $O(KN^2)$ operations. Neither the dissimilarity matrix nor the prototype coefficients are sparse and there is no way

---

[4] The cost reported in [34] for the batch relational SOM is incorrect.

to reduce this costs without introducing approximations. Notice that this cost is per iteration in both the batch and the stochastic versions of the relational SOM. This is K times larger than the Median SOM.

This large cost has motivated research on approximation techniques such as [37]. The most principled approach consists in approximating the calculation of the matrix product via the Nyström technique [50], as explored in [19].

5   Soft Topographic Mapping for Proximity Data

As pointed out in Section 3.3, if an algorithm relies on prototypes with a general possibly non metric dissimilarity, it provides only quantization and not clustering. When organized clusters are looked for, one can try to solve problem (15) directly, that is without relying on prototypes.

5.1   A Deterministic Annealing Scheme

However problem (15) is combinatorial and highly non convex. In particular, the absence of prototypes rules out standard alternating optimization schemes. Following the analysis done in the case of the dissimilarity version of the k-means in [6,26], Graepel et al. introduce in [17,18] a deterministic annealing approach to address problem (15). The approach introduces a mean field approximation which estimates by e_{ik} the effects in the criterion of problem (15) of assigning the data point x_i in cluster k. In addition, it computes soft assignments to the cluster/unit, denoted γ_{ik} for the membership of x_i to cluster k ($\gamma_{ik} \in [0,1]$ and $\sum_{k=1}^{K} \gamma_{ik} = 1$). The optimal mean field is given by

$$e_{ik} = \sum_{s=1}^{K} h_{ks} \sum_{j=1}^{N} b_{js} \left(d(x_i, x_j) - \frac{1}{2} \sum_{l=1}^{N} b_{ls} d(x_j, x_l) \right), \qquad (24)$$

where the b_{js} are given by

$$b_{js} = \frac{\sum_{k=1}^{K} \gamma_{jk} h_{ks}}{\sum_{i=1}^{N} \sum_{k=1}^{K} \gamma_{ik} h_{ks}}. \qquad (25)$$

Soft assignments are updated according to

$$\gamma_{ik} = \frac{\exp(-\beta e_{ik})}{\sum_{s=1}^{K} \exp(-\beta e_{is})}, \qquad (26)$$

where β is an annealing parameter. It plays the role of an inverse temperature and is therefore gradually increased at each step of the algorithm.

In practice, the so-called Soft Topographic Mapping for Proximity Data (STMP) is trained in an iterative batch like procedure. Given an annealing schedule (that is a series of increasing values for β) and initial random values of the mean field, the algorithm iterates evaluating equation (26), then equation (25) and finally equation (24) for a fixed value of β, until convergence. When this convergence is reached, β is increased and the iterations restart from the current value of the mean field.

Notice in equation (25) that the neighborhood function is *fixed* in this approach, whereas it is evolving with time in most SOM implementations.

5.2 Limitations of the STMP

It is well known that the quality of the results obtained by deterministic annealing are highly dependent on the annealing scheme [35]. It is particularly important to avoid missing transition phases. Graepel et al. have analyzed transition phases in the STMP in [18]. As in [35,26], the first critical temperature is related to a dominant eigenvalue of the dissimilarity matrix. As this is in general a dense matrix, the minimal cost of computing the critical temperature is $O(N^2)$. In addition, each internal iteration of the algorithm is dominated by the update of the mean field according to equation (24). The cost of a full update is in $O(N^2 K + N K^2)$. The STMP is therefore computationally intensive. It should be noted however that an approximation of the mean field update that reduces the cost to $O(N^2 K)$ is proposed in [18], leading to the same computational cost as the relational SOM.

In addition, as will appear clearly in Section 7.2, the STMP is based on prototypes, even they appear only indirectly. Therefore while it tries to optimize the clustering criterion associated to the SOM, it resorts to a similar quantization quality proxy as the relational SOM.

6 Kernel SOM

As recalled in Section 2.2, the kernel setting is easier to deal with than the dissimilarity one. Indeed the embedding into a Hilbert space $\mathcal{H}$ enables to apply any classical machine learning method to kernel data by leveraging the Euclidean structure of $\mathcal{H}$. The kernel trick allows one to implement those methods efficiently.

6.1 The Kernel Trick for the SOM

In the case of the SOM, the kernel trick is based on the same fundamental remark that enables the relational SOM (see Section 4.1): in the Batch SOM, the prototypes are linear combinations of the data points. If the initial values of the prototypes are linear combinations of the data points (and not random points), this is also the case for the stochastic/online SOM.

Then assume given a kernel k on $\mathcal{X}$, with its associated Hilbert space $\mathcal{H}$ and mapping ϕ. Implementing the Batch SOM in $\mathcal{H}$ means working on the mapped data set $(\phi(x_i))_{1 \leq i \leq N}$ with prototypes $m_k(t)$ of the form $m_k(t) = \sum_{i=1}^{N} \alpha_{ki}(t)\phi(x_i)$. Then equation (5) becomes

$$c_i(t) = \arg\min_{k \in \{1,\dots,K\}} \|\phi(x_i) - m_k(t)\|_{\mathcal{H}}^2, \qquad (27)$$

with

$$\|\phi(x_i) - m_k(t)\|_{\mathcal{H}}^2 = k(x_i, x_i) - 2\sum_{j=1}^{N} \alpha_{kj}(t)k(x_k, x_j) \qquad (28)$$

$$+ \sum_{j=1}^{N}\sum_{l=1}^{N} \alpha_{kj}(t)\alpha_{kl}(t)k(x_j, x_l).$$

Equation (28) is a typical result of the kernel trick: computing the distance between a data point and a linear combination of the data points can be done using solely the kernel function (or matrix). To our knowledge, the first use of the kernel trick in a SOM context was made in [17].

Notice that equation (6) can also been implemented without using explicitly the mapping ϕ as one needs only the coefficients of the linear combination which are given by

$$\alpha_{ki}(t+1) = \frac{h_{kc_i(t)}}{\sum_{l=1}^{N} h_{kc_l(t)}}, \qquad (29)$$

exactly as in equation (22). While the earliest kernel SOM (STMK) in [17] is optimized using deterministic annealing (as the SMTP presented in Section 5), the kernel trick enables the more traditional online SOM [31] and batch SOM [5,32,49] derived from the previous equations.

It should be noted for the sake of completeness that another kernel SOM was proposed in [2]. However, this variant assumes that $\mathcal{X}$ is a vector space and therefore is not applicable to the present setting.

6.2 Limitations of the Kernel SOM

As it is built indirectly on a Hilbert space embedding, the kernel SOM does not suffer from constrained prototypes. The stronger assumptions made on kernels compared to dissimilarities guarantee the equivalence between finding good prototypes and finding compact clusters. Kernel SOM has also both online and batch versions.

Then the main limitation of the kernel SOM is its computational cost. Indeed, as for the relational SOM, evaluating the distances in equation (28) has a $O(KN^2)$ cost. The approximation schemes proposed for the relational SOM [19,37] can be used for the kernel SOM at the cost of reduced performances in terms of data representation.

7 Equivalences between SOM Variants

It might seem at first that all the variants presented in the previous sections are quite different, both in terms of goals and algorithms. On the contrary, with the exception of the Median SOM which is very specific in some aspects, the variations between the different methods are explained by optimization strategies rather than by hypothesis on the data.

7.1 Relational and Kernel Methods are Equivalent

We have already pointed out that relational SOM and kernel SOM share the very same principle of representing prototypes by a linear combination of the data points. Both cases use the same coefficient update formulas whose structure depends only on the type of the algorithm (batch or online).

The connections are even stronger in the sense that given a kernel, the relational SOM algorithm obtained by using the dissimilarity associated to the kernel is *exactly* identical to the kernel SOM algorithm. Indeed if K is the kernel matrix, then the dissimilarity matrix is given by $D_{ij} = K_{ii} + K_{jj} - 2K_{ij}$. Then for all $\alpha \in \mathbb{R}^N$ such that $\sum_{i=1}^N \alpha_i = 1$ and for all $i \in \{1, \ldots, N\}$

$$(D\alpha)_i - \frac{1}{2}\alpha^T D\alpha = \sum_{j=1}^N D_{ij}\alpha_j - \frac{1}{2}\sum_{j=1}^N \sum_{l=1}^N \alpha_j \alpha_l D_{jl}$$

$$= \sum_{j=1}^N (K_{ii} + K_{jj} - 2K_{ij})\alpha_j - \frac{1}{2}\sum_{j=1}^N \sum_{l=1}^N \alpha_j \alpha_l (K_{jj} + K_{ll} - 2K_{jl})$$

Using $\sum_{i=1}^N \alpha_i = 1$, the first term becomes

$$\sum_{j=1}^N (K_{ii} + K_{jj} - 2K_{ij})\alpha_j = K_{ii} + \sum_{j=1}^N K_{jj}\alpha_j - 2\sum_{j=1}^N K_{ij}\alpha_j.$$

The same condition on α shows that

$$\sum_{j=1}^N \sum_{l=1}^N \alpha_j \alpha_l K_{jj} = \sum_{j=1}^N K_{jj}\alpha_j,$$

and that

$$\sum_{j=1}^N \sum_{l=1}^N \alpha_j \alpha_l K_{ll} = \sum_{l=1}^N K_{ll}\alpha_l.$$

Therefore

$$\sum_{j=1}^N \sum_{l=1}^N \alpha_j \alpha_l (K_{jj} + K_{ll} - 2K_{jl}) = 2\sum_{j=1}^N K_{jj}\alpha_j - 2\sum_{j=1}^N \sum_{l=1}^N \alpha_j \alpha_l K_{jl}.$$

Combining those equations, we end up with

$$(D\alpha)_i - \frac{1}{2}\alpha^T D\alpha = K_{ii} - 2\sum_{j=1}^N K_{ij}\alpha_j + \sum_{j=1}^N \sum_{l=1}^N \alpha_j \alpha_l K_{jl}. \tag{30}$$

The second part of this equation is exactly $\|\phi(x_i) - m\|_{\mathcal{H}}^2$ when $m = \sum_{j=1}^N \alpha_j \phi(x_j)$ as recalled in equation (28). Therefore, the best matching unit determination in the relational SOM according to equation (21) is exactly equivalent to the BMU determination in the kernel SOM according to equation (27). This shows the equivalence between the two algorithms (in both batch and online variants).

This equivalence shows that the batch relational SOM from [21] is a rediscovery of the batch kernel SOM from [32], while the online relational SOM from [34] is a rediscovery of the online kernel SOM from [31]. Results from [20] show that those rediscoveries are in fact *generalizations* of kernel SOM variants as they extend the Hilbert embedding to the more general pseudo-Euclidean embedding. In practice, there is no reason to distinguish the kernel SOM from the relational SOM.

7.2   STMP is a Prototype Based Approach

On the surface, the STMP described in Section 5 looks very different from relational/kernel approaches as it tries to address the combinatorial optimization problem (15) rather than the different problem (16) associated to the generalized median. However, as analyzed in details in [20], the STMP differs from the relational approach only by the use of deterministic annealing, not by the absence of prototypes.

A careful analysis of equations (24) and (22) clarifies this point. Indeed, let us consider $\alpha_s = (b_{js})_{1 \leq j \leq N}^T$ as the coefficient vector for a linear combination of the data points x_j embedded in the pseudo-Euclidean space associated to the dissimilarity matrix D. Then

$$\sum_{j=1}^{N} b_{js} \left(d(x_i, x_j) - \frac{1}{2} \sum_{l=1}^{N} b_{ls} d(x_j, x_l) \right) = (D\alpha_s)_i - \frac{1}{2} \alpha_s^T D\alpha_s.$$

The right hand part is the distance in the pseudo-Euclidean space between the prototype associated to α_s and x_i. Then e_{ik} in equation (24) is a weighted average of distances between x_i and each of the α_s, where the weights are given by the neighborhood function. As pointed out in [20], this can be seen as a relational extension of the assignment rule proposed by Heskes and Kappen in [25].

However, rather than using crisp assignments to a best matching unit with the lowest value of e_{ik}, the STMP uses a soft maximum strategy implemented by equation (26) to obtain assignment probabilities γ_{ik}. Those are used in turn to update the coefficients of the prototypes in equation (25).

In fact the three algorithms proposed in [17] are all based on the same deterministic annealing scheme, with an initial implementation in $\mathbb{R}^p$ (the STVQ) and two generalization in the Hilbert space associated to a kernel (STMK) and in the pseudo-Euclidean space associated to a dissimilarity (STMP). The discussion of the previous section shows that the kernel and the dissimilarity variants are strictly equivalent.

7.3   Summary

We summarizes in the following tables the variants of the SOM discussed in this paper. Table 1 maps a data type and an optimization strategy to a SOM

Table 1. Variants of the SOM

		$\mathbb{R}^p$ data	Data type Kernel	Dissimilarity
Optimization strategy	Online	online SOM	online relational SOM [31,34]	
	Batch	batch SOM	batch relational SOM [21,32]	
	Batch	NA	NA	Median SOM [27]
	Deterministic annealing	STVQ [17]	STMK [17]	STMP [17]

Table 2. Computational complexity of SOM variants for N data points, K units and in $\mathbb{R}^p$ for the classical SOM

Algorithm	Assignment cost	Prototype update cost
Batch SOM	$O(NKp)$	$O(NKp)$
Online SOM	$O(Kp)$	$O(Kp)$
Median SOM	$O(NK)$	$O(N^2 + NK^2)$
Batch relational SOM	$O(N^2K)$	$O(NK)$
Online relational SOM	$O(N^2K)$	$O(NK)$
STVQ	$O(NKp + NK^2)$	$O(NKp + NK^2)$
STMK/STMP	$O(N^2K + NK^2)$	$O(NK^2)$

variant. Relational variants include here the kernel presentation. Table 2 gives the computational costs of one iteration of the SOM variants.

8   Discussion

Even if the kernel approaches are special cases of the relational ones, we have numerous candidates for dissimilarity processing with the SOM. We discuss those variants in this section.

8.1   Median SOM

In our opinion, there is almost no reason to use the Median SOM in practice, except possibly the reduced computational burden compared to the relational SOM ($O(N^2)$ compared to $O(N^2K)$) for the dominating terms). Indeed, the Median SOM suffers from constraining the prototypes to be data points and gives in general lower performances than the relational/kernel SOM as compared to a ground truth or based on the usability of the results (see for instance [19,34,49]). The lack of interpolation capability is particularly damaging as it prevents in general to display gaps between natural clusters with u-matrix like visual representation [45,46].

For large data sets, the factor K increase in the cost of one iteration of the relational SOM compared to the median SOM could be seen as a strong argument for the latter. In our opinion, approximation techniques [19,37] are probably a better choice. This remains however to be tested as to our knowledge the effects of the Nyström approximation have only been studied extensively for the relational neural gas and the relational GTM [16,19,40].

8.2   Optimization Strategy

To our knowledge, no systematic study of the influence of the optimization strategy has been conducted for SOM variants, even in the case of numerical data. In this latter case, it is well known that the online/stochastic SOM is less sensitive to initial conditions than the batch SOM. It is also generally faster to converge and leads in general to a better overall topology preservation [14]. Similar results are observed in the dissimilarity case in [34]. It should be noted however that both analyses use only random initializations while it is well known (see e.g. [28]) that a PCA[5] based initialization gives much better results than a random one in the case of the batch SOM. It is also pointed in [28] that the neighborhood annealing schedule as some strong effects on topology preservation in the batch SOM. Therefore, in terms of the final quality of the SOM, it is not completely obvious that an online solution will provide better results than a batch one.

In addition, the relational setting negates the computational advantage of the online SOM versus the batch SOM. Indeed in the numerical case, one epoch of the online SOM (a full presentation of all the data points) has roughly the same cost as one iteration of the batch SOM. As the online SOM converges generally with a very small number of epochs, its complete computational cost is lower than the batch SOM. On the contrary, the cost of the relational SOM is dominated by the calculation of $\alpha^T D\alpha$ in equation (21). In the batch relational SOM this quantity can be computed one time per prototype and per iteration, leading to a cost of $O(N^2K)$ per iteration (this is overlooked in [34] which reports erroneously a complexity of $O(N^3K)$ per iteration). In the online version, it has also to be computed for each data point (because of the prototype update that takes place after each data point presentation). This means that one epoch of the online relational SOM costs N times more than one iteration of the batch relational SOM. We think therefore that a careful implementation of the batch relational SOM should outperform the online version, provided the initialization is conducted properly.

Comparisons of the online/batch variants with the deterministic annealing variants is missing, as far as we know. The extensive simulations conducted in [20] compare the relational neural gas to the dissimilarity deterministic annealing clustering of [6,26]. Their conclusion is the one expected from similar comparisons done on numerical data [35]: the sophisticated annealing strategy of deterministic annealing techniques leads in general to better solutions provided

---

[5] PCA initialization is easily adapted to the relational case, as it was for kernel data [41].

the critical temperatures are properly identified. This comes with a largely increased cost, not really because of the cost per iterations but rather because the algorithm comprises two loops: an inner loop for a given temperature and an outer annealing loop. Therefore the total number of iterations is in general of an order of magnitude higher than with classical batch algorithms (see also [38] for similar results in the context of a graph specific variant of the SOM principle). It should be also noted that in all deterministic variants proposed in [17], the neighborhood function is not adapted during learning. The effects of this choice on the usability of the final results remain to be studied.

To summarize, our opinion is that one should prefer a careful implementation of the batch relational SOM, paired with a PCA like algorithm for initialization and using the Nyström approximation for large data sets. Further experimental work is needed to validate this choice.

8.3 Clustering versus Quantization

As explained in Section 3.3, an algorithm that resorts (directly or indirectly) on prototypes for an arbitrary dissimilarity does in fact of form of *quantization* rather than a form of *clustering*. To our knowledge, no attempt has been made to minimize directly the prototype free criterion used in problem (15) and we can only speculate on this point.

We should first note that in the case of classical clustering, it has been shown in [9] that optimizing directly the criterion from problem (15) in its k-means simplified form gives better results than using the relational version of the k-means. While the computational burden of both approaches are comparable, the direct optimization of the pairwise dissimilarities criterion is based on a much more sophisticated algorithm which combines state-of-the-art hierarchical clustering [33] with multi-level refinement from graph clustering [24].

Assuming such a complex technique could be used to train a SOM like algorithm, one would obtain in the end a set of non empty clusters, organized according to a lattice in 2 dimensions, something similar to what can be obtained with the Median SOM. While the clusters would have a better quality, no interpolation between them would be possible, as in the Median SOM.

8.4 How Useful Are the Results?

In our personal opinion, the main interest of the SOM is to provide rich and yet readable visual representations of complex data [47,48]. Unfortunately, the visualization possibilities are quite limited in the case of dissimilarity data.

The main limitation is that for arbitrary data in an abstract space $\mathcal{X}$, one cannot assume that an element of $\mathcal{X}$ can be easily represented visually. Then even the Median SOM prototypes (which are data points) cannot be visualized. As the prototypes (in all the variants) do not have meaningful coordinates, component planes cannot be used.

In fact, the only aspects of the results that can be displayed as in the case of numerical data are dissimilarities between prototypes (in U matrix like displays

[45]) as well as numerical characteristics of the clusters (size, compactness, etc.). But as pointed out in [46], among others, this type of visualization is interesting mainly when the SOM uses a large number of units. While this is possible with the relational SOM, it implies a high computational because of the dominating $O(N^2K)$ term. The case of deterministic annealing versions of the SOM is even more problematic with the $O(NK^2)$ complexity term induced by the soft memberships.

In some situations, specific data visualization techniques can be built upon the SOM's results. For instance by clustering graph nodes via a kernel/dissimilarity SOM, one can draw a clustered graph representation, as was proposed in [5]. However, it has been shown in this case that specialized models derived from the SOM [38] or simpler dual approaches based on graph clustering and graph visualization [39] give in general better final results.

To summarize, our opinion is that the appeal of a generic dissimilarity SOM is somewhat reduced by the limited visualization opportunity it offers, compared to the traditional SOM. Further work is needed to explore whether classical visualization techniques, e.g. brushing and linking [4] could be used to provide more interesting displays based on the dissimilarity SOM.

9 Conclusion

We have reviewed in this paper the main variants of the SOM that are adapted to dissimilarity data and to kernel data. Following [20], we have shown that the variants differ more in terms of their optimisation strategy that in other aspects. We have recalled in particular that kernel variants are strictly identical to their relational counterpart. Taking into account computational aspects and known experimental results, our opinion is that the best solution is the batch relational SOM coupled with a structured initialization (PCA like) and with the Nyström approximation for large data sets and thus that we need one dissimilarity/kernel SOM variant only.

However, as discussed above, the practical usefulness of the dissimilarity SOM is reduced compared to the numerical SOM as most of the rich visual representation associated to the SOM of not available for its dissimilarity version. Without improvement in its visual outputs, it is not completely clear if the dissimilarity SOM serves a real practical purpose beyond its elegant generality and simplicity.

References

1. Ambroise, C., Govaert, G.: Analyzing dissimilarity matrices via Kohonen maps. In: Proceedings of 5th Conference of the International Federation of Classification Societies (IFCS 1996), Kobe (Japan), vol. 2, pp. 96–99 (March 1996)
2. Andras, P.: Kernel-Kohonen networks. International Journal of Neural Systems 12, 117–135 (2002)
3. Aronszajn, N.: Theory of reproducing kernels. Transactions of the American Mathematical Society 68(3), 337–404 (1950)

4. Becker, A., Cleveland, S.: Brushing scatterplots. Technometrics 29(2), 127–142 (1987)
5. Boulet, R., Jouve, B., Rossi, F., Villa, N.: Batch kernel SOM and related Laplacian methods for social network analysis. Neurocomputing 71(7-9), 1257–1273 (2008)
6. Buhmann, J.M., Hofmann, T.: A maximum entropy approach to pairwise data clustering. In: Proceedings of the International Conference on Pattern Recognition, Hebrew University, Jerusalem, Israel, vol. II, pp. 207–212. IEEE Computer Society Press (1994)
7. Conan-Guez, B., Rossi, F.: Speeding up the dissimilarity self-organizing maps by branch and bound. In: Sandoval, F., Prieto, A.G., Cabestany, J., Graña, M. (eds.) IWANN 2007. LNCS, vol. 4507, pp. 203–210. Springer, Heidelberg (2007)
8. Conan-Guez, B., Rossi, F.: Accélération des cartes auto-organisatrices sur tableau de dissimilarités par séparation et évaluation. Revue des Nouvelles Technologies de l'Information, pp. 1–16 (June 2008), RNTI-C-2 Classification: points de vue croisés. Rédacteurs invités : Mohamed Nadif et François-Xavier Jollois
9. Conan-Guez, B., Rossi, F.: Dissimilarity clustering by hierarchical multi-level refinement. In: Proceedings of the XXth European Symposium on Artificial Neural Networks, Computational Intelligence and Machine Learning (ESANN 2012), Bruges, Belgique, pp. 483–488 (March 2012)
10. Conan-Guez, B., Rossi, F., El Golli, A.: Fast algorithm and implementation of dissimilarity self-organizing maps. Neural Networks 19(6-7), 855–863 (2006)
11. Cottrell, M., Hammer, B., Hasenfuß, A., Villmann, T.: Batch and median neural gas. Neural Networks 19(6), 762–771 (2006)
12. El Golli, A., Conan-Guez, B., Rossi, F.: Self organizing map and symbolic data. Journal of Symbolic Data Analysis 2(1) (November 2004)
13. El Golli, A., Conan-Guez, B., Rossi, F.: A self organizing map for dissimilarity data. In: Banks, D., House, L., McMorris, F.R., Arabie, P., Gaul, W. (eds.) Classification, Clustering, and Data Mining Applications (Proceedings of IFCS 2004), Chicago, Illinois, pp. 61–68. IFCS, Springer (2004)
14. Fort, J.C., Letremy, P., Cottrell, M.: Advantages and drawbacks of the batch kohonen algorithm. In: Proceedings of Xth European Symposium on Artificial Neural Networks (ESANN 2002), vol. 2, pp. 223–230 (2002)
15. Gärtner, T., Lloyd, J.W., Flach, P.A.: Kernels and distances for structured data. Machine Learning 57(3), 205 232 (2004)
16. Gisbrecht, A., Mokbel, B., Schleif, F.M., Zhu, X., Hammer, B.: Linear time relational prototype based learning. Int. J. Neural Syst. 22(5) (2012)
17. Graepel, T., Burger, M., Obermayer, K.: Self-organizing maps: Generalizations and new optimization techniques. Neurocomputing 21, 173–190 (1998)
18. Graepel, T., Obermayer, K.: A stochastic self-organizing map for proximity data. Neural Computation 11(1), 139–155 (1999)
19. Hammer, B., Gisbrecht, A., Hasenfuss, A., Mokbel, B., Schleif, F.-M., Zhu, X.: Topographic mapping of dissimilarity data. In: Laaksonen, J., Honkela, T. (eds.) WSOM 2011. LNCS, vol. 6731, pp. 1–15. Springer, Heidelberg (2011)
20. Hammer, B., Hasenfuss, A.: Topographic mapping of large dissimilarity data sets. Neural Computation 22(9), 2229–2284 (2010)
21. Hammer, B., Hasenfuss, A., Rossi, F., Strickert, M.: Topographic processing of relational data. In: Proceedings of the 6th International Workshop on Self-Organizing Maps (WSOM 2007), Bielefeld, Germany (September 2007)
22. Hathaway, R.J., Bezdek, J.C.: Nerf c-means: Non-euclidean relational fuzzy clustering. Pattern Recognition 27(3), 429–437 (1994)

23. Hathaway, R.J., Davenport, J.W., Bezdek, J.C.: Relational duals of the c-means clustering algorithms. Pattern Recognition 22(2), 205–212 (1989)
24. Hendrickson, B., Leland, R.: A multilevel algorithm for partitioning graphs. In: Proceedings of the 1995 ACM/IEEE Conference on Supercomputing (CDROM), Supercomputing 1995. ACM, New York (1995), http://doi.acm.org/10.1145/224170.224228
25. Heskes, T., Kappen, B.: Error potentials for self-organization. In: Proceedings of 1993 IEEE International Conference on Neural Networks (Joint FUZZ-IEEE 1993 and ICNN 1993 [IJCNN 1993]), vol. III, pp. 1219–1223. IEEE/INNS, San Francisco (1993)
26. Hofmann, T., Buhmann, J.M.: Pairwise data clustering by deterministic annealing. IEEE Transactions on Pattern Analysis and Machine Intelligence 19(1), 1–14 (1997)
27. Kohonen, T.: Self-organizing maps of symbol strings. Technical report A42, Laboratory of computer and information science, Helsinki University of Technology, Finland (1996)
28. Kohonen, T.: Self-Organizing Maps, 3rd edn. Springer Series in Information Sciences, vol. 30. Springer (2001)
29. Kohonen, T., Somervuo, P.J.: Self-organizing maps of symbol strings. Neurocomputing 21, 19–30 (1998)
30. Kohonen, T., Somervuo, P.J.: How to make large self-organizing maps for nonvectorial data. Neural Networks 15(8), 945–952 (2002)
31. Mac Donald, D., Fyfe, C.: The kernel self organising map. In: Proceedings of 4th International Conference on Knowledge-Based Intelligence Engineering Systems and Applied Technologies, pp. 317–320 (2000)
32. Martín-Merino, M., Muñoz, A.: Extending the SOM algorithm to non-euclidean distances via the kernel trick. In: Pal, N.R., Kasabov, N., Mudi, R.K., Pal, S., Parui, S.K. (eds.) ICONIP 2004. LNCS, vol. 3316, pp. 150–157. Springer, Heidelberg (2004), http://dx.doi.org/10.1007/978-3-540-30499-9_22
33. Müllner, D.: Modern hierarchical, agglomerative clustering algorithms. Lecture Notes in Computer Science 3918(1973), 29 (2011), http://arxiv.org/abs/1109.2378
34. Olteanu, M., Villa-Vialaneix, N., Cottrell, M.: On-line relational SOM for dissimilarity data. In: Estévez, P.A., Príncipe, J.C., Zegers, P. (eds.) Advances in Self-Organizing Maps. AISC, vol. 198, pp. 13–22. Springer, Heidelberg (2013)
35. Rose, K.: Deterministic annealing for clustering, compression,classification, regression, and related optimization problems. Proceedings of the IEEE 86(11), 2210–2239 (1998)
36. Rossi, F.: Model collisions in the dissimilarity SOM. In: Proceedings of XVth European Symposium on Artificial Neural Networks (ESANN 2007), Bruges (Belgium), pp. 25–30 (April 2007)
37. Rossi, F., Hasenfuss, A., Hammer, B.: Accelerating relational clustering algorithms with sparse prototype representation. In: Proceedings of the 6th International Workshop on Self-Organizing Maps (WSOM 2007), Bielefeld, Germany (September 2007)
38. Rossi, F., Villa-Vialaneix, N.: Optimizing an organized modularity measure for topographic graph clustering: a deterministic annealing approach. Neurocomputing 73(7-9), 1142–1163 (2010)
39. Rossi, F., Villa-Vialaneix, N.: Représentation d'un grand réseau à partir d'une classification hiérarchique de ses sommets. Journal de la Société Française de Statistique 152(3), 34–65 (2011)

40. Schleif, F.-M., Gisbrecht, A.: Data analysis of (non-)metric proximities at linear costs. In: Hancock, E., Pelillo, M. (eds.) SIMBAD 2013. LNCS, vol. 7953, pp. 59–74. Springer, Heidelberg (2013)
41. Schölkopf, B., Smola, A., Müller, K.R.: Kernel principal component analysis. In: Gerstner, W., Hasler, M., Germond, A., Nicoud, J.-D. (eds.) ICANN 1997. LNCS, vol. 1327, pp. 583–588. Springer, Heidelberg (1997)
42. Shawe-Taylor, J., Cristianini, N.: Kernel Methods for Pattern Analysis. Cambridge University Press (2004)
43. Somervuo, P.J.: Self-organizing map of symbol strings with smooth symbol averaging. In: Workshop on Self-Organizing Maps (WSOM 2003), Hibikino, Kitakyushu, Japan (September 2003)
44. Somervuo, P.J.: Online algorithm for the self-organizing map of symbol strings. Neural Networks 17, 1231–1239 (2004)
45. Ultsch, A., Siemon, H.P.: Kohonen's self organizing feature maps for exploratory data analysis. In: Proceedings of International Neural Network Conference (INNC 1990), pp. 305–308 (1990)
46. Ultsch, A., Mörchen, F.: Esom-maps: tools for clustering, visualization, and classification with emergent som. Tech. Rep. 46, Department of Mathematics and Computer Science, University of Marburg, Germarny (2005)
47. Vesanto, J.: Som-based data visualization methods. Intelligent Data Analysis 3(2), 111–126 (1999)
48. Vesanto, J.: Data Exploration Process Based on the Self–Organizing Map. Ph.D. thesis, Helsinki University of Technology, Espoo (Finland) (May 2002), Acta Polytechnica Scandinavica, Mathematics and Computing Series No. 115
49. Villa, N., Rossi, F.: A comparison between dissimilarity som and kernel som for clustering the vertices of a graph. In: Proceedings of the 6th International Workshop on Self-Organizing Maps (WSOM 2007), Bielefeld, Germany (September 2007)
50. Williams, C., Seeger, M.: Using the nyström method to speed up kernel machines. In: Advances in Neural Information Processing Systems 13 (2001)

Dynamic Formation of Self-Organizing Maps

Jérémy Fix*

Supélec, IMS-MaLIS Research Group UMI-2958,
2 rue Edouard Belin, 57070 Metz, France
Jeremy.Fix@Supelec.fr

Abstract. In this paper, an original dynamical system derived from dynamic neural fields is studied in the context of the formation of topographic maps. This dynamical system overcomes limitations of the original Self-Organizing Map (SOM) model of Kohonen. Both competition and learning are driven by dynamical systems and performed continuously in time. The equations governing competition are shown to be able to reconsider dynamically their decision through a mechanism rendering the current decision unstable, which allows to avoid the use of a global reset signal.

Keywords: dynamic neural fields, self-organizing maps.

1   Introduction

Within the context of situated or embodied cognition[9], there is a growing interest in finding complex, non-linear dynamical systems supporting various functions such as perception, working memory or action. These dynamical systems would interact continuously with their environment being shaped by the stimuli they receive and producing continuously motor programs to be executed. In this paper, we focus on self-organizing maps. In self-organizing maps, there is a notion of topology and of similarity: neurons that are close together will encode similar aspects of those stimuli that they receive as input. This situation is observed for example in the visual cortex where, despite some non-linearity such as the pinwheels, there is a continuous orientation selectivity along the cortical tissue. In the context of the formation of topographically organized maps, there is one usual assumption having strong consequences on the model. It is indeed usually assumed that the network is reset when a new sample is presented to the network. If we completely develop this assumption, we can identify two situations. In the first one, between two successive stimuli, one may assume that the network receives no excitation and that this period is sufficient for the network to go back to a resting state. In the second, more constrained, situation, the stream of stimuli is continuous and there is no possibility to really separate one stimulus from the others. In such a continuous stream of inputs, it is hard to figure out a mechanism that would reset the network to put it in a more favorable state. In

---

* Jérémy Fix is in the IMS-MaLIS Research Group, Supélec (France).

T. Villmann et al. (eds.), *Advances in Self-Organizing Maps and Learning
Vector Quantization,* Advances in Intelligent Systems and Computing 295,
DOI: 10.1007/978-3-319-07695-9_2, © Springer International Publishing Switzerland 2014

this article, we advocate that such a mechanism is not necessary by introducing a dynamical neural network, relying on the dynamic neural field theory, able to *reconsider its decision* whenever it is no more appropriate. Indeed, as we will see, if the network creates a decision with a bump of activity not localized where the input is locally maximal, this decision becomes unstable. A second property, in line with the previous one, is that the prototypes of the self-organizing map are continuously learned and there is no need to wait that the decision is settled before triggering learning. Competition and learning are two processes occurring at the same time although at different time scales. The article is organized as follows: in section 2 we present the standard neural field equation of [2], explain some limitations in the context of decision making and self-organization and introduce a modified neural field equation. In section 3, we illustrate the behavior of the neural field on artificial data and finally demonstrate its ability to drive a learning process within the context of self-organizing maps. The python scripts used to run the simulations of the paper are available online[5].

2   Methods

The Dynamic Neural Field equation (DNF), as introduced in [2,11], reads :

$$\tau \frac{dV}{dt}(x,t) = -V(x,t) + \int_y w(x,y)f(V(y,t)) + I(x,t) \tag{1}$$

where $V(x,t)$ is the membrane potential at time t and position x, $I(x,t)$ is the input feeding position x at time t, τ is a time constant, $w(x,y)$ is the weight of the lateral connection between the units at position x and y and usually taken as an on-center/off-surround difference of Gaussians and f a transfer function mapping the membrane potential to the firing rates and usually taken as a Heaviside function or a sigmoid. Some properties of such a dynamical system have been formally described and others have been observed experimentally. In particular, essentially depending on the shape and extent of the lateral connections, one may observe competitive behaviors or working memory behaviors. While competitive behaviors mainly rely on inhibitory influences within the field, the memory property can be obtained with more local and excitatory weights. If one is interested in the ability of a neural field to form a decision, this equation has experimentally some drawbacks. In particular, if we suppose that the same neural field should be able to form a decision both when the input is randomly distributed (as we encounter initially with self-organizing maps, when the prototypes are drawn randomly) and when the input is made of several localized bumps of activities (e.g. when a self-organizing map actually gets organized, or when representing alternative motor plans), it appears quite difficult to tune the parameters. Intuitively, in order the neural field to form a localized decision in case of a randomly distributed input, one must introduce some lateral excitation within the field to recover a coherency in the shape of the decision absent from the input. When the inputs are already localized bumps, this lateral excitation is much less important relative to the lateral inhibition supporting the ability to decide. The lateral excitation can even have dramatic consequences in the second situation since it introduces some inertia of the decision and may prevent the neural field to reconsider the decision, in particular when the input changes. We must note that a self-organizing neural field, based on the original Amari's equation, has been proposed in [4]. However, as stated by the authors, when a new sample is presented to the network, *"the activity of the field is reset to zero"*. This is a limitation we want to overcome and that motivates the new neural field equation we propose. Indeed let us consider the dynamic neural field defined by equations (2).

$$\tau \frac{dU}{dt}(x,t) = -U(x,t) + \int_y g(x,y)I(y,t) - \int_y f(V(y,t))I(y,t) \qquad (2)$$

$$\tau \frac{dV}{dt}(x,t) = -V(x,t) + \int_y w(x,y)f(V(y,t)) + U(x,t)$$

where f is a transfer function, w is the lateral weight function and g a Gaussian. In the following, we consider the firing rates $f(V(x,t))$ as representing the decision of the neural field. If we stick to the framework of classical self-organizing maps [7], we expect this decision to be a bell-shaped bump, encoding the neighbourhood function. This equation is similar to the standard dynamic neural field equation except that an intermediate layer with the membrane potentials $U(x,t)$ is introduced. This intermediate layer combines the feedforward input $I(x,t)$ with a feedback from the output layer $V(x,t)$. The motivation behind this intermediate layer is to modulate the input feeding the V layer in order to boost the input when a decision emerges in the neural field that does not correspond to the input region that is locally maximal. This transient boost, that will be canceled if the decision switches, helps in destabilizing the decision. In the next section, we begin by showing experimentally that such a neural field is able to dynamically form a decision where the input is locally maximal. Then, introducing adaptive prototypes for defining the input, we show experimentally that such a neural field can actually self-organize in a way similar to Kohonen's self-organizing maps [7].

3   Results

3.1   Detecting Where the Input Is Locally Maximal

When self-organizing maps were introduced by [7], the author identified several processes *"1. Formation of an activity cluster in the array around the unit at which activation is maximum. 2. Adaptive change in the input weights of those units where activity was confined"*. The neural field equation we propose aims at being a neural system for the first process. The intuition is that a neural field governed by equation 2 should converge to a localized bump of firing rates in the V layer. This bump should be centered on the input region that is *locally* maximal. For clarity, we rewrite the neural field equation (3) that we consider in this section. The parameter β we introduce in the equation facilitates the tuning

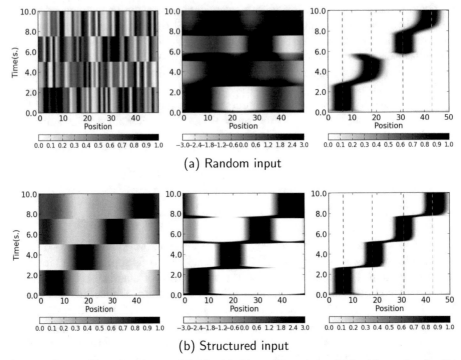

(a) Random input

(b) Structured input

Fig. 1. A one-dimensional neural field with 50 positions is excited with randomly distributed inputs (a) or with more structured inputs (b). In both situations, we plot the input $I(x,t)$ on the left, the $U(x,t)$ membrane potential on the middle and the output firing rates $f(V(x,t))$ on the right. The membrane potentials $U(x,t)$ have been saturated for the illustration to lie in $[-3,3]$ but they typically lie in a larger domain (see Fig. 2). The dashed lines shown on $f(V(x,t))$ indicate the slices plotted on Fig. 2. In all the conditions, the neural field is able to create a bump of activity localized where the input is locally maximal (in the sense of the input convolved with the Gaussian receptive field $g(x,y)$). The bump of activity becomes unstable when the input is changed and the decision is no more appropriate.

of the neural field for the latter simulations. In addition, we discretize the space in N positions and write sums rather than integrals.

$$f(x) = \frac{1}{1 + \exp(-x)}$$

$$\tau \frac{dU}{dt}(x,t) = -U(x,t) + \beta \left(\sum_y g(x,y)I(y,t) - \sum_y f(V(y,t))I(y,t) \right) \quad (3)$$

$$\tau \frac{dV}{dt}(x,t) = -V(x,t) + \sum_y w(x,y)f(V(y,t)) + U(x,t)$$

The lateral weights $w(x,y)$ are defined as an excitatory Gaussian with an inhibitory offset. The afferent weights g are defined as a Gaussian chosen to match approximately the size of the bump-like firing rates $f(V)$ the neural field is producing. Before embedding the neural field within a self-organizing process (with prototypes associated with each location and used to compute the input distribution I), we first simulate the competitive process governed by Eq. (3) with randomly distributed inputs. For clarity of the illustrations, we restrict ourselves to one dimensional neural fields. On Fig. 1 we simulate a 1D neural field with $N = 50$ spatial positions, $\tau = 0.05$, $g(x, y) = \exp(-\frac{d(x,y)^2}{2\sigma_i^2})$ with $\sigma_i = 4.7$, lateral connections $w(x, y) = A_+ \exp(-\frac{d(x,y)^2}{2\sigma_+^2}) - A_-$ with $A_+ = 1.2, \sigma_+ = 4.6, A_- = 0.9 A_+$, $\beta = 2.6$. The parameters of the neural field have been obtained following the method given in [6] with the cost function introduced in [1]. The neural field is integrated synchronously (the activities at all the spatial positions at time $t + \Delta t$ are computed with the activities at time t) with Euler and $\Delta t = 0.01$. To avoid boundary effects, the distance within the field $d(x, y)$ is cyclic. We consider two situations that will be relevant for the latter application of the neural field to self-organizing maps. In the context of self-organizing maps, we typically observe inputs that are initially random (because the prototypes are randomly sampled) and become more structured, bell-shaped, as learning goes on (because the prototypes actually get organized). These are the scenarios we simulate on Fig. 1a,b. On these figures, we plot the input $I(x, t)$ function of time on the left, the membrane potentials $U(x, t)$ on the middle and the output firing rates $f(V(x, t))$ on the right which we define as the decision of the neural field[1]. In both situations, the neural field is able to create a bell-shaped decision centered where the input is locally maximal. In case the input is changed (which we do every few iterations), and the decision is no more localized appropriately, it becomes unstable and a new decision is created after a few steps. This is this property of dynamically reconsidering its decision, what one might call automatic reset, when the input is both random or more structured, that is hard to obtain with the original Amari equation.

In order to better appreciate the behavior of the neural field, we show on Fig. 2, the membrane potentials $U(x, t)$ and output firing rates $f(V(x, t))$ at the positions where the input is locally maximal and where the decision will emerge. The transition around time 2.5s., both for random and structured inputs, is typical from the rest of the simulation. Until that time, the positions around $x = 6$ won the competition. However, as the input gets changed, these locations are no more the most excited in the input. The potentials $U(x, t)$ then begins to grow around the location $x = 18$. They grow until the current decision is destabilized and suppressed in layer $V(x, t)$. Following the suppression of the decision, a new decision appears around $x = 18$ and the $U(x, t)$ activities then come back to a resting level. The $U(x, t)$ layer really brings a boosting signal to favor the correct decision. We repeatedly mention that the decision is created where the input is locally maximal. We do not bring a clear definition of locally maximal but observed experimentally that it corresponds to the position where the input convoluted by the gaussian-like decision is maximal.

---

[1] One must note that the randomly generated inputs have been shifted in space to ensure that the field will create a bump at, successively, the positions $x = 6, 18, 31, 43$, which makes the illustrations easier to understand.

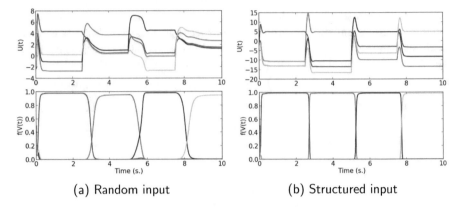

(a) Random input   (b) Structured input

Fig. 2. Slices of the membrane potentials $U(x,t)$ and firing rates $f(V(x,t))$ at the positions $x = 6, 18, 31, 43$ for the scenarios we consider on Fig. 1a,b. The spatial positions for the slices are indicated by dashed lines over the plots of $f(V(x,t))$ on Fig. 1. When a decision is not appropriately localized, the membrane potentials $U(x,t)$ get increased until the current decision is destabilized, allowing a new one to appear.

3.2 A Self-organizing Dynamic Neural Field

We now embed the neural field equation within a learning architecture similar to Kohonen Self-Organizing Maps. Indeed, we consider the dynamical system given by Eq. (4). A prototype $p(x,t) \in \mathbb{R}^2$ is associated with each position $x \in [1..N]$ in the neural field. At regular time intervals, a sample ζ, drawn from an unknown distribution $D \subseteq \mathbb{R}^2$, is presented to the field. The input $I(x,t)$ at each position is computed as a Gaussian, with standard deviation σ, of ζ centered on $p(x,t)$. Importantly, the prototypes are updated at every iteration and not just when the field has converged. Also, it is important to note that, even if the samples are presented at regular time intervals, there is no specific signal sent to the network to indicate it that a new sample is presented; the network will automatically update its decision in case it is no more appropriate. We choose $\sigma = 0.2, \tau_p = 100$ and the equations are integrated synchronously with Euler and $\Delta t = 0.01$. The learning rate is set to $\alpha = 0.001$. The other parameters are the same as in the previous simulation.

$$I(x,t) = \exp(-\frac{|p(x,t) - \zeta|^2}{\sigma^2})$$

$$\tau_p \frac{dp}{dt}(x,t) = f(V(x,t))(\zeta - p(x,t))$$

$$\tau \frac{dU}{dt}(x,t) = -U(x,t) + \beta(\sum_y g(x,y)I(y,t) - \sum_y f(V(y,t))I(y,t)) \quad (4)$$

$$\tau \frac{dV}{dt}(x,t) = -V(x,t) + \sum_y w(x,y)f(V(y,t)) + U(x,t)$$

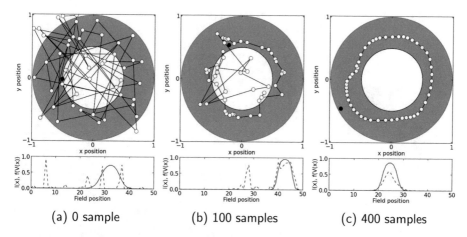

(a) 0 sample   (b) 100 samples   (c) 400 samples

Fig. 3. Simulation of the self-organizing neural fields given by Eq. 4. The samples presented to the network are drawn from a ring distribution shown in gray. The white dots represent the prototypes and are interconnected following the 1D circular topology of the neural field. The current input I (dashed line) and output firing rates $f(V)$ (solid line) are shown below each plot of the prototypes. The figures show the state of the neural field at initialization (a), after presenting 100 samples (b) and 400 samples (c). The current sample is shown as the black dot. A new sample is presented every $2s$..

For the simulation shown on Fig. 3, the samples presented to the network are drawn from a ring distribution shown as the shaded area on the figure. This distribution allows to check two important properties. In case the neural field is not able to create a decision, the prototypes will remain unchanged as the output firing rates $f(V)$ gate learning of the prototypes. In case the neural field is not able to reconsider its decision, the prototypes will converge to the center of mass of the distribution $(0,0)$. Indeed, the field would create a bump of activity and, if we suppose the decision is stuck to that position, it will always be the same prototypes that will learn whatever the input sample. These prototypes would then move to the center of mass of the distribution until they become so far away from the distribution that the associated input becomes weak enough to allow an other position in the field to win the competition. Then the same process would apply and, step by step, all the prototypes would converge to the center of mass of the distribution. In the simulation, we observe that the prototypes correctly self-organize.

As a second example, we now consider a continuous stream of inputs. The point of this simulation is to demonstrate the ability of the neural field to self-organize from a continuous stream of inputs without requiring an explicit reset signal. Indeed, we present to the network a small patch of 100×100 pixels on which a rotating bar is drawn. The bar is rotating at a constant angular speed $\omega = 10°\text{s}^{-1}$. The bar is performing 30 full turns. Examples of input shown to the neural field are displayed on Fig. 4a. The bar is always with the same contrast. On the figure, the gray levels have been adjusted for representing four consecutive

32   J. Fix

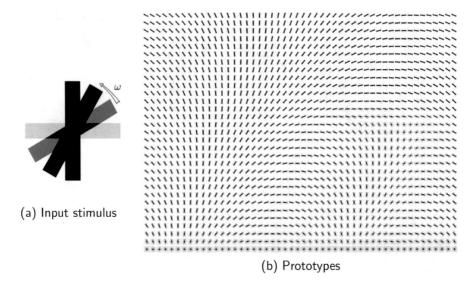

(a) Input stimulus

(b) Prototypes

Fig. 4. (a) Experimental setup: a bar rotating at a constant angular speed $\omega = 10°\text{s}^{-1}$ is drawn on a patch of 100×100 pixels. Such a patch is the input feeding all the positions of the neural field. (b) The $N = 50$ prototypes at one given time are displayed on a row. We display the prototypes after each of the 30 full turns from bottom (initial prototypes) to top (final prototypes).

stimuli. We use exactly the same neural field equations (4) than for the previous simulation. The prototypes $p(x,t)$ and input ζ are now patches of 100×100 pixels. All the parameters (time constants, learning rates, weights) are the same except the variance of the input set to $\sigma = 50$. The prototypes are initialized to the mean of 50 stimuli spanning the range $0°$ to $180°$ to which is added a uniform random noise sampled from $[-0.1, 0.1]$. We come back to the issue of initializing the prototypes in the discussion. As the neural field is evaluated synchronously, the random noise to the initial prototypes helps in breaking the symmetry which would lead to oscillating activities within the field. On Fig 4b we show the prototypes at the 50 spatial spatial positions after each full turn of the bar. The initial prototypes are shown on the bottom row and the final prototypes on the top row. Initially all the prototypes are almost identical. As learning goes on, the prototypes finally span uniformly the range of orientations from $0°$ to $180°$.

Discussion

The dynamic neural field equation we proposed for self-organizing maps in section 3.1 suffers from some limitations. One appears when the prototypes are initialized far from the unknown distribution from which the samples are drawn. This issue is linked to the fact that the neural field requires to be driven by some activities that should not be too small. Otherwise, no decision will emerge in the field. Also, in case some of the prototypes would be able to come closer to the samples, one region of the neural field would learn but not the others. The learning region would learn a mean prototype and the others would remain unchanged. A second limitation is a boundary effect when the neural field is no more laterally connected with cyclic connections. If the connections are no more cyclic, the positions on the border of the *map* are less excited and more inhibited than the positions in the middle of the map. This brings a penalty for those units on the border preventing them from adapting their prototypes. A common mechanism might solve these two issues. A constant baseline added to one of the layers might be sufficient for solving the first issue but not the second. Indeed, the boundary effect appears because of a lack of excitation or excess of inhibition on the units on the border of the map. The bonus must not be uniform in space. An alternative relies in homeostatic plasticity[10], for example by adjusting the baseline of the units depending on the difference between the mean of recent firing rates and a target firing rate. Units that never win the competition (because their prototypes are far from the samples or because these are locations on the border of the map) would receive a transient self-excitation. This transient excitation would push them to fire for everything. At some point they would win the competition and come into play for adapting their prototypes. Such an idea has already been explored in [8] where the authors used a homeostatic learning rule to modulate the lateral connections of a neural field allowing to recover activities at sites where the input has been impaired, and in [3] where the firing threshold of the units is adjusted to bring it closer to a target firing rate.

The main point for the simulations of the self-organizing dynamic neural field is that it can learn continuously; there is no separate phase for evaluating the field until convergence and for learning the prototypes. The second important property is that the neural field is able to automatically reconsider its decision and there is no need for indicating it that a new sample is presented for, for example, resetting the activities. These properties allow the neural field to learn from a continuous stream of stimuli. The dynamic neural field equation we propose can be modified to remove any dependency on space in the interactions. If the gaussian g is replaced by a Kronecker, the weight function replaced by a self-excitation and uniform inhibition from the other locations, the network is able to perform a Winner-Takes-All. One can actually demonstrate that such a network has a single fixed point corresponding to the output of a Winner-Takes-All if the transfer function is taken to be a Heaviside step function. In that respect, instead of a dynamic self-organizing map, we can obtain a dynamic k-means.

References

1. Alecu, L., Frezza-Buet, H., Alexandre, F.: Can self-organization emerge through dynamic neural fields computation? Connection Science 23(1), 1–31 (2011)
2. Amari, S.: Dynamics of Pattern Formation in Lateral-Inhibition Type Neural Fields. Biological Cybernetics 27, 77–87 (1977)

3. Bednar, J.A.: Building a mechanistic model of the development and function of the primary visual cortex. Journal of Physiology-Paris 106(5-6), 194–211 (2012)
4. Detorakis, G., Rougier, N.: A neural field model of the somatosensory cortex: formation, maintenance and reorganization of ordered topographic maps. PLoS One 7(7), e40257 (2012)
5. Fix, J.: Python source scripts for generating the illustrations (2013), http://jeremy.fix.free.fr/Simulations/dynamic_som.html (online; accessed November 5, 2013)
6. Fix, J.: Template based black-box optimization of dynamic neural fields. Neural Networks 46, 40–49 (2013)
7. Kohonen, T.: Self-organized formation of topologically correct feature maps. Biological Cybernetics 43(1), 59–69 (1982)
8. Moldakarimov, S.B., McClelland, J.L., Ermentrout, G.B.: A homeostatic rule for inhibitory synapses promotes temporal sharpening and cortical reorganization. Proceedings of the National Academy of Sciences 103(44), 16526–16531 (2006)
9. Pfeifer, R., Bongard, J.C.: How the Body Shapes the Way We Think: A New View of Intelligence (Bradford Books). The MIT Press (2006)
10. Turrigiano, G.G.: Homeostatic plasticity in neuronal networks: the more things change, the more they stay the same. Trends in Neurosciences 22(5), 221–227 (1999)
11. Wilson, H.R., Cowan, J.D.: A Mathematical Theory of the Functional Dynamics of Cortical and Thalamic Nervous Tissue. Kybernetik 13, 55–80 (1973)

MS-SOM: Magnitude Sensitive Self-Organizing Maps

Enrique Pelayo and David Buldain

Aragon Institute for Engineering Research,
University of Zaragoza
http://http://i3a.unizar.es/en

Abstract. This paper presents a new neural algorithm, MS-SOM, as an extension of SOM, that maintaining the topological representation of stimulus also introduces a second level of organization of neurons. MS-SOM units tend to focus the learning process in data space zones with high values of a user-defined magnitude function. The model is based in two mechanisms: a secondary local competition step taking into account the magnitude of each unit, and the use of a learning factor, evaluated locally, for each unit. Some results in several examples demonstrate the better performance of MS-SOM compared to SOM.

Keywords: Self-Organizing Maps, Magnitude Sensitive Competitive learning, unsupervised learning, classification, surface modelling.

1   Introduction

Soft competitive learning comprises a set of methods where more than a single neuron are adjusted with each sample pattern. These algorithms possess some features that are advantageous over hard competitive learning methods: avoiding unused ('dead') units, accelerating the learning phase, filling empty areas in the dataset space or avoiding local minima. Self Organizing Maps (SOM) [1] is one of these algorithms with the property of generating a topological representation of the input data space in a grid of neurons with low dimension. This makes SOM useful for visualizing low-dimensional views of high-dimensional data. SOM has also been used for data classification (i.e. [2], [3]).

On the other hand, Magnitude Sensitive Competitive Learning (MSCL) [4] is a hard competing algorithm which has the capability of distributing the unit centroids following any user defined magnitude.

Comparing both algorithms, the main disadvantage of SOM against MSCL is that SOM distributes unit prototypes following the data density. Magnification control methods ([5], [6]) present alternatives to SOM that try to modify the relation between data and weight vector density. However, in these kind of methods, final unit distribution is always somehow related with data density.

In this paper we describe a new algorithm, *Magnitude Sensitive Self Organizing Map (MS-SOM)*, an hybrid between MSCL and SOM, that synthesizes the

T. Villmann et al. (eds.), *Advances in Self-Organizing Maps and Learning*
Vector Quantization, Advances in Intelligent Systems and Computing 295,
DOI: 10.1007/978-3-319-07695-9_3, © Springer International Publishing Switzerland 2014

advantages of both methods. It preserves the topological representation of the input space and additionally, distributes units following a target magnitude.

The remainder of this chapter is organized as follows. Section 2 describes the proposed MS-SOM algorithm. In section 3 the new algorithm is applied to three examples: a toy example with Gaussian to show the algorithm behaviour, 3D surface modelling, and data classification. Last section presents the conclusions.

2   Magnitude Sensitive Self Organizing Maps

2.1   The Algorithm

Database Description. Dataset used in this paper $\mathcal{X}$ consists of P patterns $\boldsymbol{x}(t) = (x_1, .., x_D)(t) \in \mathbb{R}^D$.

Magnitude Definitions. the user-defined magnitude function, $MF()$, acts as an extra information for the network, forcing neurons to represent with more detail those zones of data space with higher magnitude values. We consider mainly two situations depending on the data dependency of this function: if magnitude is determined exclusively from input data, $MF(\mathcal{X})$, we define a magnitude vector, $\mathbf{m}_x$, where each component m_x corresponds to sample $\boldsymbol{x}$. Another situation occurs when magnitude function also depends on neuron parameters, $MF(i, \mathcal{X})$, then it is necessary to define for each neuron i an internal variable, $m_{w_i}(t)$ to store the value of the magnitude at that unit. These unit variables can be represented as a magnitude map, $\mathbf{M}_w(t)$, with the same dimensionality of the neural map grid. Figures 1(a) and 2(c) show examples of magnitude map.

The examples studied in next sections show both situations: 3D-surface example has a magnitude vector associated to input data, while Gaussian and classification examples present magnitude maps associated with neurons.

1. Initialization: Initial codebook $\mathcal{W}$ is formed by N prototypes $\mathbf{w}_i$ ($i = 1 \cdots N$) initialized linearly, along principal components of data, and forming a low-dimensional grid (usually 2D). For the case when magnitude depends on neurons, we need to initialize the magnitude map in $t = 0$. To carry out neural-magnitude computations along the learning process, each neuron i uses an accumulated magnitude, ρ_i, that has to be initialized too. At $t = 0$, neural-magnitude variables are initialized following:

$$m_{w_i}(0) = MF(i, \mathcal{X}) \tag{1}$$
$$\rho_i(0) = m_{w_i}(0) \tag{2}$$

2. Selection of Samples: A sample data $\boldsymbol{x}(t)$ is selected at random from $\mathcal{X}$.

3. Global Unit Competition: The index ξ of unit with minimum distance from its weights to the input data vector is selected as global winner in this first step.

$$\xi = \operatorname*{argmin}_{i \in W}(\|\boldsymbol{x}(t) - \mathbf{w}_i(t)\|) \tag{3}$$

At this point, we form the local winner set $\mathcal{S}, (\mathcal{S} \subset W)$ with the N_{grid} units belonging to the neighbourhood in the grid, of unit ξ in the MS-SOM map as:

$$\mathcal{S} = \{s_1, s_2, ..., s_{N_{grid}}\} \tag{4}$$

For example, in a two dimensional grid with hexagonal representation, N_{grid} is 7, counting the winner unit and its six closest neighbour units around.

4. Local Unit Competition: Final winner unit j is selected among the units belonging to $\mathcal{S}$, as the one that minimizes the product of its magnitude value with the distance of its weights to the input data vector, being γ the strength of the magnitude during the competition. It follows:

$$j = \operatorname*{argmin}_{s \in \mathcal{S}}(m_{w_s}(t)^\gamma \cdot \|\boldsymbol{x}(t) - \mathbf{w}_s(t)\|) \tag{5}$$

5. Winner and Magnitude Updating: Weights and magnitude are adjusted iteratively for each training sample, following:

$$m(t) = \begin{cases} m_x(t), & \text{if magnitude comes from data} \\ m_{w_j}(t), & \text{if used internal magnitude of winner unit } j \end{cases} \tag{6}$$

$$\rho_i(t+1) = \rho_i(t) + m(t) \cdot h_{ji}(t) \tag{7}$$

$$\alpha_i(t) = \left(\frac{m(t) \cdot h_{ji}(t)}{\rho_i(t+1)}\right)^\beta \tag{8}$$

$$\mathbf{w}_i(t+1) = \mathbf{w}_i(t) + \alpha_i(t)\,(\boldsymbol{x}(t) - \mathbf{w}_i(t)) \tag{9}$$

$$m_{w_i}(t+1) = \begin{cases} m_{w_i}(t) + \alpha_i(t)\,(m(t) - m_{w_i}(t)), & \text{if } m_x \text{ is used} \\ MF(i, \mathcal{X}), & \text{otherwise} \end{cases} \tag{10}$$

Equation (6) shows how magnitude for updating phase is obtained. Magnitude can be provided by data, or it could correspond to the winner neuron. In the last case, unit magnitude must be fed back to the rest of neurons for their updating phase. In the above equations $h_{ji}(t)$ is the neighbourhood kernel around the winner unit j at time t. This kernel is a function depending on the distance of map units j and i in the map grid, and m_{w_i} is the value of the magnitude at unit i. Finally, $\alpha_i(t)$ is the learning factor, and β is a scalar value between 0 and 1. Observe in eq. (10) that, if magnitude is presented as an extra input (m_x), the magnitude of the unit, m_{w_i} is updated as any other weight.

6. Stopping Condition: Training is stopped when a termination condition is reached. It may be the situation when all data samples has been presented to the MS-SOM or defined by any function that measure the training stabilization.

2.2   Analysis of the Algorithm

Competition: Competition for the Best Matching Unit (BMU) includes a second local competition step taking into account the magnitude, that forces units to move towards space regions of higher value of magnitude. Neurons with high values of magnitude become less competitive than those with low values, so these data-space zones recruit more neurons in their representations.

Learning: Learning factor $\alpha_i(t)$ for each unit depends on:

1. The value of the magnitude $m(t)$ associated to each sample data. High magnitude produces high changes in unit weights, while values near zero produces practically no learning.
2. The distance from each unit to the winner unit. The importance of this factor is modulated by the kernel function h_{ji}. High distance means lower learning.
3. The accumulated magnitude at the unit. It is related to the firing history of each unit. High accumulated magnitude means high learning up to the moment, and therefore unit becomes practically static.
4. The value of β, the forgetting factor. Using the definition of learning factor of (9), when β is equal to one, unit weights become the running weighted mean of the value of the data samples belonging to its Voronoi region, and adjacent regions (weighted according to its neighbourhood). To the contrary, lower values of β means that recent patterns have higher importance in the running weighted mean. In the limit case ($\beta = 0$), each unit would become the last presented sample: $\mathbf{w}_i(t+1) = \boldsymbol{x}(t)$

3   Application Examples

3.1   Modelling Gaussian Distributions

In this example we test the performance of a MS-SOM with four different types of magnitude functions. It is compared with a SOM.

We use a synthetic data set consisting of $P = 5000$ samples in a 2D plane ($\boldsymbol{x}(t) \in \mathbb{R}^2$) that was generated from a mixture of three Gaussian distributions with means [0,0], [3,4] and [6,0], and covariance matrix [0.1 0; 0 0.1] for all of them. The fraction of samples placed in each cluster is approximately P/3.

A SOM is trained using a Gaussian function for $h_{ji}(t)$ with ratios within [3,0.05] and a learning factor that decreases exponentially with time. Three SOMs are initialized linearly in the data space using codebooks of 40, 80 and 160 units.

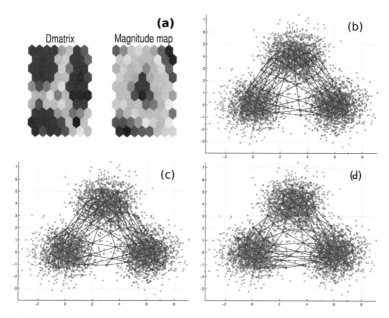

Fig. 1. *Gaussian example.* (a) D-matrix and Magnitude map of MS-SOM avoiding data mean (using MF_4 as magnitude function). (b) Trained SOM. (c) MS-SOM trained with MF_1 and (d) trained with MF_4 .

MS-SOMs have the same number of units than SOM (also uses 40, 80 and 160), uses the same initial codebooks, $h_{ji}(t)$ and a value of $\beta = 1$. We apply them to four different magnitude functions:

1. Constant value: $MF_1(i, \mathcal{X}) = 1$.
2. Distance to ordinate axis : $MF_2(i, \mathcal{X}) = abs(w_{i2})$.
3. Distance to point (0,0): $MF_3(i, \mathcal{X}) = \| \mathbf{w}_i \|$.
4. Distance to the mean of dataset: $MF_4(i, \mathcal{X}) = \| \mathbf{w}_i \quad \boldsymbol{x}_{mean} \|$.

Figure 1 shows some results for SOM and MS-SOM (with 80 units) of the grid representation over the data space. Figure 1(a) shows the corresponding D-matrix (mean distances from weights of neighbour units), and the magnitude map for one MS-SOM. Figure 1(b) shows the typical result of a trained SOM where units tend to allocate their centroids in areas with higher data density.

MS-SOM in Fig. 1(c) used a constant value for magnitude equal to one, therefore magnitude function have no effect on final training. As in the SOM case, units are centered in zones with high density. However their distribution is not so affected by the 'border effect', of the SOM representation, because the learning factor is different for each unit in MS-SOM and $\alpha_i(t)$ depends on the winning frequency of unit i.

Figure 1(d) shows a more expansive MS-SOM than using constant magnitude. Magnitude of units becomes higher as their weights are farther from mean of dataset, so units focus on these areas, even if they present low data density.

Table 1. Table shows the mean values in 100 tests of the *Weighted Mean Squared Error (WMSE)* calculated in three codebooks (sizes 40, 80 and 160) after applying SOM and MS-SOM trained with four magnitude functions. WMSE is always lower in MS-SOM independently of the magnitude function used.

Units	Algorithm	MF_1	MF_2	MF_3	MF_4
40	SOM	0.574	0.726	0.688	0.692
	MSSOM	0.474	0.550	0.540	0.536
80	SOM	0.424	0.524	0.515	0.503
	MSSOM	0.357	0.435	0.402	0.404
160	SOM	0.296	0.361	0.358	0.353
	MSSOM	0.246	0.273	0.274	0.275

We use Weighted Mean Squared Error (WMSE) as a measure of quantization quality. It is the weighted mean of the quantization squared error, where values of $m(t)$ (eq. (6)) are the weighting factor, and $\mathcal{V}_i$ the Voronoi set of i:

$$WMSE(\mathcal{X}, \mathcal{W}) = \frac{\sum\limits_{i \in \mathcal{W}} \left(\sum\limits_{x \in \mathcal{V}_i} m(t) \cdot \|\boldsymbol{x}(t) - \mathbf{w}_i(t)\| \right)}{\sum\limits_{x \in \mathcal{X}} m(t)} \tag{11}$$

Table 1 show the Weighted Mean Squared Error (WMSE) calculated in three codebooks (sizes 40, 80 and 160) after applying SOM and MS-SOM trained with four magnitude functions indicated above. Results are different for the four magnitude functions because weights in the magnitude input change depending on the selected function.

In all the cases MS-SOM surpass SOM in all the situations, getting lower weighted quantization error. It is significant that in the case of the constant one magnitude function, MS-SOM is better than SOM, because the 'border effect' is lower in MS-SOM.

3.2 Classification

The dataset in classification problems consists on P samples $\boldsymbol{x}(t) \in \mathbb{R}^D$ separated in K possible classes, $\mathcal{C} = \{C_1, C_2, \ldots, C_K\}$. Each sample has a label vector that indicates, with a binary codification, the class index of the sample (see eq.(12)). It is provided to the neural network during training, so the neuron is able to process class information for the magnitude calculation.

We will compare SOM and a MS-SOM that focus units in zones with high magnitude, associated to high miss-classification error. The process is as follows:

1. Vector data in the sample dataset are joined with the class-label vector:

$$\mathbf{x}(t) = (x_1, \ldots, x_D, c_1, \ldots, c_K) \in \mathbb{R}^{(D+K)}, .\tag{12}$$

being $c_k = 1$ if $\boldsymbol{x}(t) \in C_k$, or $c_k = 0$ if it belongs to other class. Similarly, we form an expanded weight vector for each unit i, where component ψ_{ik} acts as a mean-counter of k-class samples of class k captured by the unit:

$$\mathbf{w}_i(t) = (w_{i1}, \dots, w_{iD}, \psi_{i1}, \dots, \psi_{iK}) \in \mathbb{R}^{(D+K)}. \tag{13}$$

2. Data samples are normalized for the first D components, the only components considered during the first competition step based in distance. Components $(c_1, \dots, c_K)$ of vector $\mathbf{x}(t)$ and $(\psi_{i1}, \dots, \psi_{iK})$ of vector $\mathbf{w}_i(t)$ are masked during the global competition step, but they are updated as the other weights.

3. SOM and MS-SOM are trained using $\mathbf{x}(t)$ with data inputs selected randomly. Magnitude function for MS-SOM depends on each unit and has the following value:

$$MF(i, \mathcal{X}) = K \cdot \frac{1 - max_k(\psi_{ik})}{(K-1)} \tag{14}$$

With this definition, m_{w_i} is 0 if unit i has captured only data samples of one single class, and it is close to 1 in the situation of maximum confusion between the classes (when $max_k(\psi_{ik}) = 1/K$).

4. The final class assigned to each unit i is: $class(i) = argmax_k(\psi_{ik})$

In this classification comparative, we used three data sets: the Iris Dataset [7] and two datasets downloaded from the Proben1 library [8]. First one consists of 150 samples from three species of Iris (Setosa, Virginia and Versicolor). The second dataset presents 6 types of glasses; defined in terms of their oxide content (i.e. Na, Fe, K, etc). The Third dataset is based on patient data to decide whether a Pima Indian individual is diabetes positive or not. Number of samples, inputs and classes are specified for each problem in Table 2.

SOM and MS-SOM were trained with the same parameters, with Neigthbour radios within [3,0.05] using a Gaussian neighbouring function and a decreasing learning factor. Both neural networks received the same linear initialization, and their map sizes vary depending on the problem, as they are displayed in column *Map* of Table 2. MS-SOM uses a value of $\beta = 1$.

Table 2. Percentage of mean classification error (E) and Weighted Mean Square Error (WMSE) for SOM and MS-SOM (*MS*) obtained after training both algorithms with the three datasets. Additionally number of samples, number of inputs, classes, and map size is displayed for each problem.

Problem	Samples/Inputs	Classes	Map	E_{SOM}	E_{MS}	$WMSE_{SOM}$	$WMSE_{MS}$
Iris	150 / 4	3	[5x3]	1.2	0.2	0.643	0.556
Glass	214 / 9	6	[6x4]	24.4	13.7	1.484	1.306
Diabetes	768 / 8	2	[10x8]	7.6	4.9	1.493	1.366

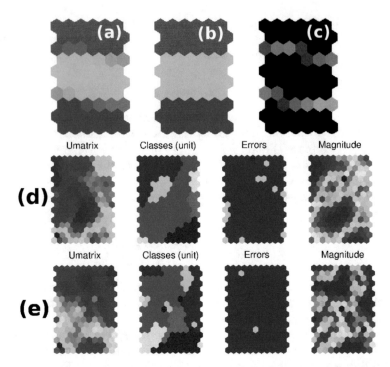

Fig. 2. *Classification regions for MS-SOM. Iris example:* (a)map with colours depending on the classes for each unit (interpolating colours mean that a unit has samples from different classes), (b) map with the final assigned class for each unit, and (c) magnitude associated to each unit (clearer grey means higher magnitude). In this figure, the map size (10x6 units) was bigger than the one used in the comparative to highlight the value of the magnitude in zones of high class confusion. *Glass example:* (d)Results of training a (17x11) grid with SOM. (e) Corresponding results of MS-SOM, that produces lower number of errors.

This table shows the percentage of mean classification errror (E) and the mean Weighted MSE (WMSE) averaged in 20 trainings with each dataset. E is the number of samples asigned to an erroneous class after each test, divided by the total number of samples in the dataset (expresed as percentage). Columns E_{SOM} and E_{MS} display classification errors for SOM and MS-SOM respectively. Columns $WMSE_{SOM}$ and $WMSE_{MS}$ are the equivalent for the Weighted Mean Square Error.

It is clear that in the three problems, MS-SOM with units focused in the limits between classes is able to distinguish more accurately the class to which each sample belongs to (it has lower error). The reason is that MS-SOM assigns less units in areas with no class-confusion (where classification error is null) while many of the units tend to be assigned to the decision regions among classes. On the other hand, WMSE reflects the quantization error, focussing in areas of high magnitude. This measure is lower in MS-SOM algorithm, what means that

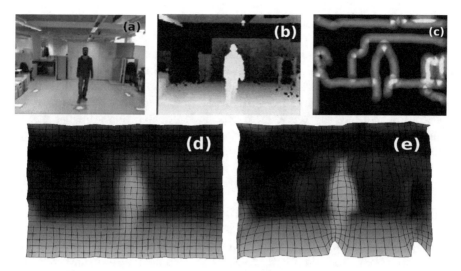

Fig. 3. *Surface modelling example.*(a) Original image. (b) 3D depth image. (c) Curvature map applying Canny. Zones with higher curvature are clearer. (d) Final surface models after training a SOM and (e) a MS-SOM following curvature.

its centroid density is higher in the decision regions, giving as result a better performance in the classification task.

Figure 2 shows a MS-SOM trained with the iris dataset: (a) Map with colours depending on the classes for each unit (interpolating colours mean that a unit has samples from different classes), (b) map with the final assigned class for each unit, and (c) magnitude associated to each unit. In the magnitude map, limits between the three classes are more clearly represented because MS-SOM tends to distribute units in the decision regions between contiguous classes.

3.3   3D Surface Modelling

In 3D computer graphics, a depth map or a 3D depth image is an image that contains information relating to the surface distances of scene objects from a viewpoint, usually represented by a grey level.

In this example we compare the performance of SOM and MS-SOM in the task of modelling a 3D surface, given a depth image downloaded from [9]. Depth information facilitates the computation of the curvature at each region of the image, which is closely related to the problem of discovering the edges in a grayscale image. An edge, detectable by a canny filter, mostly corresponds to a change in depth, and therefore it is a region with high curvature.

Top of figure 3 shows the original image, the depth gray-image (nearer is clearer) and its associated curvature values obtained with Canny filter applied to the depth image. This curvature is used as magnitude vector $\mathbf{m}_x$ associated to dataset $\mathcal{X}$.

Data samples consist in the three dimensional vectors formed by the pixel coordinates and the pixel depth. We trained both SOM and MS-SOM with the same number of units, and training parameters, including the linear codebook initialization. Figure 3 (*d* and *e*) shows the surface modelled with SOM and MS-SOM respectively. MS-SOM allocates more units than SOM in the zones corresponding to the edges in the 3D depth map, therefore three-dimensional borders are more clearly represented, and it is possible to distinguish the human figure and other details, while in the SOM representation they are mostly confused with the background.

4   Conclusions

In this paper we have presented MS-SOM, a variant of SOM, which forces units to focus in zones with high values of a user defined magnitude.

We provided three experiments, a simple Gaussian example, a surface modelling application, and the use of MS-SOM in classification problems. MS-SOM surpassed SOM in all of the experiments, so is an excellent alternative to SOM in situations were it is desirable vector quantization oriented by a magnitude.

The topological representation of stimulus naturally emerges in the biological model of lateral connectivity with excitation/inhibition in the form of Mexican hat. SOM algorithm was developed as an smart simplification of this biological model. MS-SOM introduces a second level of organization of neurons following any magnitude function. This magnitude mechanism could be simplifying other types of biological processes as, for example, a magnitude derived from a chemical difussion map. This proposition is not supported by experimental biological proofs, as we know, but we considered interesting to develop a new method that, preserving the topological behaviour, added other levels of organization with certain biological plausibility.

References

1. Kohonen, T.: Self-Organizing Maps. Springer, New York (1997)
2. Martinez, P., et al.: Hyperspectral image classification using a self-organizing map. In: Summaries of the X JPL Airborne Earth Science Workshop (2001)
3. Arias, S., Gómez, H., Prieto, F., Botón, M., Ramos, R.: Satellite image classification by self organized maps on GRID computing infrastructures. In: Proceedings of the Second EELA-2 Conference, pp. 1–11 (2009)
4. Pelayo, E., Buldain, J.D., Orrite, C.: Magnitude Sensitive Competitive Learning. Neurocomputing 112, 4–18 (2013)
5. Villmann, T., Claussen, J.C.: Magnification Control in Self-Organizing Maps and Neural Gas. Neural Computation 18, 446–469 (2006)
6. Merenyi, E., Jain, A., Villmann, T.: Explicit magnification control of self-organizing maps for 'forbidden' data. IEEE Trans. on Neural Networks 18(3), 786–797 (2007)
7. Iris Dataset, http://archive.ics.uci.edu/ml/datasets/Iris
8. Prechelt, L.: Proben1 - a set of neural network benchmark problems and benchmarking rules. Technical Report 21/94, Fac. of Informatics, Univ. Karlsruhe (1994)
9. DGait Database, http://www.cvc.uab.es/DGaitDB/Summary.html

Bagged Kernel SOM

Jérôme Mariette[1], Madalina Olteanu[2],
Julien Boelaert[2], and Nathalie Villa-Vialaneix[1,2,3]

[1] INRA, UR 0875 MIA-T, BP 52627, 31326 Castanet Tolosan cedex, France
[2] SAMM, Université Paris 1 Panthéon-Sorbonne,
90 rue de Tolbiac, 75634 Paris cedex 13, France
[3] UPVD, 52 avenue Paul Alduy, 66860 Perpignan cedex 9, France
jerome.mariette@toulouse.inra.fr, madalina.olteanu@univ-paris1.fr,
julien.boelaert@gmail.com, nathalie.villa@toulouse.inra.fr

Abstract. In a number of real-life applications, the user is interested
in analyzing non vectorial data, for which kernels are useful tools that
embed data into an (implicit) Euclidean space. However, when using
such approaches with prototype-based methods, the computational time
is related to the number of observations (because the prototypes are
expressed as convex combinations of the original data). Also, a side effect
of the method is that the interpretability of the prototypes is lost. In
the present paper, we propose to overcome these two issues by using a
bagging approach. The results are illustrated on simulated data sets and
compared to alternatives found in the literature.

1   Introduction

In a number of real-life applications, the user is interested in analyzing data that
are non described by numerical variables as is standard. For instance, in social
network analysis, the data are nodes of a graph which are described by their
relations to each others. Self-Organizing Maps (SOM) and other prototype based
algorithms have already been extended to the framework of non numerical data,
using various approaches. One of the most promising one is to rely on kernels to
map the original data into an (implicit) Euclidean space in which the standard
SOM can be used [1,2,3]. A closely related approach, called "relational SOM"
[4,5], extends this method to dissimilarity data which are pertaining to a pseudo-
Euclidean framework, as demonstrated in [4]. Further, in [6], we addressed the
issue of using several sources of (possibly non numeric) data by combining several
kernels. The combination of kernels is made optimal with a stochastic gradient
descent scheme that is included in the on-line version of the SOM algorithm.

However, while able to handle non Euclidean data, that can eventually come
from different sources, these approaches suffer from two drawbacks: as pointed
out in [7], when the data set is very large, the computational time of such approaches can be prohibitive. Indeed, prototypes are expressed as convex combinations of the original data and are thus expressed with a number of coefficients
equal to the number of observations in the data set. Also, adding an extra gradient descent step to optimize the kernel combination requires to increase the

number of iterations of the algorithm, which yields to an augmented computational time. The second drawback is emphasized in [8]: as the prototypes are expressed as a convex combination of the original data, they are no longer given as explicit representative points in the data space and the interpretability of the model is lost.

In the present paper, we propose to overcome these two issues by using a bagging approach in which only a small subset of the original data set is used. The results coming from several bags are combined to select the most representative observations that are then utilized to define the prototypes in a final map. This approach is both sparse (the resulting map is based on a small subset of observations only), fast and parallelizable, which makes it an interesting approach to analyze large samples. The rest of the paper is organized as follow: Section 2 describes the method and its relations to previous approaches in the literature. Section 3 provides the analysis of the results obtained on simulated data sets and on a real-world data set which is a graph.

2   Method

2.1   A Brief Description of the Kernel SOM Approach

Let us suppose that we are given input data, $(x_i)_{i=1,\ldots,n}$ taking values in an arbitrary space $\mathcal{G}$. When $\mathcal{G}$ is not Euclidean, a solution to handle the data set $(x_i)_i$ with standard learning algorithms is to suppose that a *kernel* is known, i.e., a function $K : \mathcal{G} \times \mathcal{G} \to \mathbb{R}$ which is symmetric ($\forall z, z' \in \mathcal{G}$, $K(z, z') = K(z', z)$) and positive ($\forall N \in \mathbb{N}$, $\forall (z_j)_{j=1,\ldots,N} \subset \mathcal{G}$ and $\forall (\alpha_j)_{j=1,\ldots,N} \subset \mathbb{R}$, $\sum_{j,j'} \alpha_j \alpha_{j'} K(z_j, z_{j'}) \geq 0$). When such conditions are fulfilled, the so-called kernel defines a dot product in an underlying Hilbert space: more precisely, there exists a Hilbert space $(\mathcal{H}, \langle ., . \rangle_\mathcal{H})$, called the *feature space*, and a function $\phi : \mathcal{G} \to \mathcal{H}$, called the *feature map*, such that

$$\forall x, x' \in \mathcal{G}, \qquad \langle \phi(x), \phi(x') \rangle_\mathcal{H} = K(x, x')$$

(see [9]). Hence using the kernel as a mean to measure similarities between data yields to implicitly rely on the Euclidean structure of $\mathcal{H}$. Many algorithms have been *kernelized*, i.e., modified to handle (possibly non vectorial) data described by a kernel. In particular, the general framework of kernel SOM is described in [1,2,3]. In this framework, as in the standard SOM, the data are clustered into a low dimensional grid made of U neurons, $\{1, \ldots, U\}$ and these neurons are related to each other by a neighborhood relationship on the grid, h. Each neuron is also represented by a prototype p_u (for some $u \in \{1, \ldots, U\}$) but unlike standard numerical SOM, this prototype does not take value in $\mathcal{G}$ but in the previously defined feature space $\mathcal{H}$. Actually, each prototype is expressed as a convex combination of the image of the input data by the feature map:

$$p_u = \sum_{i=1}^n \gamma_{ui} \phi(x_i), \qquad \text{with} \qquad \gamma_{ui} \geq 0 \text{ and } \sum_i \gamma_{ui} = 1.$$

In the on-line version of the algorithm, two steps are iteratively performed:

- **An affectation step** in which an observation x_i is picked at random and affected to its closest prototype using the distance induced by K:

$$f(x_i) = \arg\min_u \|p_u - \phi(x_i)\|_{\mathcal{H}}^2,$$

where $\|p_u - \phi(x_i)\|_{\mathcal{H}}^2 = K(x_i, x_i) - 2\sum_l \gamma_{uj} K(x_i, x_j) + \sum_{jj'} \gamma_{uj}\gamma_{uj'} K(x_j, x_{j'})$.

- **A representation step** in which the prototypes are updated according to their value at the previous step t and to the observation chosen in the previous step. A gradient descent-like step is used for this update:

$$\forall\, u = 1, \ldots, U, \qquad \gamma_u^{t+1} = \gamma_u^t + \mu(t)h^t(f(x_i), u)\left(\delta_i^n - \gamma_u^t\right)$$

where δ_i^n is the n-dimensional vector in which only entries indexed by i is non zero and equal to 1, $\mu(t) \sim 1/t$ is a vanishing coefficient and, usually, the neighborhood relationship h^t also vanishes with until being restricted to the neuron itself.

Note that this algorithm has also been extended to the case where the observations are described by several kernels, each corresponding to one particular type of data, in the *multiple Kernel SOM* algorithm [6]. In this algorithm, an additional gradient descent step is added to the algorithm to tune the respective contribution of each kernel in an optimal way.

2.2   Ensemble of SOMs

Despite their generalization properties to complex data, kernel SOM and related methods are not well-suited for large data sets since the algorithms generally require the storage of the entire Gram matrix and since the prototypes are expressed as convex combinations of the input data and thus have a very high dimension. Another important drawback of the prototype representation is that, being expressed as a very large linear correlation of the mapped input data, they are not easy to interpret. Indeed, for non vectorial data, such as e.g., nodes in a graph whose similarities can be described by several types of kernels (see [10]), there is no way to describe the prototypes in terms of an object in the input space (here, the graph). As prototypes are commonly used to decipher the clusters' meaning, one of the main interesting feature of the SOM algorithm is lost in the process, as pointed out in [8].

Several techniques have been proposed in the literature to overcome the dimensionality issues, which can be adapted to the kernel SOM framework: some are related to sparse representations and some to bootstraping and bagging. In [4], the large size of the data set is handled using "patch clustering", which is particularly suited for streaming data but can also be used to handle large dimensional data. The initial data set is randomly split into several patches, $\mathcal{P}_b$ and the algorithm processes each patch iteratively. At step b, the b-th patch is clustered until convergence. Each of the resulting prototypes is approximated by the closest P input data points. During the next step, the index set of the P-approximations of all prototypes and the index set of the next patch $\mathcal{P}_{b+1}$ are put together into the extended patch $\mathcal{P}_{b+1}^{\star}$ and the clustering process is performed on all the observations indexed by $\mathcal{P}_{b+1}^{\star}$. This is iterated until all patches are clustered. This approach leads to good clustering results, however it is not parallelizable and the algorithm may be sensitive to the order in which patches are processed. Another technique for handling large data sets is to use bootstrap and bagging. In [11], bagging is applied to a batch version of the SOM algorithm for numerical data in a semi-supervised context. The prototypes of the map trained on the first bag are initialized to lie in the first principal component and each trained map is used to initialize the subsequent map for the subsequent bag. This procedure reduces the dimensionality and improves the classification error, but it is not parallelizable. Alternatively, [12,13] propose by combining SOM based on separate bootstrap samples with a fusion of their prototypes. These approach, which can be used in parallel are however only valid if the prototypes are expressed on the same representation space, which is not directly generalizable when using kernel SOM in which prototypes are directly expressed with the bootstrap sample.

2.3   Bagged Kernel SOM

Our proposal is to use a bagging approach that is both parallelizable and sparse. Bagging uses a large number of small sub-samples, all randomly chosen, to select the most relevant observations: B subsets, $(\mathcal{S}_b)_b$ each of size $n_B \ll n$, are built, at random, within the original data set $\{x_1, \ldots, x_n\}$. Using the on-line algorithm described in [14], a map with U neurons is trained, which results in the prototypes $p_u^b = \sum_{x_i \in \mathcal{S}_b} \gamma_{ui}^b \phi(x_i)$ where ϕ is the feature map associated with the kernel K. The most representative observations are chosen as the first P largest weights for each prototype: $\forall\, u = 1, \ldots, U$,

$$\mathcal{I}_u^b := \left\{ x_i \,:\, \gamma_{ui} \text{ is one of the first } P \text{ largest weights among } (\gamma_{uj}^b)_{x_j \in \mathcal{S}_b} \right\},$$

and $\mathcal{I}_b = \cup_u \mathcal{I}_u^b$. Alternative methods to select the most interesting prototypes are reported in [8]; the one we chose is referred in this paper as the *K-convex hull* but it would be interesting to test other methods for selecting the most interesting observations.

   Then, the number of times each observation $(x_i)_{i=1,\ldots,n}$ is selected in one sub-sample is computed: $\mathcal{N}(x_i) := \sharp\{b\,:\, x_i \in \mathcal{I}_b\}$ which is finally used as a quality criterion to select the most important variables which are the first $P \times U$ observations with the largest values for $\mathcal{N}(x_i)$:

$$\mathcal{S} := \left\{ x_i : \mathcal{N}(x_i) \text{ is one of the first } PU \text{ largest numbers among } (\mathcal{N}(x_j))_{j \geq n} \right\}.$$

A final map is then trained with the selected observations in $\mathcal{S}$ which has prototypes expressed as $p_u = \sum_{x_i \in \mathcal{S}} \gamma_{ui} \phi(x_i)$. The final classification for all observations $(x_i)_{i=1,\ldots,n}$ is deduced from these prototypes by applying the standard affectation rule: $\mathcal{C}(x_i) := \arg\min_{u=1,\ldots,U} \|p_u - \phi(x_i)\|^2$ where $\|p_u - \phi(x_i)\|_{\mathcal{H}}^2$ is computed using K, as described in Section 2.1. The algorithm is described in Algorithm 1.

Algorithm 1. Multiple online kernel SOM

1: Initialize for all $i = 1, \ldots, n$, $\mathcal{N}(x_i) \leftarrow 0$
2: **for** $b = 1 \to B$ **do**
3:     **Sample** randomly with replacement n_B observations in $(x_i)_{i=1,\ldots,n}$ **return** $\mathcal{S}_b$
4:     Perform **kernel SOM** with $\mathcal{S}_b$ **return** prototypes $(p_u^b)_u \sim (\gamma_{ui}^b)_{ui}$
5:     **for** $u = 1 \to U$ **do**
6:         Select the P largest $(\gamma_{ui}^b)_{x_i \in \mathcal{S}_b}$ **return** $\mathcal{I}_u^b$ (set of the observations corresponding to the selected γ_{ui}^b)
7:     **end for**
8:     **for** $i = 1 \to n$ **do**
9:         **if** $x_i \in \cup_u \mathcal{I}_u^b$ **then**
10:            $\mathcal{N}(x_i) \leftarrow \mathcal{N}(x_i) + 1$
11:        **end if**
12:    **end for**
13: **end for**
14: Select the PU observations x_i corresponding to the largest $\mathcal{N}(x_i)$ **return** $\mathcal{S}$
15: Perform **kernel SOM** with $\mathcal{S}$ **return** prototypes $(p_u)_u \sim (\gamma_{ui})_{u=1,\ldots,U,x_i \in \mathcal{S}}$ and classification $(f(x_i))_{x_i \in \mathcal{S}}$
16: Affect $(x_i)_{x_i \notin \mathcal{S}}$ with
$$f(x_i) := \arg\min_u \|\phi(x_i) - p_u\|_{\mathcal{H}}^2$$
17: **return** final classification $(f(x_i))_{i=1,\ldots,n}$ and sparse prototypes $(p_u)_u \sim (\gamma_{ui})_{u=1,\ldots,U,x_i \in \mathcal{S}}$

Note that, strictly speaking, only the sub-kernels $\mathbf{K}_{\bar{\mathcal{S}},\mathcal{S}} := (K(x_i, x_j))_{i \notin \mathcal{S}, j \in \mathcal{S}}$ and $\mathbf{K}_{\mathcal{S}} = (K(x_j, x_j'))_{j,j' \in \mathcal{S}}$ are required to perform the final affectation step because the closest prototype does not depend on the term $K(x_i, x_i)$ and thus the affectation step for $(x_i)_{i \notin \mathcal{S}}$ can be performed by computing:

$$-2\mathbf{K}_{\bar{\mathcal{S}},\mathcal{S}}\gamma + \mathbf{1}_{|\bar{\mathcal{S}}|}\left[\text{Diag}\left(\gamma^T\mathbf{K}_{\mathcal{S}}\gamma\right)\right]^T,$$

where $\gamma = (\gamma_{ui})_{u=1,\ldots,U,i \in \mathcal{S}}$ and $\mathbf{1}_{|\bar{\mathcal{S}}|}$ if the vector with all entries equal to 1 and having length the number of elements in $\bar{\mathcal{S}} = \{x_i : x_i \notin \mathcal{S}\}$.

The complexity of the approach is $\mathcal{O}(Un_B^2 B + U^2 P)$, compared to the direct approach which has a complexity equal to $\mathcal{O}(Un^2)$. Hence, the computational time is reduced as long as $Bn_B^2 + U^2 P < n^2$ and is even more reduced if the B sub-SOMs are performed in parallel. Usually, B is chosen to be large, n_B is small compared to n and P is only a few observations to obtain sparse representations of the prototypes. However, the computational times are not directly comparable since the bagged approach can be performed in parallel, unlike the direct approach or the patch SOM.

3   Applications

Bagged Kernel SOM on Simulated Data

First, a simple simulated dataset with 5000 observations is considered. The observations are randomly generated in $[0, 1]^{20}$ and are then mapped onto a 5×5 grid using the kernel SOM algorithm with a Gaussian kernel. Several algorithms are compared with varying parameters:

- The **patch SOM** with different numbers of patches (250, 375, 500, 1 000) and different values for P (2, 5, 10, 15, 20, 30 and 50). A last kernel SOM was trained with the selected observations to make the results (based on P selected observations) comparable with those of the patch SOM;
- The **bagged SOM** with different values for n_B (5%, 7.5%, 10% and 20% of the original data set size) and for B (500 and 1000) and the same values for P as with the patch SOM;
- A **full kernel SOM** used on the whole data set and aimed at being the reference method.

Figure 1 gives the quantization and topographic [15] errors of the resulting maps versus the value of P. In this figure, two classical quality criteria for SOM results are used: the quantization error (which assesses the quality of the clustering) and the topographic error (which assesses the quality of the organization; see [15]). In some cases, the results can be even better than the full kernel SOM. Considering the bootstrap version, the performances are consistent with the full kernel SOM (for about $P \sim 5 - 10$, which corresponds to using only 250 observations at most, instead of 5000, to represent the prototypes).

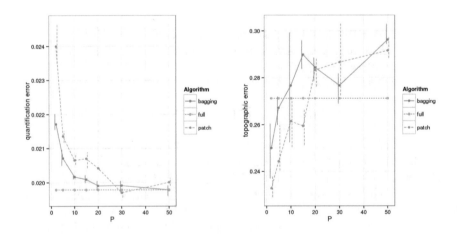

Fig. 1. Evolution of the quantization (left) and topographic (right) errors versus P. Error bars indicates the first and last quantiles and dots the average values over all simulations.

The analysis of the other parameters of the algorithm (bag size n_B and number of bootstrap samples B) does not show any particular feature. This is explained because the final clustering is obtained from the PU most representative observations and thus P has a much greater impact on the performances than, e.g., n_B.

Application to ego-facebook© Network

The bagging method is then applied to one of the ego-facebook© networks described in [16][1]. The data used in this section are the ones extracted from the network number 107: the largest connected component of the facebook© network was extracted, which contained 1 034 nodes. This section presents the comparaison of bagged SOM and standard SOM to map the nodes of the graph onto a two-dimensional grid (having sizes 10×10). As explained in [3,17], using such mappings can provide a simplified representation of the graph, which is useful for the user to help him or her understand its macro-structure before focusing more deeply on some chosen clusters. The kernel used to compute similarities between nodes in the facebook© network was the *commute time kernel*, [18]. If the graph is denoted by $\mathcal{G} = (V, E, W)$, with $V = \{x_1, \ldots, x_n\}$ the set of vertices, E the set of edges which is a subset of $V \times V$ and W a weight matrix (a symmetric matrix with positive entries and null diagonal), the commute time kernel is the generalized inverse of the graph Laplacian, L, which is: $L_{ij} = \begin{cases} d_i & \text{if } i = j \\ -W_{ij} & \text{otherwise} \end{cases}$ where $d_i = \sum_j W_{ij}$ is the degree of node x_i. As explained in [19], the Laplacian is closely related to the graph structure and thus, it is not surprising that a number of kernel has been derived from this matrix [10]. As shown in [18], the commute kernel yields to a simple similarity interpretation because it computes the average time needed for a random walk on the graph to reach a node from another one.

Different approaches were compared: (i) the standard kernel SOM (on-line version), using all available data; (ii) the bagged kernel SOM, as described in Section 2, with $B = 1000$ bootstrap sample, $n_B = 200$ in each sample and $P = 3$ observations selected per prototype and (iii) a standard kernel SOM trained with an equivalent number of randomly chosen observations. The relevance of the results was assessed using different quality measures. Some quality measures were related to the quality of the map (quantification error and topographic error) and some were related to a ground truth: some of the nodes have been indeed labeled by users to belong to one "list" (as named by facebook©). We confronted these groups to the clusters obtained on the map calculating (i) the average node purity (i.e., the mean over all clusters of the maximal proportion of one list in a given cluster; only individuals belonging to one list were used to compute this quality measure) and (ii) the normalized mutual information [20] (also restricted to individuals belonging to one list only) and also to the graph structure using the modularity [21], which is a standard quality measure for node clustering.

---

[1] available at http://snap.stanford.edu/data/egonets-Facebook.html

Table 1. Quality measures for different versions of kernel SOM (standard using all data, bagged, standard using randomly selected data) on facebook© data

	Quantification Error (×100)	Topographic Error	Node Purity	Normalized Mutual Information	Modularity
bagged K-SOM	7.66	4.35	89.65	70.10	0.47
full K-SOM	9.06	5.22	86.53	53.79	0.34
random K-SOM	8.08	6.09	87.26	60.79	0.40

The results are summarized in Table 1. Surprisingly, the maps trained with a reduced number of samples (bagged K-SOM and random K-SOM) obtain better quality measures than the map trained with all samples. Using a bootstraping approach to select the relevant observations also significantly improves all quality measures as compared to a random choice with the same number of observations. The results obtain with the bagged SOM are displayed in Figures 2 and 3 (from, respectively, the map and the network points of view). They show that the nodes are mainly dispatched into three big clusters, which correspond each to approximately only one "list", as defined by the user. The results provided with the K-SOM using all the data tend to provide smaller communities and to scatter the biggest lists on the map. Using this approach, it is however hard to conclude if interpretability has been increased (i.e., if the selected observations used for training are representative of their cluster) as, in Figure 3, they do not seem to have a particularly high degree or centrality.

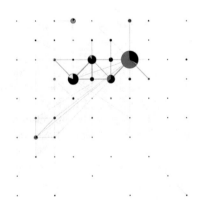

Fig. 2. Simplified representation of the facebook© network projected on the map resulting from bagged K-SOM. The circle sizes are proportional to the number of nodes classified in the cluster and the edge width are proportional to the number of edges between the nodes in the two clusters. Colors correspond to the proportion of user-defined lists (black is used for "no list").

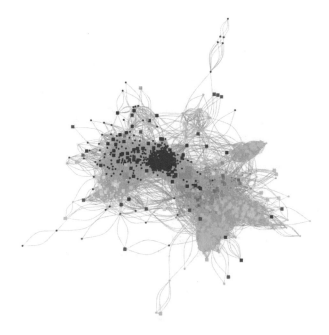

Fig. 3. The facebook© network represented with a force-directed placement algorithm [22]. Colors represent the clusters on the map and selected nodes used to train the map are represented by squares (instead of circles)

4   Conclusion

This paper presents a parallelizable bagged approach which results in a reduced computational time and a sparse representation of prototypes for kernel SOM. The simulations show good performances and only a small loss of accuracy which is compensated by a faster computational time. Obtained prototypes are also easier to interpret, as based on a smaller number of observations.

References

1. Mac Donald, D., Fyfe, C.: The kernel self organising map. In: Proceedings of 4th International Conference on Knowledge-Based Intelligence Engineering Systems and Applied Technologies, pp. 317–320 (2000)
2. Lau, K., Yin, H., Hubbard, S.: Kernel self-organising maps for classification. Neurocomputing 69, 2033–2040 (2006)
3. Boulet, R., Jouve, B., Rossi, F., Villa, N.: Batch kernel SOM and related laplacian methods for social network analysis. Neurocomputing 71(7-9), 1257–1273 (2008)
4. Hammer, B., Hasenfuss, A.: Topographic mapping of large dissimilarity data sets. Neural Computation 22(9), 2229–2284 (2010)

5. Olteanu, M., Villa-Vialaneix, N.: On-line relational and multiple relational SOM. Neurocomputing (2013) (forthcoming)
6. Olteanu, M., Villa-Vialaneix, N., Cierco-Ayrolles, C.: Multiple kernel self-organizing maps. In: Verleysen, M. (ed.) XXIst European Symposium on Artificial Neural Networks, Computational Intelligence and Machine Learning (ESANN), Bruges, Belgium, pp. 83–88. i6doc.com (2013)
7. Massoni, S., Olteanu, M., Villa-Vialaneix, N.: Which distance use when extracting typologies in sequence analysis? An application to school to work transitions. In: International Work Conference on Artificial Neural Networks (IWANN 2013), Puerto de la Cruz, Tenerife (2013)
8. Hofmann, D., Hammer, B.: Sparse approximations for kernel learning vector quantization. In: Verleysen, M. (ed.) Proceedings of XXIst European Symposium on Artificial Neural Networks, Computational Intelligence and Machine Learning (ESANN), Bruges, Belgium, pp. 549–554. i6doc.com (2013)
9. Aronszajn, N.: Theory of reproducing kernels. Transactions of the American Mathematical Society 68(3), 337–404 (1950)
10. Smola, A.J., Kondor, R.: Kernels and regularization on graphs. In: Schölkopf, B., Warmuth, M.K. (eds.) COLT/Kernel 2003. LNCS (LNAI), vol. 2777, pp. 144–158. Springer, Heidelberg (2003)
11. Petrakieva, L., Fyfe, C.: Bagging and bumping self organising maps. Computing and Information Systems Journal 9, 69–77 (2003)
12. Vrusias, B., Vomvoridis, L., Gillam, L.: Distributing SOM ensemble training using grid middleware. In: Proceedings of IEEE International Joint Conference on Neural Networks (IJCNN 2007), pp. 2712–2717 (2007)
13. Baruque, B., Corchado, E.: Fusion methods for unsupervised learning ensembles. SCI, vol. 322. Springer, Heidelberg (2011)
14. Villa, N., Rossi, F.: A comparison between dissimilarity SOM and kernel SOM for clustering the vertices of a graph. In: 6th International Workshop on Self-Organizing Maps (WSOM), Bielefield, Germany, Neuroinformatics Group, Bielefield University (2007)
15. Polzlbauer, G.: Survey and comparison of quality measures for self-organizing maps. In: Paralic, J., Polzlbauer, G., Rauber, A. (eds.) Proceedings of the Fifth Workshop on Data Analysis (WDA 2004), Sliezsky dom, Vysoke Tatry, Slovakia, pp. 67–82. Elfa Academic Press (2004)
16. McAuley, J., Leskovec, J.: Learning to discover social circles in ego networks. In: NIPS Workshop on Social Network and Social Media Analysis (2012)
17. Rossi, F., Villa-Vialaneix, N.: Optimizing an organized modularity measure for topographic graph clustering: a deterministic annealing approach. Neurocomputing 73(7-9), 1142–1163 (2010)
18. Fouss, F., Pirotte, A., Renders, J., Saerens, M.: Random-walk computation of similarities between nodes of a graph, with application to collaborative recommendation. IEEE Transactions on Knowledge and Data Engineering 19(3), 355–369 (2007)
19. von Luxburg, U.: A tutorial on spectral clustering. Statistics and Computing 17(4), 395–416 (2007)
20. Danon, L., Diaz-Guilera, A., Duch, J., Arenas, A.: Comparing community structure identification. Journal of Statistical Mechanics, P09008 (2005)
21. Newman, M., Girvan, M.: Finding and evaluating community structure in networks. Physical Review, E 69, 026113 (2004)
22. Fruchterman, T., Reingold, B.: Graph drawing by force-directed placement. Software, Practice and Experience 21, 1129–1164 (1991)

Probability Ridges and Distortion Flows: Visualizing Multivariate Time Series Using a Variational Bayesian Manifold Learning Method

Alessandra Tosi[1], Iván Olier[2], and Alfredo Vellido[1,*]

[1] Dept. de Llenguatges i Sistemes Informàtics - Universitat Politècnica de Catalunya
Barcelona, 08034, Spain
[2] Manchester Institute of Biotechnology - The University of Manchester,
M1 7DN, UK

Abstract. Time-dependent natural phenomena and artificial processes can often be quantitatively expressed as multivariate time series (MTS). As in any other process of knowledge extraction from data, the analyst can benefit from the exploration of the characteristics of MTS through data visualization. This visualization often becomes difficult to interpret when MTS are modelled using nonlinear techniques. Despite their flexibility, nonlinear models can be rendered useless if such interpretability is lacking. In this brief paper, we model MTS using Variational Bayesian Generative Topographic Mapping Through Time (VB-GTM-TT), a variational Bayesian variant of a constrained hidden Markov model of the manifold learning family defined for MTS visualization. We aim to increase its interpretability by taking advantage of two results of the probabilistic definition of the model: the explicit estimation of probabilities of transition between states described in the visualization space and the quantification of the nonlinear mapping distortion.

Keywords: Multivariate time series, Nonlinear dimensionality reduction, Mapping distortion, Magnification Factors, Visualization, Generative Topographic Mapping, Variational Bayesian methods.

1 Introduction

Most applied analysis of MTS involves, in one way or another, problems with specific targets such as prediction, forecasting, or anomaly detection. A less explored avenue of research is the exploratory analysis of MTS using machine learning and computational intelligence methods [1].

Data exploration may be a key stage in knowledge extraction from MTS using complex nonlinear methods, as it opens the door to their interpretability [2]. As in any other process of knowledge extraction from data, the analyst could benefit from the exploration of the characteristics of MTS consisting of a high number of individual series through their visualization [3]. The direct

---

* This research was partially funded by MINECO research project TIN2012-31377.

T. Villmann et al. (eds.), *Advances in Self-Organizing Maps and Learning Vector Quantization,* Advances in Intelligent Systems and Computing 295,
DOI: 10.1007/978-3-319-07695-9_5, © Springer International Publishing Switzerland 2014

visualization of such high-dimensional data, though, can easily be beyond the interpretation capabilities of human experts. Therefore, the exploration of MTS can be assisted by dimensionality reduction (DR) methods. In particular, the visualization of MTS using nonlinear DR (NLDR) methods [4] can provide the expert with inductive reasoning tools as a means to hypothesis generation [3,5]. Visualization can thus facilitate interpretation, which is paramount given that NLDR methods can be rendered useless in practice if interpretability is lacking.

In this brief paper, we merge two strands of previous research on data visualization. The first one involves the visualization of MTS using Statistical Machine Learning (SML) NLDR methods [6]. The second tackles one of the main interpretability bottlenecks of NLDR techniques: the difficulty of expressing the nonlinear mapping distortion they introduce in the data visualization space in an intuitive manner. Specifically, we attempt to increase the interpretability of the Variational Bayesian Generative Topographic Mapping Through Time (VB-GTM-TT), a variational Bayesian variant of a constrained hidden Markov model (HMM) [7] of the manifold learning family, defined for MTS visualisation [8]. For this, we use two results of the probabilistic definition of the model: the explicit estimation of probabilities of transition between states described in the visualization space and the quantification of the distortion introduced by the nonlinear mapping of the MTS in the form of Magnification Factors (MF).

Note that this paper does not address the assessment of the quality of the mapping as such. In fact, the proposed visualization strategies are meant to be independent from it. Although VB-GTM-TT is used here for illustration (as a method that, even if prone to limitations such as local minima, has been shown to perform robustly in the presence of noise), the proposed approach could be extended to alternative MTS DR models for which distortion and probability of state transition (or some approximations to them) were quantifiable.

2   Methods

2.1   Variational Bayesian GTM Through Time

The Generative Topographic Mapping (GTM: [9]) is a NLDR latent variable model of the manifold learning family. It can be seen as a mixture of distributions whose centres are constrained to lay on an intrinsically low-dimensional space. Given that the generative model specifies a mapping from latent space to observed data space, such latent space can be used for data visualization when its dimensionality is 1 or 2. Unless regularization is included, the GTM is prone to overfitting. Adaptive regularization for GTM was proposed in [10].

The GTM was redefined as a constrained HMM in [6]. The resulting GTM Through Time (GTM-TT) can be considered as a GTM model in which the latent states are linked by transition probabilities, in a similar fashion to HMM. This model, even if useful for MTS clustering and visualization, did not implement any regularization process.

Recently, the GTM was reformulated within a variational full Bayesian framework in [11], which was extended to the analysis of MTS in [8]. The result was the VB-GTM-TT: a model that integrates regularization explicitly and provides adaptive optimization of most of the model parameters involved. Assuming a sequence of N hidden states $\mathbf{Z} = \{z_1, z_2, \ldots, z_n, \ldots, z_N\}$ and the observed MTS $\mathbf{X} = \{\mathbf{x}_1, \mathbf{x}_2, \ldots, \mathbf{x}_n, \ldots, \mathbf{x}_N\}$, the complete-data likelihood for VB-GTM-TT is given by:

$$p(\mathbf{Z}, \mathbf{X}|\boldsymbol{\Theta}) = p(z_1) \prod_{n=2}^{N} p(z_n|z_{n-1}) \prod_{n=1}^{N} p(\mathbf{x}_n|z_n). \tag{1}$$

The model parameters are $\boldsymbol{\Theta} = (\pi, \mathbf{A}, \mathbf{Y}, \beta)$, where $\pi = \{\pi_j\} : \pi_j = p(z_1 = j)$ are the initial state probabilities; $\mathbf{A} = \{a_{ij}\} : a_{ij} = p(z_n = j|z_{n-1} = i)$ are the transition state probabilities; and

$$\{\mathbf{Y}, \beta\} : p(\mathbf{x}_n|z_n = j) = \left(\frac{\beta}{2\pi}\right)^{D/2} \exp\left(-\frac{\beta}{2}\|\mathbf{x}_n - \mathbf{y}_j\|^2\right)$$

are the emission probabilities, which are controlled by spherical Gaussian distributions with common inverse variance β and a matrix $\mathbf{Y}$ of K centroids $\mathbf{y}_j$, $1 \le j \le K$. They can be considered as hidden variables and integrated out to describe the marginal likelihood as:

$$p(\mathbf{Z}, \mathbf{X}) = \int p(\boldsymbol{\Theta}) p(\mathbf{Z}, \mathbf{X}|\boldsymbol{\Theta}) d\boldsymbol{\Theta},$$
$$\text{where } \boldsymbol{\Theta} = (\pi, \mathbf{A}, \mathbf{Y}, \beta). \tag{2}$$

VB-GTM-TT assumes its parameters to be independent, so that $p(\boldsymbol{\Theta}) = p(\pi)p(\mathbf{A})p(\mathbf{Y})p(\beta)$, where the set of prior distributions $p(\boldsymbol{\Theta})$ are defined as:

$$p(\pi) = \mathtt{Dir}(\{\pi_1, \ldots, \pi_K\}|\nu)$$
$$p(\mathbf{A}) = \prod_{j=1}^{K} \mathtt{Dir}(\{a_{j1}, \ldots, a_{jK}\}|\lambda)$$
$$p(\mathbf{Y}) = \left[(2\pi)^K |\mathbf{C}|\right]^{-D/2} \prod_{d=1}^{D} \exp\left(-\frac{1}{2}\mathbf{y}_{(d)}^T \mathbf{C}^{-1} \mathbf{y}_{(d)}\right)$$
$$p(\beta) = \Gamma(\beta|d_\beta, s_\beta).$$

Here, $\mathtt{Dir}(\cdot)$ represents the Dirichlet distribution and $\Gamma(\cdot)$ is the Gamma distribution. The vector ν, the matrix λ and the scalars d_β and s_β correspond to the hyperparameters of the model which are fixed $a\ priori$. The prior over the parameter $\mathbf{Y}$ defines the mapping from the hidden states to the data space as a Gaussian Process (GP), where $\mathbf{y}_{(d)}$ is each of the row vectors (centroids) of the matrix $\mathbf{Y}$ and $\mathbf{C}$ is a matrix where each element is defined by the covariance function as:

$$C_{i,j} = c\,(\mathbf{u}_i, \mathbf{u}_j) = \exp\left(-\frac{\|\mathbf{u}_i - \mathbf{u}_j\|^2}{2\alpha^2}\right), \quad i, j = 1 \ldots K. \tag{3}$$

The α parameter controls the flexibility of the mapping from the latent space to the data space. The vector $\mathbf{u}_j$, $j = 1 \ldots K$ corresponds to the state j in a latent space of usually lower dimension than that of the data space (for MTS visualization purposes). Thus, a topography over the states is defined by the GP as in the standard GTM. The VB-GTM-TT is optimized using variational approximation techniques. A more detailed description of the VB-GTM-TT and its formulation is provided in [8,12].

2.2   Distortion Measures and Local Metrics

NLDR techniques usually attempt to minimize the unavoidable distortion they introduce in the projection of the high-dimensional data from the observed space onto lower-dimensional spaces. For a more faithful interpretation of models, many distortion measures have been proposed and adapted to visualization techniques for different methods. While reducing dimensionality, different levels of local mapping distortion are generated, leading to a loss of information that we aim to recover, to some extent, to improve the interpretability of the model.

An interesting approach, proposed in [13] for GTM (and extended to Self Organizing Maps), is the calculation of the MFs. The concept of magnification has been applied to manifold learning methods in order to quantify the distortion due to the embedding of a manifold in a high-dimensional space. Importantly, the distortion caused by the mapping can be explicitly computed in a continuous way over the low-dimensional latent space of visual representation.

From the theory of differential geometry, we can describe the local geometry of a q-dimensional differential manifold through the mapping $\xi \mapsto \zeta^j(\xi)$ between the set of curvilinear coordinates system to the set of rectangular Cartesian coordinates systems defined in the high dimensional space. The Jacobian of this transformation can be written as:

$$J = \left(\frac{\partial \zeta^i}{\partial \xi^j}\right)_{i,j}. \tag{4}$$

Every point on the manifold has local geometrical properties which are given by its metric tensor $\mathbf{g}_{i,j}$, which is defined by:

$$\mathbf{g}_{i,j} = \delta_{i,j} \frac{\partial \zeta}{\partial \xi} \frac{\partial \zeta}{\partial \xi}, \tag{5}$$

where δ is the Kronecker delta. From (4) and (5), it follows that $\mid J \mid = \mid \mathbf{g} \mid^{\frac{1}{2}}$.

2.3 Magnification Factors for VB-GTM-TT

As stated in [13], the MF can be explicitly computed for the batch-SOM and GTM. In this paper, we provide the calculation of the MF for the VB-GTM-TT model. For this, we first consider the jointly Gaussian random variables:

$$\begin{bmatrix} \mathbf{y} \\ \mathbf{y}_* \end{bmatrix} \sim \mathcal{N} \left(\mathbf{0}, \begin{bmatrix} C & C_{(*,\cdot)} \\ C_{(\cdot,*)} & C_{(*,*)} \end{bmatrix} \right), \tag{6}$$

where y_* is a test point and $C_{(\cdot,\cdot)}$ is the covariance matrix defined according to (3). Due to the properties of Gaussian distributions, we can explicitly write the posterior probability as follows:

$$\mathbf{y}_* | \mathbf{u}_*, \mathbf{U}, \mathbf{Y}, \theta \sim \mathcal{N} \left(\mathbf{C}_{(*,\cdot)} \mathbf{C}^{-1} \mathbf{Y}, \mathbf{C}_{(*,*)} - \mathbf{C}_{(*,\cdot)} \mathbf{C}^{-1} \mathbf{C}_{(\cdot,*)} \right). \tag{7}$$

The Jacobian J of this mapping can be obtained computing the derivatives of $\langle (\mathbf{y}_* | \mathbf{u}_*, \mathbf{U}, \mathbf{Y}, \theta) \rangle$ with respect to u, using:

$$\frac{\partial c_{(*,j)}}{\partial \mathbf{u}_*^l} = \frac{1}{\alpha^2} (u_*^l - u_j^l) \exp \left(-\frac{\|\mathbf{u}_* - \mathbf{u}_j\|^2}{2\alpha^2} \right), \quad l = 1 \ldots q, j = 1 \ldots K, \tag{8}$$

being q the dimension of the latent space. As a result, the MF is calculated as:

$$\mu_* = det^{-\frac{1}{2}} \left(J J^T \right) \tag{9}$$

The MF does not only provide us with a quantification of the local mapping distortion that separates areas of the visual map which have undergone much compression or stretching from those which have not; it also tells us about data sparsity: the model distorts the most in areas which are mostly empty of data and the least in densely populated areas. For this reason, the MF has been used as an indicator of the existence of data clusters and the boundaries between those clusters [14]. For MTS, we would expect the time series to flow over time through areas of low MF mostly when the MTS evolve slowly, whereas fast transitions between MTS regimes might require crossing areas of higher distortion.

2.4 Cumulative State Transition Probabilities

Another metric that might help improving the interpretability of the mapping is the likelihood for a state to be transited by any of the potential trajectories through states. Again, this can explicitly be quantified, for each state j defined by VB-GTM-TT, as the estimated cumulative state transition probability (CSTP) defined as the sum of the probabilities of transition from all states to it:

$$\mathcal{CSTP}_j = \sum_{i=1}^{K} a_{ij}. \tag{10}$$

We would expect the MTS trajectory to happen through areas of high $CSTP$, because these should be areas of highly likely transition. As such, the $CSTP$

plays the opposite role to MF, because the areas of large manifold stretching (high MF) should mostly be areas that the MTS is unlikely to cross (low $CSTP$).

3   Experiments and Discussion

3.1   Materials and Experimental Setup

We illustrate the proposed MTS visualization using two different datasets. The first is an *artificial* 3-variate time series, with $1,000$ time points. The second set is the *Shuttle-data* from Space Shuttle mission STS-57[1]: a time series consisting of 1,000 points described by 6 features. This data set has previously been used for cluster detection in [15].

3.2   MTS Visualization

The considered MTS are particularly suitable for the illustration of the proposed visualization techniques due to the nature of their regimes and transitions periods. The *artificial* dataset (displayed in Fig.(1), top-row, left) is characterized by two intervals with regular regimes, divided by a sudden transition at point 700. The VB-GTM-TT model was trained over a 8×8, $2-D$ grid of hidden states and each of the MTS points was mapped by VB-GTM-TT to a particular state in the grid. The result of this mapping assignment is shown in Fig.(1) (top-row, right). Before point 700, the periodicity of the data is well-captured by the roughly circular structure of populated states. The sudden transition to a higher-amplitude periodic interval is also neatly visualized.

On the other hand, *Shuttle_Data* presents four periods of little variability A-C-D-E and one period of high (quasi-periodic) variability B, which are separated by sudden transitions, as evidenced by their display in Fig.(2) (top row, left). The VB-GTM-TT model was trained over a 13×13 grid of hidden states and each of the MTS points was mapped by VB-GTM-TT to a particular state in the grid, as shown in Fig.(2) (top row, right). There is a clear interpretation for this state membership mapping, as the *Shuttle-data* trajectory is confined to a limited number of its states (a common characteristic of VB-GTM-TT mappings, in which over-complexity is penalized). Only a few of them are relatively big: these are mostly stationary states with little MTS change in intervals C, D and E. The quasi-periodic interval B evolves slowly through a cloud of states on the top-left and center of the map.

The MFs were computed for *artificial* and *Shuttle-data* and represented in Figs.1 and 2 (bottom, right) through color maps over the grid of hidden states. For both datasets, it might seem at first sight that the MTS cross through areas of high MF (high distortion), a behaviour that would refute the hypothesis that the densely data populated areas correspond to low mapping distortion. In fact,

---

[1] Which can be requested from www.cs.ucr.edu/~eamonn.

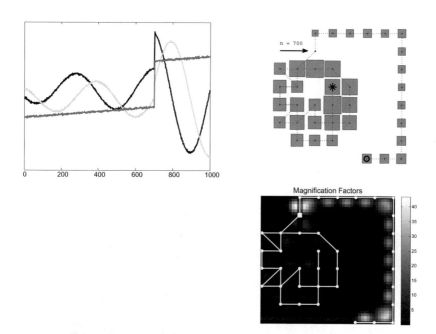

Fig. 1. Top row, left: Artificial dataset: a 3-variate time series, characterized by a sudden transition at $n = 700$. Top row, right: State-membership map of VB-GTM-TT, with a 8×8 grid of hidden states represented as squares, whose relative size is proportional to the time data points assigned to them; the starting point of the MTS is represented as a star and the ending point is represented as a circle. The sudden transition point is signaled by an arrow. Bottom row, right: MF gray-shade color map, represented in the VB-GTM-TT latent space visualization grid. The trajectory of the MTS over the map is displayed as a white solid line.

this is not the case: the MTS mostly flows over time through *channels* of low distortion surrounded by borders of high distortion. These borders seem to act as barriers that compel the MTS to follow a given trajectory. In fact, these barriers are only breached (with the MTS moving briskly towards higher MF) in sudden transitions between regimes. These can clearly be seen for *Shuttle-data* if we plot the value of MF over time, as in Fig.(2) (bottom row, left): MF narrow spikes of varying magnitude (particularly strong in the transition from B to C) appear in the transitions between time intervals. These spikes take values well over the mean MF of the map. This result suggests that the evolution of the MF over time could directly be used to detect sudden regime transitions in MTS.

The $CSTP$ maps in Fig.3 are very consistent with their MF counterparts, and complement them. Alternatively displayed as $3 - D$ maps over the grid of hidden states, they provide an intuitive illustration of the previously described

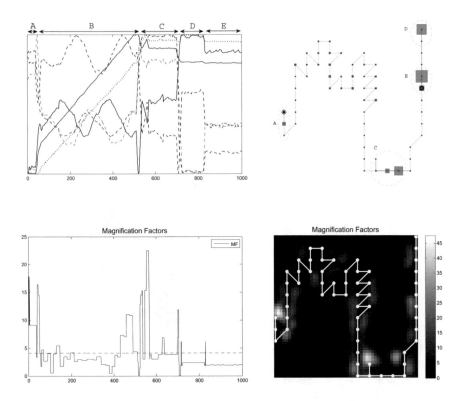

Fig. 2. Top left: Plot of *Shuttle-data*; the five intervals or regimes separated by sudden transitions are identified as A, B, C, D and E. Top right: State-membership map generated by VB-GTM-TT, with a 13×13 grid of hidden states represented as squares; the relative size of these squares is proportional to the time data points assigned to them; the starting point of the MTS is represented as a star, the ending point as a circle. Bottom Left: The Magnification Factors as a function of time, including the mean MF over all states (represented as a dashed line); narrow peaks of distortion are detected precisely in the areas of sudden transitions. Bottom Right: MF gray-shade color map, represented in the VB-GTM-TT latent space visualization grid; white areas correspond to high distortion.

behaviour. Following a geographical representation visual metaphor, the MTS can be seen to flow across cumulative state transition probability *ridges*, where rapid transitions between regimes see the MTS moving through relatively lower-valued *depressions* in those *ridges*. An opposite graphical metaphor could be used for the MF distortion, with the MTS flowing through its *valleys*, that is, across areas of the map characterized by low MF values.

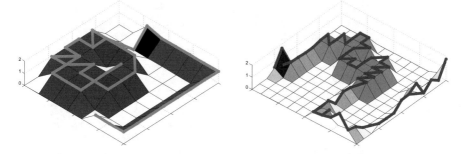

Fig. 3. A $3 - D$ representation for the CSTP plots. The values in the vertical axis correspond to the CSTP values over the latent space. Left: *artificial* data; right: *Shuttle-data*.

4   Conclusions

Data visualization can be of great assistance in knowledge extraction processes. High dimensionality is always a barrier for visualization. In the case of MTS, this is compounded by their i.i.d. nature, because the search for patterns over time is often relevant in their study. Dimensionality reduction can make visualization operative for high-dimensional MTS. The use of NLDR methods to this purpose poses a challenge of model interpretability due to the existence of locally-varying distortion.

In this study, we have proposed two methods to improve interpretability for VB-GTM-TT, a manifold learning NLDR method. The model mapping distortion has been explicitly quantified in the latent space continuum and the probabilistic nature of the method has allowed us to define a cumulative probability of state transition. The reported preliminary experiments have shown that both metrics can provide interesting insights that enhance the low-dimensional visualization of the MTS provided by the model.

This exploration approach is quite flexible and could be extended to other dimensionality reduction models for MTS analysis, provided their local distortion can be quantified. Examples of this may include Gaussian process latent variable models (GP-LVM, [16]), Gaussian process dynamical models (GPDM, [18], [17]) or temporal Laplacian eigenmaps ([19]). It could also be extended to alternative visual display methods, such as the recently proposed cartograms[2] [20],[21] and topographic maps [22]. Future research will also investigate the VB-GTM-TT generalization capabilities and its use for prediction.

References

1. Fu, T.C.: A Review on Time Series Data Mining. Engineering Applications of Artificial Intelligence 24(1), 164–181 (2011)
2. Vellido, A., Martín-Guerrero, J.D., Lisboa, P.J.G.: Making Machine Learning Models Interpretable. In: ESANN 2012, pp. 163–172. d-Side Pub. (2012)

---

[2] www.worldmapper.org/

3. Vellido, A., Martín, J.D., Rossi, F., Lisboa, P.J.G.: Seeing is Believing: The Importance of Visualization in Real-World Machine Learning Applications. In: ESANN 2011, pp. 219–226. d-Side Pub. (2011)
4. Lee, J.A., Verleysen, M.: Nonlinear Dimensionality Reduction. Springer (2007)
5. Van Belle, V.: Lisboa, P.: Research Directions in Interpretable Machine Learning Models. In: ESANN 2013, pp. 533–541. i6doc.com Pub. (2013)
6. Bishop, C.M., Hinton, G.E., Strachan, I.G.D.: GTM Through Time. In: Fifth International Conference on Artificial Neural Networks, pp. 111–116 (1997)
7. Rabiner, L.R.: A tutorial on hidden Markov models and selected applications in speech recognition. Proceedings of the IEEE 77(2), 257–286 (1989)
8. Olier, I., Vellido: A Variational Formulation for GTM Through Time. In: International Joint Conference on Neural Networks (IJCNN 2008), pp. 517-522 (2008)
9. Bishop, C.M., Svensén, M., Williams, C.K.I.: GTM: The Generative Topographic Mapping. Neural Computation 10, 215–234 (1998)
10. Bishop, C.M., Svensén, M., Williams, C.K.I.: Developments of the Generative Topographic Mapping. Neurocomputing 21(1), 203–224 (1998)
11. Olier, I., Vellido, A.: Variational Bayesian Generative Topographic Mapping. Journal of Mathematical Modelling and Algorithms 7(4), 371–387 (2008)
12. Olier, I., Amengual, J., Vellido, A.: A Variational Bayesian Approach for the Robust Estimation of Cortical Silent Periods from EMG Time Series of Brain Stroke Patients. Neurocomputing 74(9), 1301–1314 (2011)
13. Bishop, C.M., Svensén, M., Williams, C.K.I.: Magnification Factors for the SOM and GTM Algorithms. In: Proceedings of the 1997 Workshop on Self-Organizing Maps (WSOM), pp. 333–338 (1997)
14. Tosi, A., Vellido, A.: Robust Cartogram Visualization of Outliers in Manifold Learning. In: ESANN 2013, pp. 555–560. i6doc.com Pub. (2013)
15. Lin, J., Vlachos, M., Keogh, E., Gunopulos, D.: Iterative Incremental Clustering of Time Series. In: Bertino, E., Christodoulakis, S., Plexousakis, D., Christophides, V., Koubarakis, M., Böhm, K. (eds.) EDBT 2004. LNCS, vol. 2992, pp. 106–122. Springer, Heidelberg (2004)
16. Lawrence, N.: Probabilistic Non-Linear Principal Component Analysis with Gaussian Process Latent Variable Models. The Journal of Machine Learning Research 6, 1783–1816 (2005)
17. Damianou, A.C., Titsias, M.K., Lawrence, N.D.: Variational Gaussian Process Dynamical Systems. In: Advances in Neural Information Processing Systems, NIPS (2011)
18. Wang, J.M., Fleet, D.J., Hertzmann, A.: Gaussian Process Dynamical Models for Human Motion. IEEE Transactions on Pattern Analysis and Machine Intelligence 30(2), 283–298 (2008)
19. Lewandowski, M., Martínez-del-Rincón, J., Makris, D., Nebel, J.C.: Temporal Extension of Laplacian Eigenmaps for Unsupervised Dimensionality Reduction of Time Series. In: 20th International Conference on Pattern Recognition (ICPR), pp. 161–164. IEEE (2013)
20. Tosi, A., Vellido, A.: Cartogram Representation of the Batch-SOM Magnification Factor. In: ESANN 2012, pp. 203–208 (2012)
21. Vellido, A., García, D., Nebot, À.: Cartogram Visualization for Nonlinear Manifold Learning Models. Data Mining and Knowledge Discovery 27(1), 22–54 (2013)
22. Gianniotis, N.: Interpretable magnification factors for topographic maps of high dimensional and structured data. In: IEEE CIDM, pp. 238–245 (2013)

Short Review of Dimensionality Reduction Methods Based on Stochastic Neighbour Embedding

Diego H. Peluffo-Ordóñez[1,*], John A. Lee[1,2], and Michel Verleysen[1]

[1] Machine Learning Group - ICTEAM,
Université catholique de Louvain, Machine Learning Group - ICTEAM, Place du Levant 3,
B-1348 Louvain-la-Neuve, Belgium
[2] Molecular Imaging Radiotherapy and Oncology - IREC,
Université catholique de Louvain, Avenue Hippocrate 55, B-1200 Bruxelles, Belgium
{diego.peluffo,john.lee,michel.verleysen}@uclouvain.be

Abstract. Dimensionality reduction methods aimed at preserving the data topology have shown to be suitable for reaching high-quality embedded data. In particular, those based on divergences such as stochastic neighbour embedding (SNE). The big advantage of SNE and its variants is that the neighbor preservation is done by optimizing the similarities in both high- and low-dimensional space. This work presents a brief review of SNE-based methods. Also, a comparative analysis of the considered methods is provided, which is done on important aspects such as algorithm implementation, relationship between methods, and performance. The aim of this paper is to investigate recent alternatives to SNE as well as to provide substantial results and discussion to compare them.

Keywords: Dimensionality reduction, divergences, similarity, stochastic neighbor embedding.

1 Introduction

For pattern recognition and data mining tasks involving high dimensional data sets, dimensionality reduction (DR) is a key tool. The aim of DR approaches is to extract lower dimensional, relevant information from high-dimensional input data, so that the performance of a pattern recognition system might be improved. As well, the data visualization will become more intelligible. Among the classical DR approaches, we may mention principal component analysis (PCA) and classical multidimensional scaling (CMDS), which are respectively based on variance and distance preservation criteria [1]. Nowadays, the focus of DR approaches relies on more developed criteria, which are aimed at preserving the data topology. In particular, the data topology is involved within the formulation through pairwise similarities between data points. Therefore, these approaches can be readily understood from a graph-theory point of view such that the data are represented by a non-directed and weighted graph, in which data points represent the nodes, and a non-negative similarity (also affinity) matrix holds the

---

* J.A. Lee is a Research Associate with the FRS-FNRS (Belgian National Scientific Research Fund). This work is funded by FRS-FNRS (Belgian National Scientific Research Fund) project 7.0175.13 DRedVis.

pairwise edge weights. The pioneer methods incorporating similarities are Laplacian eigenmaps [2] and locally linear embedding [3], which are spectral approaches. More recently, given the fact that the rows of the normalized similarity matrix can be seen as probability distributions, methods based on divergences have emerged. Due to the probabilistic connotation, the most representative method is so named stochastic neighbour embedding (SNE) [4]. SNE and its variants have shown to be suitable for getting high-quality embedding data, since they preserve similarities in both low- and high-dimensional space during the optimization process. As alternatives to SNE, enhanced versions have been proposed. In [5, 6], a mixture of divergences is proposed. Additionally, an improved gradient to speed up the procedure is also introduced in [6]. Another approach, which consists of simplifying the SNE's formulation, is introduced in [7]. Such simpler version is founded on the same principle as elastic network [8] and it is solved by an approximate gradient following the direction of an underlying eigenvalue problem [9].

In this work, we present a short review of recent alternatives to SNE. A comparative analysis is done regarding some key aspects, namely: algorithm implementation, performance, and links between methods. For comparison purposes, we also evaluate a classic technique (CMDS), as well as a spectral approach (Laplacian eigenmaps – LE). Experiments are carried out over third conventional databases: an artificial spherical shell, the COIL-20 image bank [10], and a subset of the MNIST image bank [11]. To quantify the performance of studied methods, an improved version of the average agreement rate is used, as described in [6]. Experimentally, we show the relationship between the divergence-based methods with the similarity preservation. The grounds and reasonings provided here may encourage new researches on any of the issues presented in this work, as well as the conclusions and discussions may facilitate users to select a method according to the compromise between complexity and performance.

The outline of this paper is as follows: Section 2 explains the studied methods and discusses in detail algorithm implementation issues and the links between methods. Experimental results and discussion are shown in Section 3. Finally, Section 4 draws the final remarks and conclusions.

2   Alternatives to Stochastic Neighbor Embedding

The DR problem is to embeded a high dimensional data matrix $Y = [y_i]_{1 \le i \le N}$ into a low-dimensional, latent data matrix $X = [x_i]_{1 \le i \le N}$, such that the relevant information is preserved. Denote $y_i \in \mathbb{R}^D$ and $x_i \in \mathbb{R}^d$ $(d < D)$ as the i-th data point from the high- and low-dimensional space. To cope with this problem, stochastic neighbor embedding (SNE) [4] minimizes the information divergence D between two distributions $P_n = [p_{nm}]_{1 \le m \le N}$ and $Q_n = [q_{nm}]_{1 \le m \le N}$ associated with the n-th point from observed and latent data, respectively. Then, using the Kullback-Leibler directed divergence D_{KL}, the SNE objective function is in the form:

$$E_{SNE}(X) = \sum_{n=1}^{N} D_{KL}(P_n \| Q_n) = \sum_{n,m=1}^{N} p_{nm} \log \frac{p_{nm}}{q_{nm}}. \tag{1}$$

Defining $\delta_{nm} = \|y_n - y_m\|^2$ and $d_{nm} = \|x_n - x_m\|^2$, distributions P_n and Q_n can be chosen as generalized, normalized nonsymmetric affinities in the form

$$p_{nm} = \frac{\exp\left(-\frac{1}{2}\delta_{nm}^2/\sigma_n^2\right)}{\sum\limits_{n \neq m'} \exp\left(-\frac{1}{2}\delta_{nm'}^2/\sigma_n^2\right)}, \quad \text{and} \quad q_{nm} = \frac{\exp\left(-\frac{1}{2}d_{nm}^2/\pi_n^2\right)}{\sum\limits_{n \neq m'} \exp\left(-\frac{1}{2}d_{nm'}^2/\pi_n^2\right)}, \tag{2}$$

with $q_{nn} = 0$ and $p_{nn} = 0$.

Symmetric SNE: A symmetric version of SNE (SSNE) can be achieved by selecting full normalized affinities which can readily be obtained by slightly expressions in (2). In this case, rather than a restricted sum, all entries must be summed on the denominator in order to enforce that all normalized entries sum to 1. This can be done by guaranteeing that $1_N^\top Q 1_N = 1_N^\top P 1_N = 1$.

t-SNE: SNE-based methods suffer from reaching distorted and overlapped latent space, when d is smaller than the intrinsic dimension [7]. To cope with this issue, another variant raised, which is named *t*-SNE and consists of selecting the Q_n as a *t*-distributed sequence [5].

Jensen-Shanon embedding: In [12], it is proposed a mixture by adding a regularization parameter β to balance *precision* and *recall* so: $(1 - \beta)\,D_{KL}(P_n\|Q_n) + \beta\,D_{KL}(Q_n\|P_n)$. Similarly, in [6], a novel approach is introduced which mixes the divergences as $D_{KL}^\beta = (1-\beta)\,D_{KL}(P_n\|S_n) + \beta\,D_{KL}(Q_n\|S_n)$, where S_n is a distribution following the same mixture rule so that $S_n = (1 - \beta)P_n + \beta Q_n$. This divergence is used in the so-called Jensen-Shannon embedding (JSE), which aims then to minimize $E_{JSE} = \sum_{n=1}^N D_{KL}^\beta(Q_n\|S_n)$ [6].

Elastic embedding Another alternative to SNE is introduced in [7], which is called elastic embedding (EE). EE is aimed to optimize:

$$E_{EE}(X|\lambda) = \sum_{n,m=1}^N w_{nm}^+ d_{nm}^2 + \lambda \sum_{n,m=1}^N w_{nm}^- \exp(d_{nm}^2) = E_{EE}^+(X) + \lambda E_{EE}^-(X). \tag{3}$$

Briefly put, this method attempts to involve the two objectives that SNE fulfills but in a simpler way. To this end, which is the key of this method, two graphs are used. Then, we have two kind of weighting coefficients w_{nm}^+ and w_{nm}^- being the entries of attractive W^+ and repulsive W^- affinity matrices, respectively. Both of them are positive semi-definite matrices. For simplicity, full graphs affinities are considered: $w_{nm}^- = \|y_n - y_m\|^2$ and $w_{nm}^+ = \exp(-\frac{1}{2}\delta_n^2/\sigma^2)$. From Eq. (3), the gradient of E_{EE} can be written as:

$$G(X|\lambda) = 4X(L^+ - \lambda \widetilde{L}^-) = 4XL, \tag{4}$$

where $\widetilde{w}_{nm}^- = w_{nm}^- \exp(-d_{nm}^2)$, $w_{nm} = w_{nm}^+ - \lambda \widetilde{w}_{nm}^-$, and their corresponding Laplacians $\widetilde{L} = \widetilde{D} - \widetilde{W}$ and $L = D - W$. Likewise, L^+ is the non-normalized Laplacian and thus $L^+ = D^+ - W^+$. In [7], to carry out the search for the suboptimal embedded solution X, a gradient descent algorithm is used, which is powered via the spectral direction (SD) proposed in [9].

Following are discussed in detail some implementation issues in Section 2.1 as well as the links between methods in Section 2.2.

2.1 Implementation and Algorithms

In this section, we discuss about two recent implementations, here called: spectral direction and full gradient.

Implementation via spectral direction: Methods such as EE, SNE and SSNE can be implemented in a fast fashion via a SD-based gradient descent search [7]. We denote the n-th embedded data point at iteration r as $\boldsymbol{x}_n[r] = \boldsymbol{x}_n[r-1] + \alpha[r]\boldsymbol{\varrho}_n[r]$. SD is aimed at determining the optimal direction $\boldsymbol{\varrho}_n[r]$ by incorporating a partial-Hessian strategy within the gradient descent heuristic [9]. Then, by design, Hessian is heavily exploited which is advantageous for subsequent developments since it can be be computed fast and has the suitable property to be positive semi-definite. As an intuitive condition, sought direction must hold that $\boldsymbol{B}[r]\boldsymbol{\varrho}_n[r] = -\boldsymbol{g}_n$, being $\boldsymbol{g}_n$ the column n of $\boldsymbol{G}(\boldsymbol{X}|\lambda)$ and $\boldsymbol{B}[r]$ any positive semi-definite matrix. SD consists of calculating the gradient of $E_{\mathrm{EE}}(\boldsymbol{X}|\lambda)$ following the direction of an underlying convex function which arises when $\lambda = 0$. Such a function is in fact the attractive part $E_{EE}^+(\boldsymbol{X}) = E_{\mathrm{EE}}(\boldsymbol{X}|0)$, whose Hessian is $\nabla^2 E_{EE}^+(\boldsymbol{X}) = 4\boldsymbol{L}^+$ being evidently positive semi-definite. As a matter of fact, possible alternatives for selecting $\boldsymbol{B}[r]$ span from null perplexity to $k = N$ (full graph) which match respectively with degree $\boldsymbol{D}^+$ and Laplacian $\boldsymbol{L}^+$ [9].

Moreover, the calculation of step $\alpha[r]$ is powered by a backtracking line search [13] following the updating rule $\alpha_l[r] = \rho\alpha_{l-1}[r]$ for a user-provided constant ρ. Gathering the spectral directions in matrix $\mathcal{P} \in \mathbb{R}^{d \times N}$, per each iteration, output embedded data can be calculated as $\boldsymbol{X}^* = \boldsymbol{X} + \alpha_l[r]\mathcal{P}$ under the convergence criterion given by $E_{\mathrm{EE}}(\boldsymbol{X} + \alpha_l[r]\mathcal{P})|\lambda) > E_{\mathrm{EE}}(\boldsymbol{X}|\lambda) + c\alpha_l[r] \operatorname{tr}(\mathcal{P}\boldsymbol{G}(\boldsymbol{X}|\lambda))$, where c is a small positive value. Steps for performing EE with backtracking line search are summarized in Algorithm 1. Within this framework, SNE and its variants can be alternatively implemented. To do so, the cost function of the method to be run should take place in $E(\boldsymbol{X})$. The gradient is the same for SNE-like methods, since the suboptimal solution is sough via a spectral direction.

Also, the calculation of SD is speeded up by using Cholesky decomposition. Namely, rather than calculating matrix directly with $\mathcal{P} = -\boldsymbol{G}(\boldsymbol{X}|\lambda)(\boldsymbol{B})^{-1}$ (which is $O(N^3 D)$ when using conventional Gaussian-Jordan elimination), two solve triangular systems in the form $\boldsymbol{R}^\top \boldsymbol{R} \operatorname{vec}(\boldsymbol{P}) = -\operatorname{vec}(\boldsymbol{G})$ are solved, where $\boldsymbol{R}$ is the upper triangular matrix resulting from the Cholesky decomposition of $\boldsymbol{B} \otimes \boldsymbol{I}_d$. Latter calculation can be done in $O(N^2 d)$ with standard linear algebra routines. In addition, computation of $\boldsymbol{R}$ needs to be done only once at first iteration and its complexity is $O(\frac{1}{3}N)$.

Implementation via a full gradient and Hessian: In [6], the search is done by using a full gradient calculated over the whole cost function (no approximations are done). In this case, the search is done via $\boldsymbol{x}_n[r] = \boldsymbol{x}_n[r-1] + \mu_n[r]\nabla E$, where $\mu_n[r]$ is an adaptive step size dependent on the Hessian. Given the nature of divergences, doing so can increase the complexity. Even more when using a mixture of divergences ($E = E_{\mathrm{JSE}}$), calculation of gradient and Hessian may be more expensive. Nonetheless, the advantage of this implementation is that scaling is considered in both high and low dimensional space. This provides a more modulated gradient and then a better tracking of the local structure of data during the optimization process.

Algorithm 1. SNE via SD

Input: Affinity matrices $\boldsymbol{W}^+$ and $\boldsymbol{W}^-$, N_{iter}, ϵ, λ, $\boldsymbol{X}$, $r = 1$
Compute the graph Laplacian $\boldsymbol{L}^+$
Compute the objective function $E(\boldsymbol{X})$ (3)
Set $\delta(\boldsymbol{X}^*, \boldsymbol{X}) \geq \epsilon$
while $\delta(\boldsymbol{X}^*, \boldsymbol{X}) \geq \epsilon$ **do**
    Calculate the gradient $\boldsymbol{G}(\boldsymbol{X}|\lambda)$ using Eq. (4)
    Calculate spectral direction matrix: $\mathcal{P} = -\boldsymbol{G}(\boldsymbol{X}|\lambda)(\boldsymbol{L}^+)^{-1}$

    Backtracking line search for estimating $\boldsymbol{X}^*$
    Set c, ρ, and α_0
    Initialize $l = 1$
    while $E(\boldsymbol{X} + \alpha\mathcal{P}|\lambda) > E(\boldsymbol{X}|\lambda) + c\alpha_l\,\text{tr}(\mathcal{P}\boldsymbol{G}(\boldsymbol{X}|\lambda))$ **do**
        $\alpha_l = \rho\alpha_{l-1}$
        Calculate $E(\boldsymbol{X} + \alpha_l\mathcal{P}|\lambda)$
        Increase l by 1
    end while

    Estimate $\boldsymbol{X}^*$ as: $\boldsymbol{X}^* = \boldsymbol{X} + \alpha_l\mathcal{P}$
    $\delta(\boldsymbol{X}^*, \boldsymbol{X}) = \|\boldsymbol{X}^* - \boldsymbol{X}\|_F$
    Update $\boldsymbol{X} = \boldsymbol{X}^*$
end while
Output: Embedded data $\boldsymbol{X}$

2.2 Links between Methods

Relation between SNE and EE: Eliminating independent terms from $\boldsymbol{X}$, Equation (1) can be expanded as

$$E_{\text{SNE}}(\boldsymbol{X}) = \sum_{n,m=1}^{N} p_{nm}\|\boldsymbol{x}_n - \boldsymbol{x}_m\|^2 + \sum_{n=1}^{N} \log \sum_{n \neq m} \exp(\|\boldsymbol{x}_n - \boldsymbol{x}_m\|^2). \tag{5}$$

Hence we can appreciate that by omitting the log operator and adding a homotopy parameter λ, E_{SNE} becomes the EE's cost function. Furthermore, EE is a variant of the elastic network applied to solve the traveling salesman problem as explained in [8].

Relation between SNE and LE: Laplacian Eigenmaps (LE) introduced in [2] is a popular approach for DR. This approach is spectral and is aimed at minimizing local distances. The LE's cost function can be written as $\sum_{n,m=1}^{N} w_{nm}\|\boldsymbol{x}_n - \boldsymbol{x}_m\|$, where $\boldsymbol{W} = [w_{nm}]_{1 \leq n \leq N}$ is the similarity matrix and $\|\cdot\|$ stands for Euclidean distance. Alternatively, we can express LE's formulation as

$$E_{\text{LE}}(\boldsymbol{X}) = \text{tr}(\boldsymbol{X}\boldsymbol{L}\boldsymbol{X}^\top) \text{ s. t. } \boldsymbol{X}\boldsymbol{D}\boldsymbol{X}^\top = \boldsymbol{I}_d, \quad \boldsymbol{X}\boldsymbol{D}\boldsymbol{1}_N = \boldsymbol{0}_d, \tag{6}$$

where $\boldsymbol{D} = \text{Diag}(\boldsymbol{W}\boldsymbol{1}_N)$ is the degree matrix and $\boldsymbol{L}$ is the graph Laplacian matrix given by $\boldsymbol{L} = \boldsymbol{D} - \boldsymbol{W}$. LE's constraints facilitates the solution leading to a generalized eigenvalue problem. Along this line, the embedded data is then the d smallest vector eigenvectors of normalized Laplacian $D^{-1/2}LD^{-1/2}$. This formulation is also useful to determine underline data clusters within input data [14]. Recalling Equation (5), it is noticeable that, doing as in diffusion maps [15] which means using the normalized affinities so that $p_{nm} = w_{nm}$, the right hand side of the Equation is the same as the LE objective function.

Relation between EE and LE: This relationship is quite similar to that when comparing SNE with EE. However, it is worth mentioning that by setting $\lambda = 0$, EE does not reach the same embedding as LE, since the optimization is different. EE's embedding is determined through a search and that of LE comes from a spectral decomposition under orthonormality assumptions.

3   Experiments and Results

Following the experiments to compare the DR methods are described. First, the considered data sets and the methods to be compared are mentioned. Also, the parameter settings to carry out the DR procedure as well as the performance measure are described. Finally, obtained results and discussion are drawn.

Data sets and methods: Experiments are carried out over three conventional data sets. The first data set is an artificial spherical shell ($N = 1500$ data points and $D = 3$). The second data set is the COIL-20 image bank [10], which contains 72 gray-level images representing 20 different objects ($N = 1440$ data points –20 objects in 72 poses/angles– with $D = 128^2$). The third data set is a randomly selected subset of the MNIST image bank [11], which is formed by 6000 gray-level images of each of the 10 digits ($N = 1500$ data points –150 instances for all 10 digits– and $D = 24^2$). Figure 1 depicts examples of the considered data sets.

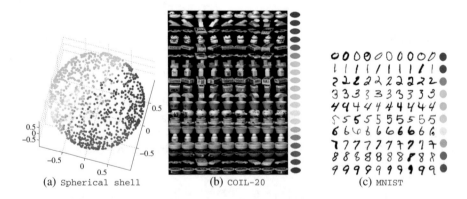

(a) Spherical shell   (b) COIL-20   (c) MNIST

Fig. 1. The three considered data sets. To carry out the DR procedure, images from COIL-20 and MNIST data sets are vectorized.

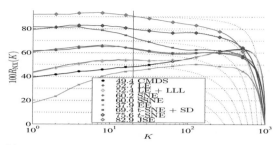

(a) $R_{NX}(K)$ for all considered methods. The value of AUC is shown in the legend besides the method's name.

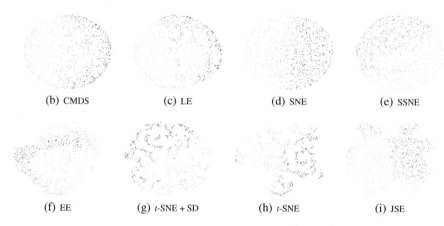

| (b) CMDS | (c) LE | (d) SNE | (e) SSNE |

| (f) EE | (g) t-SNE + SD | (h) t-SNE | (i) JSE |

Fig. 2. Results for Spherical shell. Results are shown regarding the quality measure $R_{NX}(K)$. The curves and their AUC (a) for all considered methods are depicted, as well as the embedding data (b)-(j).

Methods to be compared: We consider the SNE-like methods, namely: classical SNE, SSNE, t-SNE, EE, t-SNE via spectral direction (t-SNE + SD), and JSE. Also, we evaluate a representative classical technique, which is a CMDS; and a spectral technique being LE.

Performance measure and parameter settings: To quantify the performance of studied methods, the scaled version of the average agreement rate $R_{NX}(K)$ introduced in [6] is used, which is ranged within the interval $[0, 1]$. Since $R_{NX}(K)$ is calculated at each perplexity value from 2 to $N - 1$, a numerical indicator of the overall performance can be obtained by calculating its area under the curve (AUC). EE, t-SNE+SD and SSNE are implemented via a spectral direction procedure. Meanwhile, SNE, t-SNE and JSE are implemented via a full gradient scheme. Both SD and full gradient implementations involve a backtracking line search.

To form the similarity matrices, given a perplexity parameter K, the relative bandwidth parameter σ_n is estimated regarding its distribution P_n so that the entropy over neighbors of such distribution is approximately $\log K$. This is done by a binary search

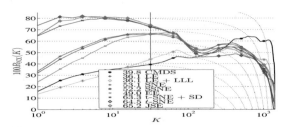

(a) $R_{NX}(K)$ for all considered methods. The value of AUC is shown in the legend besides the method's name.

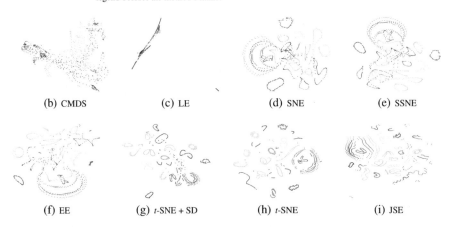

(b) CMDS   (c) LE   (d) SNE   (e) SSNE

(f) EE   (g) t-SNE + SD   (h) t-SNE   (i) JSE

Fig. 3. Results and obtained embedding data for `COIL-20`

as explained in [7]. The homotopy parameter for EE is set $\lambda = 100$. Regularization parameter β for JSE is set to be $1/2$. For all methods, input data is embedded into a 2-dimensional space, then $d = 2$. The number of neighbors is established as $K = 30$. The rest of free parameter are $\epsilon = 10^{-3}$, $c = 0.1$, $\rho = 0.8$, and $\alpha_0 = 1$.

Results and discussion: Overall results for `Sphere`, `COIL` and `MNIST` regarding AUC $R_{NX}(K)$ are respectively shown in Figures 2, 3 and 4. As well, the resultant embedded spaces reached by each method are depicted.

    For all considered databases, SNE-like methods perform a better embedding preserving smaller neighbours (local structure) in comparison the other methods. We can notice that SNE, SSNE and EE have a similar performance. In this case, SD makes that SNE and EE behave as a symmetrized version due to the strong assumption done over the gradient calculation. On the contrary, t-SNE + SD performs a better embedding since t-distributed probabilities may improve the separation of underline clusters despite that the gradient is biased to be that of the related, quadratic and symmetric form. Indeed, t-SNE + SD accomplishes a similar $R_{NX}(K)$ shape and AUC in comparison with t-SNE. JSE outperforms the remaining considered methods due to both the divergence type, and the identical similarity definition in the high-dimensional and low-dimensional space.

    As another important observation from this work, we notice that the spectral methods (LE and CMDS), in general, attempt to preserve the global structure (larger neighbors).

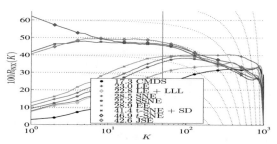

(a) $R_{NX}(K)$ for all considered methods. The value of AUC is shown in the legend besides the method's name.

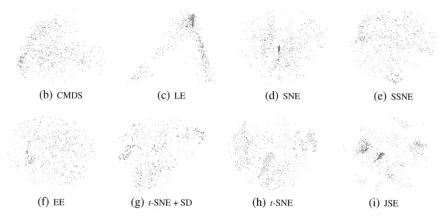

(b) CMDS   (c) LE   (d) SNE   (e) SSNE

(f) EE   (g) t-SNE + SD   (h) t-SNE   (i) JSE

Fig. 4. Results and obtained embedding data for MNIST

Particularly, CMDS exhibiting a pronounced peak on large neighbors. Then, we can claim that SNE based methods are better at preserving local structure, meanwhile those based on spectral analysis preserve the global structure.

4   Conclusions

This work reviews recent dimensionality reduction methods based on divergences. In particular, stochastic neighbor embedding and its improved variants. We provide a short comparative analysis involving key aspects such as relations between methods, algorithm implementation, and performance. Empirically, we demonstrate that methods using normalized similarities as probabilities and optimizing divergences reach better embedding by preserving the local structure of data. This is the case of SNE and its variants, in which the similarities are optimized in both high- and -low dimensional spaces. Meanwhile, spectral methods like multidimensional scaling and Laplacian eigenmaps are better at preserving global structure.

Discussion and results given here may facilitate users to choose a method seeking a good trade-off between performance and complexity.

References

1. Borg, I.: Modern multidimensional scaling: Theory and applications. Springer (2005)
2. Belkin, M., Niyogi, P.: Laplacian eigenmaps for dimensionality reduction and data representation. Neural Computation 15(6), 1373–1396 (2003)
3. Roweis, S.T., Saul, L.K.: Nonlinear dimensionality reduction by locally linear embedding. Science 290(5500), 2323–2326 (2000)
4. Hinton, G.E., Roweis, S.T.: Stochastic neighbor embedding. In: Advances in Neural Information Processing Systems, pp. 833–840 (2002)
5. Van der Maaten, L., Hinton, G.: Visualizing data using t-sne. Journal of Machine Learning Research 9(2579-2605), 85 (2008)
6. Lee, J.A., Renard, E., Bernard, G., Dupont, P., Verleysen, M.: Type 1 and 2 mixtures of kullback-leibler divergences as cost functions in dimensionality reduction based on similarity preservation. Neurocomputing (2013)
7. Carreira-Perpinán, M.A.: The elastic embedding algorithm for dimensionality reduction. In: ICML, vol. 10, pp. 167–174 (2010)
8. Durbin, R., Szeliski, R., Yuille, A.: An analysis of the elastic net approach to the traveling salesman problem. Neural Computation 1(3), 348–358 (1989)
9. Vladymyrov, M., Carreira-Perpiñán, M.Á.: Partial-hessian strategies for fast learning of nonlinear embeddings. CoRR, abs/1206.4646 (2012)
10. Nene, S.A., Nayar, S.K., Murase, H.: Columbia object image library (coil-20). Dept. Comput. Sci., Columbia Univ., New York, 62 (1996),
http://www.cs.columbia.edu/CAVE/coil-20.html
11. LeCun, Y., Bottou, L., Bengio, Y., Haffner, P.: Gradient-based learning applied to document recognition. Proceedings of the IEEE 86(11), 2278–2324 (1998)
12. Venna, J., Peltonen, J., Nybo, K., Aidos, H., Kaski, S.: Information retrieval perspective to nonlinear dimensionality reduction for data visualization. The Journal of Machine Learning Research 11, 451–490 (2010)
13. Nocedal, J., Wright, S.: Numerical optimization. Series in operations research and financial engineering. Springer, New York (2006)
14. Yu, S.X., Shi, J.: Multiclass spectral clustering. In: Proceedings of the Ninth IEEE International Conference on Computer Vision, pp. 313–319. IEEE (2003)
15. Singer, A., Wu, H.-T.: Vector diffusion maps and the connection Laplacian. Communications on Pure and Applied Mathematics 65(8), 1067–1144 (2012)

Part II
Prototype Based Classification

Attention Based Classification Learning
in GLVQ and Asymmetric Misclassification Assessment

Marika Kaden[1,*], W. Hermann[2], and Thomas Villmann[1]

[1] University of Applied Sciences Mittweida - Dept. of Mathematics
Mittweida, Germany
[2] Department of Neurology, Paracelsus Hospital Zwickau,
Zwickau, Germany

Abstract. The general aim in classification learning by supervised training is to achieve a high classification performance, frequently judged in terms of classification accuracy. A powerful method is the generalized learning vector quantizer, which realizes a gradient based optimization scheme based on a cost function approximating the usual symmetric misclassification rate. In this paper we investigate a modification of this approach taking into account asymmetric misclassification penalties to reflect structural knowledge of external experts about the data, as it is frequently the case for instance in medicine. Further we also discuss the weighting of importance for the considered classes in the classification problem. We show that both aspects can be seen as a kind of attention based learning strategy.

1 Introduction

The standard classification learning task consists of decision learning based on a labeled data set, the training data are provided together with the class assignments for predefined classes. If the data are available as data vectors, one powerful strategy for those tasks is prototype based classification learning. Several approaches belong to this kind of classifier models including k-nearest neighbor (k-NN), support vector machines (SVMs) or learning vector quantization (LVQ). While in k-NN and LVQ the learning scheme is heuristically motivated, SVM learning is based on a convex optimization task with well defined convex cost function to be minimized. A modification of LVQ leads to the *generalized* LVQ (GLVQ), which also optimizes an error function based on a smooth approximation of the classification accuracy during classification learning. The GLVQ model widely keeps the intuitive but heuristically motivated learning scheme and interpretability of the resulting classification model while applying stochastic gradient descent learning as optimization method. It turns out that GLVQ or variant thereof are a powerful prototype based classifiers with performance comparable to other advanced classifiers like SVMs.

Cost function based classifiers have the advantage of a precisely defined objective compared to heuristic classification learning schemes. However, although the training data maybe properly labeled, the information might be not sufficient to learn an accurate

* Supported by the European Social Fund (ESF), Saxony.

T. Villmann et al. (eds.), *Advances in Self-Organizing Maps and Learning
Vector Quantization*, Advances in Intelligent Systems and Computing 295,
DOI: 10.1007/978-3-319-07695-9_7, © Springer International Publishing Switzerland 2014

classification. Thus, additional information is required to improve classification. Otherwise, auxiliary information could provide knowledge about importance about correct class separation. For example, in medicine it is frequently more important to distinguish between healthy and infected persons than precise differentiation of all subtypes of a considered illness, i.e. it might be more important to differentiate between main classes than to detect precisely all sub-classes.

In this paper we will deal with those problems involving auxiliary information or additional requirements specified in advance. We show how these informations/requirements can be fed into the GLVQ classifier model by means of an attention based learning scheme. Attention based learning weights the influence of considered data vectors according to their available auxiliary knowledge. Finally, this idea leads to a modification of the GLVQ cost function. In this paper, we particularly focus on weighting of class importances and asymmetric error assessment.

The outline of the paper is as follows: First, we briefly describe standard GLVQ. Then we give suggestions how to integrate auxiliary knowledge into this GLVQ model. After presenting the framework we give beside an illustrative example, real-life medical application, which are followed by concluding remarks.

2   Generalized Learning Vector Quantization and Its Modifications

We assume that the data points $\mathbf{v} \in V \subset \mathbb{R}^n$ with their labels $c(\mathbf{v}) \in \mathcal{C} = \{1, 2, \ldots, C\}$ are given. The cardinality of V is N_V. The GLVQ is a prototype based classifier, i.e. we suppose the set W of prototypes $\mathbf{w} \in W \subset \mathbb{R}^n$ with their labels $y(\mathbf{w}) \in \mathcal{C}$ forming the set $Y \subseteq \mathcal{C}$. The objective to be minimized by the GLVQ method

$$E_{GLVQ} = \sum_{\mathbf{v} \in V} f_\Theta(\mu_W(\mathbf{v}))$$ (1)

as the cost function with the *squashing function* f_Θ and the *classifier* function

$$\mu_W(\mathbf{v}) = \frac{d^+(\mathbf{v}) - d^-(\mathbf{v})}{d^+(\mathbf{v}) + d^-(\mathbf{v})},$$ (2)

where $d^+(\mathbf{v}) = d(\mathbf{v}, \mathbf{w}^+)$ is the distance or dissimilarity between the best matching prototype $\mathbf{w}^+$ for the data point $\mathbf{v}$ of the same class and the data point itself [18]. Otherwise, $d^-(\mathbf{v}) = d(\mathbf{v}, \mathbf{w}^-)$ is the distance between $\mathbf{v}$ and the closest, wrong labeled prototype $\mathbf{w}^-$, i.e. $c(\mathbf{v}) \neq y(\mathbf{w}^-)$. Obviously, the classifier function $\mu_W(\mathbf{v})$ from (2) is negative, iff the data point is correct classified, i.e. it is valid $d^+(\mathbf{v}) \leq d^-(\mathbf{v})$. Because of the normalization term $d^+(\mathbf{v}) + d^-(\mathbf{v})$ the range of $\mu_W(\mathbf{v})$ is $[-1, 1]$. The classifier function becomes negative if $\mathbf{v}$ is correctly classified.

Furthermore, the parametrized squashing function f_Θ has to be monotonically increasing. In this paper, we use a common choice: the sigmoid squashing function

$$f_\Theta(x) = \frac{1}{1 + e^{-\Theta \cdot x}}.$$ (3)

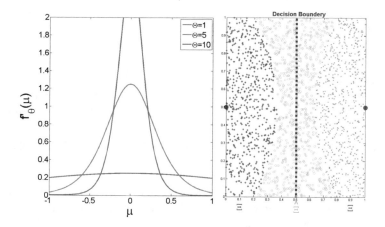

Fig. 1. Left: Derivative of the sigmoid function; right: visualization of the active set $\hat{\Xi}$ (green) with the two classes $\star$ and $+$ and their belonging prototypes ($\bullet$, $\bullet$)

with parameter $\Theta \geq 0$ determining their slope and $0 < f_\Theta(\mu) < 1$ is always valid. For large values $\Theta \gg 0$ the cost function E_{GLVQ} approximatly counts the number of misclassifications [12].

If the dissimilarity function $d(\mathbf{v}, \mathbf{w})$ is supposed to be differentiable with respect to the second argument, the minimization of the cost function (1) can be done by stochastic gradient descent. The resulting prototype update is

$$\mathbf{w}^\pm \leftarrow \mathbf{w}^\pm - \alpha \frac{\partial E_{GLVQ}}{\partial \mathbf{w}^\pm} \qquad (4)$$

with the learning rate $0 < \alpha \ll 1$ and the derivatives

$$\frac{\partial E_{GLVQ}}{\partial \mathbf{w}^\pm} = f'_\Theta(\mu_W(\mathbf{v})) \cdot \frac{\pm d^\mp(\mathbf{v})}{(d^+(\mathbf{v}) + d^-(\mathbf{v}))^2} \cdot \frac{\partial d^\pm(\mathbf{v})}{\partial \mathbf{w}^\pm} . \qquad (5)$$

This gradient scales with with the derivative $f'_\Theta(\mu_W(\mathbf{v}))$ of the squashing function. High values for Θ effect a significant change only for those data and prototypes, which generate a small value absolute value $|\mu_W(\mathbf{v})|$, see Fig. 1.

Thus only data points nearby the decision border contribute to the prototype update. Therefore, the subset

$$\hat{\Xi} = \{\mathbf{v} \in V | f'_\Theta(\mu_W(\mathbf{v})) > \epsilon\} \qquad (6)$$

of all training data points is denoted as *active set*, where $\epsilon > 0$ is a a predefined threshold value. The active set is also visualized in Fig.1.

The most common dissimilarity d is the squared Euclidean distance $d_2(\mathbf{v}, \mathbf{w}) = (\mathbf{v} - \mathbf{w})^2$. Other choices may be divergences, correlations or kernel distances [21,20,22]. Minkowski p-distances

$$d_p(\mathbf{v}, \mathbf{w}) = \sqrt[p]{\sum_{k=1}^{n} |v_k - w_k|^p}$$

with $1 \leq p \leq \infty$ became popular also for $p \neq 2$ [2,13,14]. Weighting of the data dimensions or correlations between them leads to distances like

$$d_\Omega(\mathbf{v}, \mathbf{w}) = (\Omega(\mathbf{v} - \mathbf{w}))^2 \tag{7}$$

with the mapping matrix Ω [5,19]. Here, the matrix $\Lambda = \Omega^T \Omega$ can also be interpreted as a *classification correlation matrix*, i.e. it describes correlations supporting the class separation if the mapping matrix Ω is also adapted during classification learning.

3 Integration of Auxiliary Knowledge into the Cost Function of the GLVQ

In this section we think about integration of auxiliary knowledge into GLVQ learning while keeping the basic structure of the cost function. In particular, we consider the a multiplicative function φ such that

$$E_{\varphi-GLVQ} = \sum_{\mathbf{v} \in V} \varphi(\Phi_V, \Psi_W, \mathbf{v}, c(\mathbf{v}), W, Y) \cdot f_\Theta(\mu_W(\mathbf{v})) \tag{8}$$

is the new cost function. The formal weighting function φ should reflect the auxiliary information where Φ_V and Ψ_W are free parameters regarding the knowledge about the data space V and the prototypes W, respectively. The specific structure of the function φ should reflect the goal of the classification and should also depend on the auxiliary knowledge about the classes/data points. In the following we present two possibilities.

3.1 Class Priors

For many classification problems, the importance of the several classes may differ depending on the focus. For example, in medicine, it might be much more critical if ill people are classified to be healthy. Otherwise, a healthy but misclassified person may be treated by pharmacy causing heavy side-effects. For the latter case, the false positive rate is an appropriate accuracy measure, whereas for the first case the true positive rate should be high. Both values are contained in the confusion matrix. Optimization of statistical measures based on the confusion matrix in GLVQ learning was recently proposed in [11].

Here we present an alternative way introducing different class priors $\beta_k \geq 0$ for each class $k \in C$ defining the class prior function

$$\beta(c(\mathbf{v})) = \begin{cases} \beta_1 & \text{, if } c(\mathbf{v}) = 1 \\ \beta_2 & \text{, if } c(\mathbf{v}) = 2 \\ \vdots \\ \beta_c & \text{, if } c(\mathbf{v}) = C \end{cases} \tag{9}$$

for a given training data vector $\mathbf{v}$ depending on the class membership. Thus the class priority function represents the auxiliary expert knowledge. In this way, the weighting function φ from (8) simply becomes

$$\varphi(\Phi_V, \Psi_W, \mathbf{v}, c(\mathbf{v}), W, Y) = \beta(c(\mathbf{v})) \tag{10}$$

abbreviated as φ_β. The respective prototype updates are obtained as

$$\Delta\mathbf{w}^\pm \sim \beta(c(\mathbf{v}))\cdot f'_\Theta(\mu_W(\mathbf{v})) \cdot \frac{\pm d^\mp(\mathbf{v})}{(d^+(\mathbf{v}) + d^-(\mathbf{v}))^2} \cdot \frac{\partial d^\pm(\mathbf{v})}{\partial \mathbf{w}^\pm} . \quad (11)$$

realizing a kind of attention based learning [3].

Of course, the usual classification accuracy is not an appropriate assessment measure of such a classifier. Instead, the *weighted* accuracy

$$wacc_\beta(V) = \frac{C}{N_V \cdot \sum_{j=1}^C \beta_j} \sum_{\mathbf{v}\in V} \beta(c(\mathbf{v})) \cdot \delta_{c(\mathbf{v}),y(\mathbf{w}_{s(\mathbf{v})})} \quad (12)$$

could serve alternatively, where $\delta_{i,j}$ is the Kronecker symbol and

$$s(\mathbf{v}) = arg\min_k \left(d\left(\mathbf{v}, \mathbf{w}\right) \right) \quad (13)$$

is the index of the best matching prototype regardless the class label also denoted as the (overall) winner.

3.2 Asymmetric Misclassification Assessment

Misclassifications by a classifier can be interpreted as costs. However, different misclassification may generate different costs. For example, we can think about (main) classes, which are further divided into sub-classes: In this scenario might be more costly to fail the main classes than violate the sub-classification. Those problems frequently arise in medicine when for a disease several clinical subtypes may be distinguished.

Generally, the non-diagonal elements of the respective confusion matrix have different impact on the overall performance of the classifier system. We collect these impact weights in a real matrix $\Gamma_0 \in \mathbb{R}^{C\times C}$ with non-negative elements $\gamma(j,i)$. The diagonal elements $\gamma(i,i)$ are set to zero. We emphasize at this point that we do not assume Γ_0 to be symmetric and denote this matrix as an asymmetric misclassification assessment matrix (AMAM).

Especially, on problems with several classes, individual ones are more important or misclassifications between several classes are more critical, respectively. In the latter case we observe an asymmetric error assessment. To consider such different weighting of misclassifications, the φ-function from (8) can be constructed as

$$\varphi(\Phi_V, \Psi_W, \mathbf{v}, c(\mathbf{v}), W, Y) = \gamma(y(\mathbf{w}^-), c(\mathbf{v})) \quad (14)$$

with the matrix elements $\gamma(y(\mathbf{w}^-), i)$ collected in the matrix $\Gamma_0 \subset \mathbb{R}^{C\times C}$ with $i \in \mathcal{C}$. Thus, the matrix Γ_0 regulates the penalizing for misclassifications of class i to class $y(\mathbf{w}^-)$, i.e. it reflects the relations between the classes. This knowledge has to be provided in advance, for example by an expert. In this way, an asymmetric confusion matrix is generated by the trained classifier. Again, the standard accuracy is not useful to evaluate the GLVQ with AMAM. Hence, we modify the weighted accuracy from (12) to

$$werr_{\Gamma_0}(V) = \sum_{\mathbf{v}\in V} \gamma(y(\mathbf{w}_{s(\mathbf{v})}), c(\mathbf{v})) \cdot \left(1 - \delta_{c(\mathbf{v}),y(\mathbf{w}_{s(\mathbf{v})})}\right) \quad (15)$$

denoted as weighted error rate with respect to Γ_0.

We immediately observe that we could interpret the learning rule resulting from (14) again as an attention based learning but now using the AMAM Γ_0. Thus we can combine this asymmetric assessment method with the above class attention based approach: For this purpose, we define the diagonal elements to be as $\gamma(i,i) = \beta(i)$ using the class weights $\beta(i)$ from (9) yielding the overall expert knowledge matrix Γ, which now serves in the attention based learning.

If only two classes C_+ (positive) and C_- (negative) are under observation, the asymmetric model (14) becomes equivalent to a 2×2 contingency table with the confusion matrix $\mathbf{C}$ Tab.1 as known from statistics. It has been shown in [10] that respective clas-

Table 1. Confusion matrix $\mathbf{C}$ of the two-class problem with classes C_+ (positive) and C_- (negative): TP - true positives, FP - false positives, TN - true negatives, FN - false negatives, $N_\pm$ - number of positive/negative data, $\widehat{N}_+$ - number of predicted positive/negative data

labels		true		
		C_+	C_-	
predicted	C_+	TP	FP	$\widehat{N}_+$
	C_-	FN	TN	$\widehat{N}_-$
		N_+	N_-	N

sification schemes based on classification evaluation quantities like *precision* and *recall* or the widely applied F_γ-*measure* developed by C.J. VAN RIJSBERGEN [16] can be easily plugged into prototype based classification according to GLVQ using the border sensitive learning model.

4   Border Sensitive GLVQ as a Kind of Auxiliary Knowledge Integration

A natural extension of the above introduced general border sensitive learning, which could be motivated also by auxiliary information, is a localized border sensitivity. For this purpose, the class-relation dependent sensitivity matrix $\mathbf{\Theta} = (\theta_{i,j})$ is introduced. Each matrix element $\theta_{i,j} = \frac{1}{s_{i,j}} \geq 0$ specifies the sensitivity between the classes i and j. High values $s_{i,j} \geq 0$ correspond to small local active sets

$$\hat{\Xi}_{i,j} = \{\mathbf{v} \in V | f'_{\theta_{i,j}}(\mu_W(\mathbf{v})) > \epsilon\} \tag{16}$$

between these classes whereas low values signalize more insensitive behavior. Thereby, we define the localized squashing function

$$f_{\theta_{i,j}}(x) = \frac{1}{1 + e^{-\theta_{i,j} \cdot x}}$$

in analogy to (3). The respectively modified cost function reads as

$$E_\Theta(W) = \sum_{\mathbf{v}} f_{\theta_{y(\mathbf{w}^-),c(\mathbf{v})}}(\mu_W(\mathbf{v})) \tag{17}$$

emphasizing the sensitivity between the correct data class $c(\mathbf{v})$ and the class $y(\mathbf{w}^-)$ of the best matching prototype $\mathbf{w}^-$ with incorrect class label.

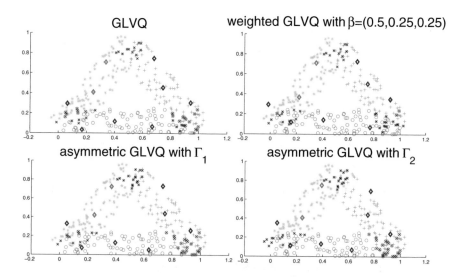

Fig. 2. Visualization of the results for the △-data. The three classes are: 1- blue ○, 2- red +, 3 - green *. The prototype positions are marked by ◇. For the interpretation of the simulations, see the text.

5   Numerical Experiments

In this chapter we report first numerical experiments. These, are of course not suffi- cient to illustrate alls aspects of GLVQ based on statistical measures. However, they give at least numerical evidence for the provided framework and should inspire future considerations and applications.

5.1   Illustrative Example for Artificial Data

The first numerical experiments are based on an artificial but illustrative example: We consider a three-class problem in $\mathbb{R}^2$. Each of the three classes is an uniform distribution in a rectangular shaped area. These areas form together a triangle with overlapping regions in the corners, as one can see in Fig.2.

We denote this dataset as △-data.

In the first experiment we apply the GLVQ variant with class priors as introduced in Sec.3.1. The class priors were set to be $\beta = (\frac{1}{2}, \frac{1}{4}, \frac{1}{4})$. For comparison, we trained a standard GLVQ with 3 prototypes per class, which results the confusion matrix depicted in Tab.2.

The resulting overall accuracy is 85.7% whereas the weighted accuracy (12) is 85.2%. If we now give more importance to the first class as imposed by the vector β and apply the weighted prototype update (11) according to the cost function $E_{\varphi-GLVQ}$ from (8) with φ_β. As we can conclude from Tab.2, the algorithm follows nicely the attention based learning increasing the accuracy of the first class compared to the GLVQ-result.

Table 2. Confusion matrix for the $\triangle$-data obtained by the unweighted (left) and weighted (right) GLVQ with the class prior vector $\beta = (\frac{1}{2}, \frac{1}{4}, \frac{1}{4})$ and 3 prototypes for each class, the latter one realizing an attention based learning with emphasis given to class $\circ$

		real		
		$\circ$	$+$	$*$
predicted	$\circ$	84 %	5 %	6 %
	$+$	7 %	86 %	7 %
	$*$	9 %	9 %	87 %

β		real		
		$\circ$	$+$	$*$
predicted	$\circ$	92 %	8 %	18 %
	$+$	6 %	87 %	12 %
	$*$	2 %	5 %	70 %

Table 3. Confusion matrices for the $\triangle$-data obtained by the asymmetric GLVQ according to the AMAMs Γ_1 and Γ_2, respectively, using 3 prototypes for each class

Γ_1		real		
		$\circ$	$+$	$*$
predicted	$\circ$	84 %	6 %	5 %
	$+$	5 %	88 %	8 %
	$*$	11 %	6 %	87 %

Γ_2		real		
		$\circ$	$+$	$*$
predicted	$\circ$	73 %	11 %	0 %
	$+$	2 %	89 %	16 %
	$*$	25 %	0 %	84 %

Further, the prototypes of class $\circ$ are more spread occupying better the overlapping regions whereas the respective prototypes of the other classes are pushed away, see Fig.2. Yet, the overall accuracy is slightly decreased to 83% but the weighted accuracy (12) ends up to 85.3%.

The second experiment is dedicated to the GLVQ with asymmetric misclassification assessment as proposed in Sec. 3.2. In particular, we consider two asymmetric misclassification assessment matrices (AMAMs)

$$\Gamma_1 = \begin{pmatrix} 1 & 1 & 2 \\ 2 & 1 & 1 \\ 1 & 2 & 1 \end{pmatrix} \text{ and } \Gamma_2 = \begin{pmatrix} 1 & 1 & 3 \\ 3 & 1 & 1 \\ 1 & 3 & 1 \end{pmatrix}$$

with the higher level of asymmetry for Γ_2. The achieved confusion matrices are depicted in Tab. 3. Again, we used 3 prototypes per class, which now try to reflect the asymmetric assessment of misclassifications moving them more into the regarding corners of the data triangle, see Fig. 2. As we can expect, the accuracy decreases with increasing asymmetry - we achieved 86% and 85%, respectively. However, the emphasized misclassifications are reduced compared to the standard GLVQ solution from Tab.2 as it is the aim of this kind of asymmetric attention based learning.

5.2 Real World Application – A Medical Diagnosis System

In this real world application we consider a medical data set of electro-physiological data. Each data vector describes an electro-physiological impairment profile (EIP) of a patient suffering from Wilson's disease (WD,[6,8]). Wilson's Disease is an autosomal disorder of copper metabolism in the liver, which leads to a disturbed copper metabolism in several extra-pyramidal motor brain regions [1]. The EIPs are 7-dimensional data vectors giving the electro-physiological response to a clinical test system for sensoric and motor evoked potentials as described in [9] and [7], respectively. The resulting dataset consist of overall 74 patient profiles and 48 records from healthy volunteers (V). The detailed class distribution of the four classes is

class	PS	PP+MT	NN	V
number	34	22	18	48

using the above mentioned fusion of PP and MT.

One can distinguish between two phases during the course of WD: the non-neurological phase (NN) in the beginning and later on followed by the neurological phase with manifested neurological symptoms. For the latter case one can further differentiate between pseudo-sclerotic (PS) and preudo-parkinsonian (PP) cases according to fine-motoric disturbances caused by the neurological impairments. Frequently, a merged type (MT) between them is considered, which can be seen neurologically belonging to PP. However, it is generally difficult to detect WD if it is in the non-neurological phase without detailed and costly medical investigations accompanying the reasonable cheap electro-physiological investigations. Otherwise, early detection is mandatory to prevent fast degeneration by pharmaceutical treatment, i.e. although a clear distinction between the classes is desired, most important is a confident diagnosis of the disease. Further, the drug treatment depends on the phase (NN or N). This medical expert knowledge can be roughly modeled by the AMAM

$$\Gamma = \begin{pmatrix} 1 & 1 & 5 & 5 \\ 1 & 1 & 5 & 5 \\ 5 & 5 & 1 & 1 \\ 10 & 10 & 10 & 1 \end{pmatrix}$$

to be utilized in GLVQ learning for classification.

We applied both asymmetric GLVQ using the above Γ-matrix and standard GLVQ. For both simulations we used two prototypes per class. The data vectors were z-transformed in advance as usually done in pattern recognition [4]. We report results obtained by an eight-fold cross-validation procedure. They are displayed in Tab.4.

We observe a clear improvement of the class distinction according to the desired diagnostic behavior. Particularly, the false-classification of NN-persons to the class V is drastically reduced although false-classifications still remain.

Besides the confusion matrices, we also calculated the weighted error $werr_\Gamma(V)$ from Sec. 3.2 for both results. For standard GLVQ we achieved an averaged error $werr_\Gamma(V) = 0.48$ whereas the asymmetric approach yields $werr_\Gamma(V) = 0.40$, which is underlying the obtained improvement of classification behavior. Thus, asymmetric GLVQ is able to reflect medical expert knowledge seriously.

Table 4. Confusion matrices for the WD-data obtained by standard GLVQ (left) and GLVQ with AMAM Γ (right) using 2 prototypes per class

		real				real			
		PS	PP+MT	NN	V	PS	PP+MT	NN	V
predicted	PS	59.4 %	12.5 %	0 %	0 %	56.3 %	18.8 %	0 %	0 %
	PP+MT	25.0 %	50.0 %	0 %	0 %	28.1 %	37.5 %	6.3 %	4.2 %
	NN	9.3 %	6.3 %	31.3 %	0 %	12.5 %	31.2 %	56.3 %	12.5 %
	V	6.3 %	31.2 %	68.7 %	100 %	3.1 %	12.5 %	37.4 %	83.3 %

6 Summary and Future Work

In this contribution we discuss several approaches to integrate expert knowledge into GLVQ. This leads to modified cost functions to be minimized. These cost functions can be seen as weighted variants of the original GLVQ cost function, which can be related to the attention-based learning paradigm. In particular, class priors as well as asymmetric misclassification assessment were considered. Further, general statistical evaluation measures as replacement for the usual cost function were discussed.

For an artificial illustrative data example the new approaches are successfully demonstrated. A real world example from medical diagnosis systems emphasize the practical application aspect, which, in fact, was inspiring this work in very early beginning. Future work should comprise other application scenarios as well as theoretical aspects like stability of learning, relevance learning and other.

References

[1] Barthel, H., Villmann, T., Hermann, W., Hesse, S., Kühn, H.-J., Wagner, A., Kluge, R.: Different patterns of brain glucose consumption in Wilsons disease. Zeitschrift für Gastroenterologie 39, 241 (2001)

[2] Biehl, M., Kästner, M., Lange, M., Villmann, T.: Non-Euclidean principal component analysis and Oja's learning rule – theoretical aspects. In: Estevez, P.A., Principe, J.C., Zegers, P. (eds.) Advances in Self-Organizing Maps. AISC, vol. 198, pp. 23–34. Springer, Heidelberg (2013)

[3] Der, R., Herrmann, M.: Attention based partitioning. In: der Meer, M.V. (ed.) Bericht Des Status–Seminar Des BMFT Neuroinformatik, pp. 441–446. DLR, Berlin (1992)

[4] Duda, R., Hart, P.: Pattern Classification and Scene Analysis. Wiley, NY (1973)

[5] Hammer, B., Villmann, T.: Generalized relevance learning vector quantization. Neural Networks 15(8-9), 1059–1068 (2002)

[6] Hermann, W., Caca, K., Eggers, B., Villmann, T., Clark, D., Berr, F., Wagner, A.: Genotype correlation with fine motor symptoms in patients with Wilson's disease. European Neurology 48, 97–101 (2002)

[7] Hermann, W., Günther, P., Kühn, H.-J., Schneider, J., Eichelkraut, S., Villmann, T., Strecker, K., Schwarz, J., Wagner, A.: FAEP und Morphometrie des Mesenzephalons bei Morbus Wilson. Aktuelle Neurologie 34(10), 547–554 (2007)

[8] Hermann, W., Villmann, T., Grahmann, F., Kühn, H., Wagner, A.: Investigation of fine motoric disturbances in Wilson's disease. Neurological Sciences 23(6), 279–285 (2003)

[9] Hermann, W., Villmann, T., Wagner, A.: Elektrophysiologisches Schädigungsprofil von Patienten mit einem Morbus Wilson'. Der Nervenarzt 74(10), 881–887 (2003)

[10] Kaden, M., Hermann, W., Villmann, T.: Optimization of general statistical accuracy measures for classification based on learning vector quantization. In: Verleysen, M. (ed.) Proc. of European Symposium on Artificial Neural Networks, Computational Intelligence and Machine Learning (ESANN 2014), Louvain-La-Neuve, Belgium (page accepted, 2014), i6doc.com

[11] Kaden, M., Villmann, T.: A framework for optimization of statistical classification measures based on generalized learning vector quantization. Machine Learning Reports, 7(MLR-02-2013), 69–76 (2013), http://www.techfak.uni-bielefeld.de/fschleif/mlr/mlr_02_2013.pdf, ISSN:1865-3960

[12] Kästner, M., Riedel, M., Strickert, M., Hermann, W., Villmann, T.: Border-sensitive learning in kernelized learning vector quantization. In: Rojas, I., Joya, G., Gabestany, J. (eds.) IWANN 2013, Part I. LNCS, vol. 7902, pp. 357–366. Springer, Heidelberg (2013)

[13] Lange, M., Biehl, M., Villmann, T.: Non-Euclidean principal component analysis by Hebbian learning. Neurocomputing (page in press, 2014)

[14] Lange, M., Villmann, T.: Derivatives of l_p-norms and their approximations. Machine Learning Reports 7(MLR-04-2013), 43–59 (2013), http://www.techfak.uni-bielefeld.de/~fschleif/mlr/mlr_04_013.pdf, ISSN:1865-3960

[15] Matthews, B.: Comparison of the predicted and observed secondary structure of T4 phage Iysozyme. Biochimica et Biophysica Acta 405, 442–451 (1975)

[16] Rijsbergen, C.: Information Retrieval, 2nd edn. Butterworths, London (1979)

[17] Sachs, L.: Angewandte Statistik, 7th edn. Springer (1992)

[18] Sato, A.S., Yamada, K.: Generalized learning vector quantization. In: Tesauro, G., Touretzky, D., Leen, T. (eds.) Advances in Neural Information Processing Systems, vol. 7, pp. 423–429. MIT Press (1995)

[19] Schneider, P., Hammer, B., Biehl, M.: Adaptive relevance matrices in learning vector quantization. Neural Computation 21, 3532–3561 (2009)

[20] Strickert, M., Schleif, F.-M., Seiffert, U., Villmann, T.: Derivatives of Pearson correlation for gradient-based analysis of biomedical data. Inteligencia Artificial, Revista Iberoamericana de Inteligencia Artificial (37), 37–44 (2008)

[21] Villmann, T., Haase, S.: Divergence based vector quantization. Neural Computation 23(5), 1343–1392 (2011)

[22] Villmann, T., Haase, S., Kaden, M.: Kernelized vector quantization in gradient-descent learning. Neurocomputing (page in press, 2014)

Visualization and Classification of DNA Sequences Using Pareto Learning Self Organizing Maps Based on Frequency and Correlation Coefficient

Hiroshi Dozono

Department of Advanced Fusion, Saga University,
1 Honjyo Saga 840-8502 Japan
hiro@dna.ec.saga-u.ac.jp

Abstract. Next-generation sequencing techniques produce an enormous amount of sequence data. Analyzing these sequences requires an efficient method that can handle large amounts of data. Self-organizing maps (SOMs), which use the frequencies of N-tuples, can categorize sets of DNA sequences with unsupervised learning. In this study, SOM using correlation coefficients among nucleotides was proposed, and its performance was examined in the experiments through mapping experiments of the genome sequences of several species and classification experiments using Pareto learning SOMs.

1 Introduction

Next-generation sequencing [1] produces large amounts of sequence data that are applied to many areas of genome science. Meta-genome and comparative genome analyses are examples of such applications. Meta-genome analysis uses mixtures of genomes from a group of species for analysis of the composition of species or expressed sequences. Comparative genome analysis uses the sequenced genome data of a group of species to analyze evolutionary relationships or species diversity. Both applications require a global comparison of DNA sequences among species.

Self organizing maps(SOMs)[2] are often used for the global comparison of DNA sequences. SOMs are neural networks that use the architecture of feedforward networks and train the network with an unsupervised learning method. A set of input vectors is given to the network, and SOM extracts the features of the input vectors on two-dimensional maps according to vector similarity.

The frequencies of N-tuples, which denote the occurrence of each N-tuple for a fixed N, are effective for global comparison, and we proposed an analysis of DNA sequences with an SOM by using the vectors of N-tuple frequencies as input vectors [3]. For large-scale data, the use of these frequencies as feature SOM vectors is effective, and it is also applied to the analysis of IP-packet traffic For large scale data, it is effective to use the frequencies as feature vector of SOM,

T. Villmann et al. (eds.), *Advances in Self-Organizing Maps and Learning
Vector Quantization*, Advances in Intelligent Systems and Computing 295,
DOI: 10.1007/978-3-319-07695-9_8, © Springer International Publishing Switzerland 2014

and it is also applied to the analysis of the traffic of IP-packets [4]. In a previous study [3], the relationships among the genomes of species were visualized on the basis of frequencies of N-tuples. Further research proceeded using this method.

Herein, we propose another preprocessing method on the basis of correlation coefficients (CCs) of the occurrences of each nucleotide in a DNA sequence. All combinations between 2 nucleotides A-A, A-C, A-G, A-T, C-A, ..., T-G, T-T, CCs of the occurrences in the sequences are calculated by shifting 1 of the sequences in 1 to N. For 1 to N shifts, the number of CCs is $4^2 \times N$. CCs are arranged in vectors and used as input vectors for SOM, which determines the global features of the DNA sequences.

Furthermore, we apply Pareto learning SOMs (P-SOMs) [5] to visualize and classify DNA sequences. P-SOMs use a multi-modal vector composed of multiple vectors, including the category vector that denotes the class of the vector for supervised learning. The category vector operates cooperatively with the original input vectors to improve visualization and classification. P-SOMs were examined in the benchmark data set iris [5] and applied to the authentication method for behavior biometrics [6] and IP-packet traffic analysis [4].

2   Pareto Learning Self Organizing Map (P-SOM)

2.1   Pareto Learning SOM for Multi Modal Vector

2.2   P-SOM for Multi-modal Vectors

A multi-modal vector $(\{x_1\}, \{x_2\}, \ldots, \{x_n\})$ is a vector composed of multiple vectors and attributes. For example, keystroke timing and key typing intensity are the features used for authentication with key typing features. In multi-modal vectors, each vector and attribute is described in a different unit and scale, and the availability for the classification may be different. Conventional SOMs can learn multi-modal vectors by using a simply concatenated vector $(x_1, x_2, \ldots, x_n)$ or a concatenated vector with weight values $(w_1 x_1, w_2 x_2, \ldots, w_n x_n)$ as the input vector. When weight values are excluded, the map is dominated by largely scaled vectors and easily affected by unreliable vectors. A map using weight values depends heavily on these values, making the selection of optimal weight values difficult.

P-SOM makes direct use of a multi-modal vector $x = (\{x_1\}, \{x_2\}, \ldots, \{x_n\})$ as an input vector based on Pareto optimality. For each vector, x_i, the objective function is defined as $f_i(x, U^{jk}) = |x_i - m_i^{ij}|$ for unit U^{jk} on the map, where $m^{ij} = (\{m_1^{jk}\}, \{m_2^{jk}\}, \ldots, \{m_n^{jk}\})$ is the vector associated with U^{jk}. The Pareto winner set $P(x)$ for an input vector x is the set of the units U^{jk} that are Pareto optimal according to the object functions $f_i(x, U^{jk})$. Thus, P-SOM is a multi-winner SOM and all units in $P(x)$ and their neighbors are updated simultaneously.

The algorithm of P-SOM is as follows.

P-SOM Algorithm

1. Initialization of the map
   Initialize the vector $\mathbf{m^{ij}}$ which are assigned to unit U^{ij} on the map using the 1st and 2nd principal components as base vectors of 2-dimensional map.
2. Batch learning phase
   (1) Clear all learning buffer of units U^{ij}.
   (2) For each vector x^i, search for the pareto optimal set of the units $P = \{U_p^{ab}\}$. U_p^{ab} is an element of pareto optimal set P, if for all units $U_{kl} \in P - U_p^{ab}$, existing h such that $e_h^{ab} \leq e_h^{kl}$ where

$$e_h^{kl} = |\mathbf{x_h^i} - \mathbf{m_h^{kl}}| \tag{1}$$

   (3) Add x^i to the learning buffer of all units $U_p^{ab} \in P$.
3. Batch update phase
   For each unit U^{ij} update the associated vector $\mathbf{m^{ij}}$ using the weighted average of the vectors recorded in the buffer of U^{ij} and its neighboring units as follows.
   (1)For all vectors x recorded in the buffer of U^{ij} and its neighboring units in distance $d \leq Sn$, calculate weighted sum $\mathbf{S}$ of the updates and the sum of weight values W.

$$\mathbf{S} = \mathbf{S} + \eta fn(d)(\mathbf{x} - \mathbf{m^{i'j'}}) \tag{2}$$
$$W = W + fn(d) \tag{3}$$

   where $U^{i'j'}$s are neighbors of U^{ij} including U^{ij} itself, η is learning rate, $fn(d)$ is the neighborhood function which becomes 1 for d=0 and decrease with increment of d.
   (2) Set the vector $\mathbf{m^{ij}} = \mathbf{m^{ij}} + \mathbf{S}/W$.

Repeat 2. and 3. with decreasing the size of neighbors Sn for pre-defined iterations.

Fig.1 shows the differences in the SOM and P-SOM algorithms.

In the update phase, the units in the overlapped neighbors are updated more strongly, and this phase plays an important role in the integration of multimodal vectors. P-SOM is scale free because all vectors in $\boldsymbol{x}$ are handled evenly independently to the scales of $\boldsymbol{x}_i$

P-SOM can integrate any kind of vector. Thus, the category vector c^i can be introduced as an independent vector for each input vector to P-SOM.

$$\acute{\mathbf{x}}^i = (\boldsymbol{x^i}, \boldsymbol{c^i}) \tag{4}$$
$$c_j^i = \begin{cases} 1 & \boldsymbol{x^i} \in C_j \\ 0 & otherwise \end{cases} \tag{5}$$

The category vector is also used to search the Pareto winner set, and it attracts the input vectors in the same category that correspond closely on the map with the original input vector $\boldsymbol{x}$. The category of the given test vector $\boldsymbol{x}_t$ is determined as $argmax\{ \sum_{U^{ij} \in P(\boldsymbol{x}_t)} c_k^{ij} \}$ where $P(\boldsymbol{x}_t)$ is the Pareto optimal set of units for $\boldsymbol{x}_t$.

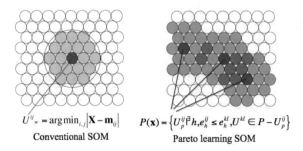

$U^{ij}_w = \arg\min_{i,j} |X - m_{ij}|$

Conventional SOM

$P(\mathbf{x}) = \{U^{ij}_p | h_{,}e^{ij}_h \le e^{kl}_h, U^{kl} \in P - U^{ij}_p\}$

Pareto learning SOM

Fig. 1. Differences between the self-organizing map (SOM) and Pareto learning SOM (P-SOM) algorithms

3 Analysis of DNA Sequences Using SOM

This section explains the preprocessing methods for effective extraction of DNA sequencw features.

3.1 Frequency of DNA Sequences

The frequency of N-tuples in DNA sequences is defined as the number of N-tuples in the sequence. Fig. 2 shows an example of the frequency for $N = 2$. Long

Sequence	AGTAATCTCTAATCT															
Frequency	AA	AC	AG	AT	CA	CC	CG	CT	GA	GC	GG	GT	TA	TC	TG
	1	0	1	2	0	0	0	3	0	0	0	1	1	2	0

Fig. 2. Frequency of the 2-tuple of a DNA sequence

sequences are divided into segments of constant length to enlarge the number of learning vectors. SOMs, which uses the frequency of DNA sequences as the input vector, can reportedly visualize the relationship of the genomes of different species for $N = 4$ and $N = 5$ [4]. However, the dimension of the frequency vector becomes 4^N. Thus, for large N values, the size of the input vector becomes very large.

3.2 Correlation Coefficient(CC)s of the Nucleotides in DNA Sequences

A DNA sequence is the sequence of the characters 'A','G','T', and 'C', thus, it is meaningless to calculate the CC directly for the sequence. A DNA sequence is converted to 4 binary sequences that represent the occurrences of every nucleotides 'A', 'G', 'T', and 'C'. For all combinations of the occurrence sequences,

$\rho_{n1,n2}(i)$, which is CC between the first occurrence sequence of nucleotides $n1$ and the sequence that shifts N nucleotides from the second occurrence sequence of nucleotides $n2$ are calculated for i=1 to N. Fig.3 shows the example of the calculation of CCs. These CCs are concatenated in a vector, and used as the input

```
        ACGCTACTAG
A   1000010010     ρ_AA(n) CC between A and n-shifted A
C   0101001000     ρ_AC(n) CC between A and n-shifted C
G   0010000001          :
T   0000100100     ρ_TT(n) CC between T and n-shifted T
```

Fig. 3. Correlation Coefficients of DNA sequence

vector for the SOM. Calculating CCs requires the scanning of the sequences 16 times, and has huge computational costs for long sequences. Using the following equation, all CCs of between 2 sequences of nucleotides, $S^1 = s_1^1 s_2^1 \cdots s_L^1$ and $S^2 = s_1^2 s_2^2 \cdots s_L^2$, can be calculated with 1 pass scan.

$$C_1 = \begin{cases} 1 & s_k^1 = n1 \\ 0 & s_i^1 \neq n1 \end{cases} \tag{6}$$

$$C_2 = \begin{cases} 1 & s_i^2 = n2 \\ 0 & s_i^2 \neq n2 \end{cases} \tag{7}$$

$$\sigma_{n1,n2} = \sum_{i=1}^{L} (C_1 - m_{n1})(C_2 - m_{n2}) \tag{8}$$

$$\sigma_{n1,n1} = \sum_{i=1}^{L} (C_1 - m_{n1})^2 \tag{9}$$

$$\sigma_{n1,n2} = \sum_{i=1}^{L} (C_2 - m_{n2})^2 \tag{10}$$

$$\rho_{n1,n2} = \frac{\sigma_{n1,n2}}{\sigma_{n1,n1}\sigma_{n1.n2}} \tag{11}$$

where m_{n1} and m_{n2} are the averages of the occurrence sequences for nucleotides $n1$ and $n2$ respectively.

Compared with the dimensions of the frequency vector, the dimension of the vector is small. It is $16 \times N$ for the concatenated vector of 1 to N shifts.

3.3 Experimental Results

The purpose of applying SOM for the analysis of DNA sequences is visualization. This subsection gives the experimental results of visualization of the relations between DNA sequences based on frequencies and CCs. We used two sets od

DNA sequences. The first set comprised the DNA sequences of 6 species registered to the pathway of amino acid metabolism in the Kyoto Encyclopedia of Genes and Genomes database. These species are colored as shown in Table 1.

Table 1. Species used in the experiments **Table 2.** Pathways used in the experiments

Genome name	Description	Color
hsa	homo sapience	red
cfa	dog	blue
mmu	mouse	green
dme	fruit fly	yellow
eco	E-Coli	magenta
osa	rice	cyan

Pathway name	Color
amino acid metabolism	red
cell growth and death	blue
metabolism of complex carbohydrates	green
metabolism of complex lipids	yellow
nucleotide metabolism	magenta
transration	cyan
transcription	white

The second set comprised the DNA sequences of 6 pathways of homo sapience. Gene sequences registered to multiple pathways were removed from the set. In this paper, The pathways are colored as shown in Table 2.

In both sets, the sequences which are longer than 1000 were segmented to the sequences with a length of 1000. The total number of the segments was 7148 for the species set, and 1135 for the pathway set.

The parameters of SOM was given as follows.

- map size: 128×64
- learning rate: from 0.8 to 0.1
- update method: batch update
- neighborhood function: gausian function
- iteration of learning: 50

Fig.4 shows the map of frequencies of 4-tuples. he length of the vector is $4^4 = 256$. Each color dot on the map represents the fragment of the sequence colored

Fig. 4. Map of the frequencies of 4-tuples in the DNA sequences of 6 species

as shown in Table.1. Sequences of dme, eco and osa were clustered separately. Sequences of hsa, cfa and mmu were loosely clustered because they are mammals. Fig.5 shows the map of CC of 1 to 4 shifts. The length of the vector is $16 \times 4 = 64$. The topologies of these maps are similar, and the clarity of the clusters is almost

Fig. 5. Map of CC of 1 to 4 shifts in the DNA sequences of 6 species

the same. When the number of shifts and length of tuples is decreased, as shown in Fig.6 to Fig.9, CCs show better clustering results than those of frequencies.

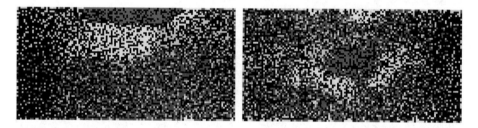

Fig. 6. Map of the frequencies of 3-tuples **Fig. 7.** Map of the frequencies of 2-tuples
L=64   L−16

Considering the length of the vector(L), CCs represented the features of DNA sequences more effectively than the frequencies of N-tuples did.

Fig.10 and Fig.11 show the maps of the frequencies of 4-tuples and CCs of 1 to 4 shifts using the pathway set respectively.

When the pathway set was used, the sequences were not clustered clearly. However, each color showed the shading in the specific area on the map, which was considered loosely clustered.

As an additional experiment, Fig.12 shows the maps of CCs using the input data of 7 different virus genomes. Some virus genomes are fragmented in some regions, however they are clustered as the species set.

Fig. 8. Map of the CCs of 1 and 2 shifts L=32

Fig. 9. Map of the CC of 1 shift L=16

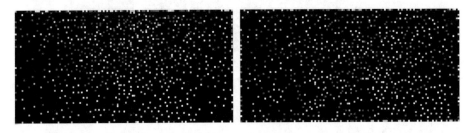

Fig. 10. Map of the frequencies of 4-tuples of the pathway set

Fig. 11. Map of the CCs of 1 to 4 shifts of the pathway set

4   Analysis of DNA Sequences Using Pareto Learning SOM(P-SOM)

We analyzed the DNA sequences using P-SOM. P-SOM can learn input vectors both in unsupervised learning mode without using category vectors for learning and in supervised learning mode with using category vectors. In supervised learning mode, category vectors cooperate with the original input vectors to organize the map. A vector of 16 CCs for each shift is used as an element of multi-modal input vectors to the P-SOM.

Fig.13 and Fig.14 show the maps of the CCs of 1 to 4 shifts using the species set and the pathway set as input vectors. The maps are torus maps. In Fig.13, the species are clustered, as seen in the results of the conventional SOM, and the mammals are clustered more strongly than those in the conventional SOM because of the supervised learning feature of the P-SOM. In Fig.14, the pathways are also more clearly clustered than those of the conventional SOM.

For the classification experiment, a randomly selected 70 % of the sequences were used for learning, and 30 % of the sequences were used for the test. CCs and frequencies of N-tuples were used as input vectors, and the experiments using conventional SOM were conducted for comparison. Table 3 shows the results for the species set. In this table, CC-N denotes the CC of 1 to N shifts, and F-N

Fig. 12. Map of CC of 1 to 4 shifts of 7 virus genome

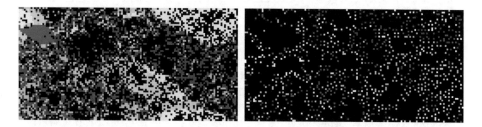

Fig. 13. Map of the CCs of 1 to 4 shifts by **Fig. 14.** Map of the CC of 1 to 4 shifts using using the species set as the input vectors for the pathway set as the input vectors for P-P-SOM   SOM

Table 3. Rates of successful classification of the species set

Input vector	CC-2	CC-4	CC-2	CC-4	F-4	F-4
Length	32	64	32	64	256	256
Method	P-SOM	P-SOM	SOM	SOM	P-SOM	SOM
Learned sequences	0.832	0.831	0.920	0.916	0.980	0.915
Test Sequences	0.609	0.643	0.593	0.599	0.624	0.629

denotes the frequency of N-tuples. For the learned sequences, the P-SOM using frequency as the input vector performed best, and for the test sequences, the P-SOM using CCs performed best. Table 4 shows the rates of successful classification for each species. As expected, the rates for mammals are poor because they were loosely clustered on the map. The sequences from cfa(dog) may be miss classified to hsa and mmu. The accuracy seems to be low as the classifier, because the coding regions of mammals include common genes. For the virus genome set, the accuracies for learned sequences and test sequences were 0.980 and 0.864 respectively.

Table 5 shows the classification results for the pathway set. For both of the learned sequences and the test sequences, P-SOMs using CCs of 1 to 4 shifts performed best. For the learned sequences, almost all sequences were successfully classified, however for the test sequences, less than one-fourth of the sequences were classified, because the map was very complicated. It is considered to be

Table 4. Rates of successful classification for each species

name	Learned sequences	Test sequences
hsa	0.803	0.521
cfa	0.564	0.121
mmu	0.809	0.618
dme	0.967	0.962
eco	0.994	0.990
osa	0.910	0.876

Table 5. Rates of successful classification of pathway set

Input vector	CC-4	CC-4	F-4	F-4
Length	64	64	256	256
Method	P-SOM	SOM	P-SOM	SOM
Learned sequences	0.999	0.985	0.836	0.938
Test Sequences	0.240	0.208	0.214	0.195

difficult to classify the genes from different pathway sets of single organism using the features of frequencies of N-tuples or CC of sequences.

5 Conclusion

We proposed a preprocessing method for DNA sequences by using correlation coefficients of the occurrence of the nucleotides. Using this method, the clustering results of the sequences were nearly compatible with those obtained using the frequencies of the N-tuples despite the difference in the length of input vectors. The correlation coefficients are considered a more effective method for preprocessing DNA sequences.

Pareto learning SOM method is applied to the classification of DNA sequences by using correlation coefficients and frequencies as input vectors. Pareto learning SOM using CC as the input vector shows good performance for classification compared with that obtained with conventional SOMs, and frequencies. Correlation coefficients are effective as indexes for classifiertion.

In the future studys, we must apply this method to additional types sequence data, such as coding region and non-coding region, and to large data sets such as whole genomea. For such experiments, we must improve the computational costs of P-SOMs, which are 5 times more than those of conventional SOMs.

References

1. illuminal, An Introduction to Next-Generation Sequencing Technology, `http://www.illumina.com/Documents/products/Illumina_Sequencing_Introduction.pdf`
2. Kohonen, T.: Self Organizing Maps. Springer (2001)
3. Abe, T., Ikekura, T., et al.: Informatics for unreveiling hidden genome signatures. Genome Res. 13, 693–702
4. Dozono, H., Kabashima, T., et al.: Visualization of the Packet Flows using Self Organizing Maps. WSEAS Transactions on Information Science & Applications 7(1), 132–141 (2010)
5. Sorjamaa, A., Corona, F., Miche, Y., Merlin, P., Maillet, B., Séverin, E., Lendasse, A.: Sparse Linear Combination of SOMs for Data Imputation: Application to Financial Database. In: Príncipe, J.C., Miikkulainen, R. (eds.) WSOM 2009. LNCS, vol. 5629, pp. 290–297. Springer, Heidelberg (2009)
6. Dozono, H., Nakakuni, M.: Application of Supervised Pareto Learning Self Organizing Maps to Multi-modal Biometric Authentication. IPSJ Journal 49(9), 3028–3037 (2008) (in Japanese)

Probabilistic Prototype Classification Using t-norms

Tina Geweniger, Frank-Michael Schleif, and Thomas Villmann

Computational Intelligence Group, University of Applied Sciences Mittweida,
Technikumplatz 17, 09648 Mittweida, Germany

Abstract. We introduce a generalization of Multivariate Robust Soft Learning
Vector Quantization. The approach is a probabilistic classifier and can deal with
vectorial class labelings for the training data and the prototypes. It employs t-
norms, known from fuzzy learning and fuzzy set theory, in the class label assignments, leading to a more flexible model with respect to domain requirements. We
present experiments to demonstrate the extended algorithm in practice.

1 Motivation

Uncertainty is a general effect of most datasets and should not be neglected during
learning. In this article we focus on classification problems where the uncertainty occurs in the data as well as in the label information. Both aspects have been addressed
before in the field of probabilistic learning or using fuzzy sets [1,2]. However, often the
obtained models are quite complex or lack sufficient flexibility to integrate additional
expert knowledge. Recently, a multivariate formulation of Robust Soft Learning Vector
Quantization (MRSLVQ) was proposed in [1] and independently recovered in [3], providing an interpretable prototype based model. Another alternative can be found in [4]
for a so called *fuzzification* of Soft Nearest Prototype Classification for fuzzy labeled
data and prototypes.

Prototypes are compact representations of a larger set of points, like the mean of a
set of points, and partition the data space e. g. into disjunct regions. They can easily be
inspected by experts and summarize large complex data sets.

All these models share many positive aspects of prototype based learning [5], such
as metric adaptation [6], kernelization [5] or the processing of dissimilarity data [7].
Here we will focus on the MRSLVQ, although the presented concepts can be transferred to other approaches straight forward. The original formulation of MRSLVQ uses
a multiplicative assignment rule for the class label assignments or fuzzy label membership which we will be replaced by the more generic concept of $t-$norms in this work.
$t-$norms occur in the field of fuzzy logic to model boolean set operations for decision
rules. It is assumed that the conjunction $\wedge$ is interpreted by a triangular norm ($t-$norm
for short) and the disjunction $\vee$ is interpreted by a triangular co-norm ($t-$co-norm).
In the considered classification task we like to express that the label of its closest prototype is consistent with the label of a data point which can be modeled by a logical
conjunction and which is approximated by a corresponding $t-$norm.

In the following, first we will give a brief overview of MRSLVQ and review different
t-norms. Subsequently we extend MRSLVQ by t-norms and show the effectiveness of
the approach for two datasets with unsafe label information.

T. Villmann et al. (eds.), *Advances in Self-Organizing Maps and Learning*
Vector Quantization, Advances in Intelligent Systems and Computing 295,
DOI: 10.1007/978-3-319-07695-9_9, © Springer International Publishing Switzerland 2014

2   Multivariate Robust Soft LVQ

The Robust Soft LVQ algorithm (RSLVQ) was introduced in [8] as a probabilistic prototype classifier. It is assumed that the probability density $p(v)$ of the data points $v \in \mathbb{R}^d$,
with d being the data dimensionality, can be described by a Gaussian mixture model.
Every component of the mixture is assumed to generate data which belongs to only one
of the N_C classes. The classification itself is based on a winner takes all scheme. The
probability density of all the data points is given by

$$p(v|W) = \sum_{k=1}^{N_C} \sum_{j:y_j=k}^{N_P} p(v|j)P(j) \tag{1}$$

where $W = \{(w_j, y_j)\}_{j=1}^{N_P}$ is the set of N_P labeled prototype vectors $w_j \in \mathbb{R}^d$ and
their assigned crisp class labels y_j. $P(j)$ stands for the probability that data points are
generated by component j of the mixture and is commonly set to an identical value
for all the prototypes. The conditional density $p(v|j)$, which describes the probability
that component j is generating a particular data point v, is a function of the prototype
w_j itself. The density $p(v|j)$ can be chosen to have the normalized exponential form
$p(v|j) = K(j) \cdot e^{f(v, w_j, \sigma_j^2)}$ where $K(j)$ is the normalization constant and the hyper
parameter σ_j^2 the width of component j.

The aim of RSLVQ is to place the prototypes such that a given data set is classified
as accurately as possible. Therefore the likelihood ratio

$$L = \prod_{i=1}^{N_V} L(v_i, c_i) , \qquad \text{with } L(v_i, c_i) = \frac{p(v_i, c_i|W)}{p(v_i|W)} \tag{2}$$

where N_V is the number of data points, has to be maximized. The ratio is built up of the
particular probability density $p(v_i, c_i|W)$, that data point v_i is generated by a mixture
component of the correct class c_i

$$p(v_i, c_i|W) = \sum_{j:y_j=c_i} p(v_i|j)P(j) \tag{3}$$

with the total probability density $p(v_i|W)$

$$p(v_i|W) = \sum_j p(v_i|j)P(j). \tag{4}$$

The cost function is given as

$$E_{RSLVQ} = \sum_{i=1}^{N_V} \log \left(\frac{p(v_i, c_i|W)}{p(v_i|W)} \right). \tag{5}$$

with learning rules as presented in [8].

While Robust Soft Learning Vector Quantization is very effective, it is only applicable for crisp labeled training data. An extension of this approach based on a vectorial
adaption scheme for handling fuzzy labeled training data was presented in [9] leading
to the following modifications:

2.1   Cost Function

The assumption of fuzzy labeled data points requires an adaption of the original RSLVQ algorithm. The originally crisp class label c_i for training data point v_i becomes a N_C-dimensional vector c_i of assignment probabilities with $\sum_{k=1}^{N_C} c_i^k = 1$ and $c_i^k \geq 0$. For RSLVQ, each prototype w_j describes exactly one class. Now we relax this condition and allow the prototypes to be (partial) representatives for different classes. Analogously to the notation for the data points, the class memberships of the prototypes are now expressed in vector notation yielding y_j with $\sum_{k=1}^{N_C} y_j^k = 1$ and $y_j^k \geq 0$. The classification of untrained data is still based on a winner takes all scheme. Taking the fuzzy class assignments of the data points into account, the particular probability density $p(v_i, c_i|W)$ with crisp data labels c_i specified in equation (3) changes to

$$p(v_i, c_i|W) = \sum_{k=1}^{N_C} c_i^k \sum_{j=1}^{N_P} y_j^k \cdot p(v_i|j)P(j) \qquad (6)$$

where $p(v_i, c_i|W)$ now is the particular probability density that data point v_i is generated by the mixture components referred to by c_i. Thereby, due to the factor c_i^k only a fraction of the sum of the respective probability densities is taken into account. The factor y_j^k ensures that only those prototypes are accounted for, which actually are representatives for the respective class.

The total probability density $p(v_i|W)$ (4)

$$p(v_i|W) = \sum_j p(v_i|j)P(j)$$

is the probability that data point v_i is generated by *any* prototype. It is the sum over all prototypes independent of matching class assignments and, therefore, does not change.

The cost function of the Multivariate RSLVQ (MRSLVQ) can now be defined as

$$E_{MRSLVQ} = \sum_{i=1}^{N_V} \log\left(\frac{p(v_i, c_i|W)}{p(v_i|W)}\right). \qquad (7)$$

2.2   Derivation of Learning Rules

In order to optimize the classification, the cost function (7) has to be minimized, which can be done by a stochastic gradient descent.

Considering an universal parameter Θ with $\Theta \neq v_i$ a general update rule can be derived:

$$\frac{\partial \log \frac{p(v_i, c_i|W)}{p(v_i|W)}}{\partial \Theta_j} = (P_{c_i}(j|v_i) - P(j|v_i))\left(\frac{1}{K(j)}\frac{\partial K(j)}{\partial \Theta_j} + \frac{\partial f(v_i, w_j, \sigma_j^2))}{\partial \Theta_j}\right). \qquad (8)$$

The terms $P_{c_i}(j|v_i)$ and $P(j|v_i)$ in (8), which are assignment probabilities, yield:

$$P_{c_i}(j|v_i) = \frac{\sum_k c_i^k y_j^k P(j)K(j)e^{f(v_i,w_j,\sigma_j^2,\lambda_j)}}{p(v_i,c_i|W)} \qquad (9)$$

$$P(j|v_i) = \frac{P(j)K(j)e^{f(v_i,w_j,\sigma_j^2,\lambda_j)}}{p(v_i|W)} \qquad (10)$$

$P_{c_i}(j|v_i)$ is the assignment probability of v_i to component j taking the partial class assignments of the data points and the prototypes into account. $P(j|v_i)$ is the assignment probability of v_i to component j independent of the class membership.

Assuming the special case of a Gaussian mixture model with $P(j) = 1/N_P \forall j$, the similarity function is set to $f(v_i, w_j, \sigma_j^2) = \frac{d(v_i,w_j)}{2\sigma_j^2}$. Thereby, $d(v_i, w_j)$ is the distance between data point v_i and prototype w_j, and $K(j)$ a normalization constant which can be set to $K(j) = (2\pi\sigma_j^2)^{(-N/2)}$.

The original RSLVQ algorithm uses the squared Euclidean distance as dissimilarity measure. In the following the update rules for the prototypes w_j and a hyper parameter σ_j^2 employing a general distance are derived. Afterwards the update rules based on specific distance measures are given.

To obtain the update rules for specific, cost function relevant parameters, Θ_j has to be substituted.

Updating the prototypes w
Replacing Θ_j in (8) by the prototype w_j yields

$$\frac{\partial \log \frac{p(v_i,c_i|W)}{p(v_i|W)}}{\partial w_j} = (P_{c_i}(j|v_i) - P(j|v_i)) \left(\frac{1}{2\sigma_j^2} \frac{\partial d(v_i,w_j)}{\partial w_j} \right). \qquad (11)$$

Updating the prototype labels y
Analogously, the update rule for the fuzzy prototype labels y_j is obtained as

$$\frac{\partial \log \frac{p(v_i,c_i|W)}{p(v_i|W)}}{\partial y_j} = \left(\frac{c_j}{P_{c_i}(j|v_i)} - \frac{1}{p(v_i|W)} \right) (P(j)p(v|j)). \qquad (12)$$

3   T-Norms

T-norms are a generalization of the triangular inequality of metrics and were introduced by Menger [10]. They can also be used as generalizations of the Boolean logic conjunctive 'AND' operator to multi-valued logic. Applied in fuzzy logic t-norms represent the union of fuzzy sets. Its dual operation t-co-norm analogously refers to the 'OR' operator and can be used to represent the intersection of fuzzy sets. T-norms are widely used in fuzzy set theory with multiple applications [11,12,13]. Recently, $t-$norms have also been analyzed in alternative frameworks [14,15], motivating their use in general classification methods as shown here.

3.1 Definition of General t-norms

A t-norm is a dual function $\mathsf{T} : [0, 1] \times [0, 1] \rightarrow [0, 1]$ to generalize the triangle inequality of ordinary metric spaces and has the following properties:

1. Commutativity $\mathsf{T}(a, b) = \mathsf{T}(b, a)$

2. Monotonicity $\mathsf{T}(a, b) \leq \mathsf{T}(c, d)$, if $a \leq c$ and $b \leq d$

3. Associativity $\mathsf{T}(a, \mathsf{T}(b, c)) = \mathsf{T}(\mathsf{T}(a, b), c)$

4. Identity $\mathsf{T}(a, 1) = a$

According to this definition, the values of t-norms are only specified on the corner points of a unit square and along the edges. In the middle area the values are restricted to the range $[0, 1]$. Therefore, there exist a variety of different t-norms. In the following a short listing of common t-norms (some of them parametrized) is given. Selected plots based on the unit square are provided in Fig. 1:

Minimum/Zadeh t-norm $\qquad \mathsf{T}_{min}(a, b) = \min(a, b)$

Product/Probabilistic t-norm $\mathsf{T}_{prod}(a, b) = a \cdot b$

Łukasiewicz t-norm $\qquad \mathsf{T}_{luka}(a, b) = \max(a + b - 1, 0)$

Drastic t-norm $\qquad \mathsf{T}_{drastic}(a, b) = \begin{cases} a \text{ if } b = 1 \\ b \text{ if } a = 1 \\ 0 \text{ otherwise} \end{cases}$

Hamacher t-norm $\qquad \mathsf{T}_{ham}(a, b) = \frac{ab}{\gamma + (1-\gamma)(a+b-ab)}$ with $\gamma > 0$

Weber t-norm $\qquad \mathsf{T}_{weber}(a, b) = \max(\frac{a+b-1+\gamma ab}{1+\gamma}, 0)$ with $\gamma > -1$

Yager t-norm $\qquad \mathsf{T}_{yager}(a, b) = \max(1 - ((1-a)^{\gamma} + (1-b)^{\gamma})^{\frac{1}{\gamma}}, 0)$
with $\gamma > 0$

Aczel-Alsina t-norm $\qquad \mathsf{T}_{ucz}(a, b) = \exp(-((-\log(u))^{\gamma} + (-\log(b))^{\gamma})^{\frac{1}{\gamma}})$
with $0 < \gamma < \infty$

In accordance to the analysis provided in [15] we focus on the Product t-norm, the Hamacher t-norm, and the Aczel-Alsina t-norm. These three t-norms permit easy differentiation, avoiding further approximation steps as necessary in case of t-norms involving $\max$ operators [16]. Further, the Product t-norm was used implicitly in the original version of MRSLVQ as will be clarified in the next section.

Note that these three t-norms are related to each other:

- for $\gamma = 1$ the Hamacher and the Aczel-Alsina t-norms are equivalent to the non-parametrized Product or Probabilistic t-norm

$$\mathsf{T}_{prod}(a, b) \equiv \mathsf{T}_{ham}(a, b) \equiv \mathsf{T}_{acz}(a, b)$$

- for $\gamma \rightarrow +\infty$ (Hamacher t-norm) respectively $\gamma \rightarrow 0$ (Aczel-Alsina t-norm) the Drastic t-norm is approximated

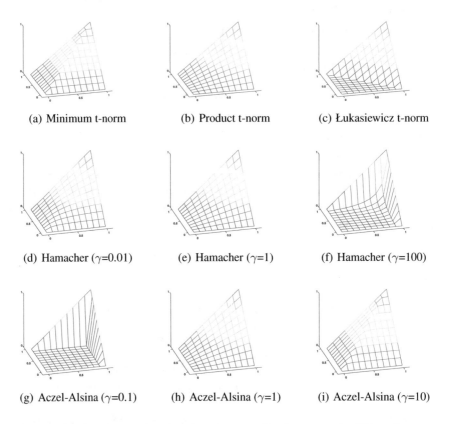

(a) Minimum t-norm   (b) Product t-norm   (c) Łukasiewicz t-norm

(d) Hamacher (γ=0.01)   (e) Hamacher (γ=1)   (f) Hamacher (γ=100)

(g) Aczel-Alsina (γ=0.1)   (h) Aczel-Alsina (γ=1)   (i) Aczel-Alsina (γ=10)

Fig. 1. Plots of various t-norms based on the unit square. For the parametrized Hamacher t-norm and the Aczel-Alsina t-norm three different values for the parameter γ are given.

4   Integrating t-norms in MRSLVQ

The large number of different (parametric) t-norms is due to different domain specific interpretations of the (dis-)similarity of multivariate vectors compared by a t-norm. Here we consider the (dis-)similarity between multivariate label vectors. In the MRSLVQ the authors made implicit use of the Probabilistic respectively Product t-norm in (6) by taking the fuzzy labels of the prototypes into the inner sum. Replacing the probabilistic t-norm in (6) we get:

$$p(\boldsymbol{v}_i, \boldsymbol{c}_i|W) = \sum_{k=1}^{N_C} \sum_{j=1}^{N_P} \top(c_i^k, y_j^k, \tau) \cdot p(\boldsymbol{v}_i|j)P(j) \tag{13}$$

with $\top(c_i^k, y_j^k, \tau)$ being a t-norm as defined before with a potential parameter τ. Due to the generalization to any t-norm the update of the prototype positions and the prototype labels has to be changed. Accordingly we replace in the equation of the assignment probabilities (9) the product of the fuzzy label assignments by a t-norm:

$$P_{c_i}(j|v_i) = \frac{\sum_k \top(c_i^k, y_j^k, \tau)P(j)K(j)e^{f(v_i, w_j, \sigma_j^2, \lambda_j)}}{p(v_i, c_i|W)} \tag{14}$$

This substitution also has to be considered in the prototype update of Eq. (11).

For the update of the fuzzy prototype labels y_j the gradient of the t-norm with respect to the prototype label $\frac{\partial \top(c_i^k, y_j^k, \tau)}{\partial y_j}$ has to be taken into account yielding the general form

$$\frac{\partial \log \frac{p(v_i, c_i|W)}{p(v_i|W)}}{\partial y_j} = \left(\frac{\frac{\partial \top(c_i^k, y_j^k, \tau)}{\partial y_j}}{P_{c_i}(j|v_i)} - \frac{1}{p(v_i|W)} \right) (P(j)p(v|j)) \tag{15}$$

By replacing $\top(c_i^k, y_j^k, \tau)$ by a specific t-norm the particular update rule is obtained. For example the Product t-norm yields

$$\frac{\partial \log \frac{p(v_i, c_i|W)}{p(v_i|W)}}{\partial y_j} = \left(\frac{c_j}{P_{c_i}(j|v_i)} - \frac{1}{p(v_i|W)} \right) (P(j)p(v|j)). \tag{16}$$

which is equivalent with update rule (12) as expected.

It would also be possible to update parameters of the t-norm by providing the corresponding gradients similar as for the metric adaptation or the σ learning, see e. g. [17]. For simplicity we will specify t-norm parameters using a grid search on an independent test set. In the following we focus on the before chosen parametrized t-norms Hamacher t-norm and Aczel-Alsina t-norm and provide experiments for different datasets taken from the life science domain. We compare with the approach using the standard Probabilistic t-norm, which is identical with the original MRSLVQ.

5   Experiments

We now apply the priorly derived approach to two datasets with multivariate labels. We chose the Hamacher t-norm and the Aczel-Alsina t-norm due to their easy differentiability. We show the effectiveness for a classification task and compare the results to the standard MRSLVQ approach based on the implicitly implemented Probabilistic or Product t-norm. Potential parameters of the t-norms have been optimized using a grid search on an independent test set. Using the optimized parameters the model performance was evaluated on the remaining data in a 10-fold cross-validation.

5.1   Overlapping Gaussian Distributions

The first data set is a simulated one consisting of two overlapping Gaussian distributions. 1000 samples are drawn randomly mixed from the two distributions. The mixing coefficients are used as fuzzy labels. Applying the Aczel-Alsina t-norm, the grid search for the optimal parameter γ reveals improvements for $\gamma \geq 0.2$ compared to the standard MRSLVQ (see Fig. 2a). These improvements are measured in terms of training accuracy on randomly selected training data. The performance test was conducted on separate test data with $\gamma = 0.5$ reaching a test accuracy of 85.77% which is slightly better than standard MRSLVQ (see Tab. 1). The Hamacher t-norm turned out to be a less effective. First, the range for the grid search for the optimal parameter γ has to be enlarged to show any effect ($0.001 \leq \gamma \leq 10000$), and second, the classification accuracy of standard MRSLVQ cannot be reached (see Tab. 1).

5.2   Barley Grain Plant Data

The second dataset are images of serial transverse sections of barley grains at different developmental stages. Developing barley grains consist of three genetically different tissue types: the diploid maternal tissues, the filial triploid endosperm, and the diploid embryo. Because of their functionality, cells of a fully differentiated tissue show differences in cell shape and water content and accumulate different compounds. Based on those characteristics, scientists experienced in histology are able to identify and to label differentiated tissues within a given section of a developing grain (segmentation). However, differentiating cells lack these characteristics. Because differentiation occurs along gradients, especially borders between different tissue types of developing grains often consist of differentiating cells, which cannot be identified as belonging to one or the other tissue type. Thus, fuzzy processing is highly desirable. However, since (training) examples, manually labeled by a biological expert, are costly and rarely available, one is interested in automatic classification based on a small training subset of the whole data set. In our example, the training set consists of 4418 data points (vectors) whereas the whole transverse section of the image contains 616×986 samples, which finally have to be classified and visualized as an image for immediate interpretation by biologists. The data vectors are 22-dimensional, the number of classes is $N_c = 11$. Using standard MRSLVQ based on the Product t-norm to classify the plant data yields a classification accuracy of 64.16% (see Tab. 1). Before testing our derived method the optimal parameter values were obtained again by a grid search using a training dataset and comparing the training accuracy to the standard MRSLVQ training accuracy. The plot of the accuracies obtained by the Aczel-Alsina t-norm is depicted in Fig. 2b. Interestingly, the parameter value yielding a slightly better classification accuracy than MRSLVQ is exactly that value, for which Azcel-Alsina t-norm and Product t-norm are equivalent. But nevertheless, as observed before for the Gaussian dataset, applying Aczel-Alsina t-norm with $\gamma = 1.0$ yields an improvement. For the current dataset this improvement amounts

Table 1. Average classification accuracy for the Gaussian dataset and the Barley grain plant data. Note that the non-parametric Product or Probabilistic t-norm is equivalent to the standard MRSLVQ model.

	Gaussian distributions		Barley grain plant data	
	class. acc.	γ	class. acc.	γ
Probabilistic/Product t-norm	0.8517 ± 0.0388	–	0.6416 ± 0.0317	–
Hamacher t-norm	0.8257 ± 0.0530	0.01	0.6664 ± 0.0464	0.01
Aczel-Alsina t-norm	0.8577 ± 0.0429	0.5	0.7857 ± 0.0317	1.0

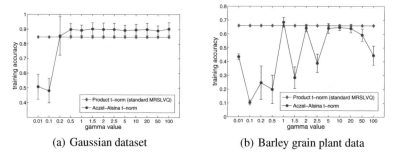

(a) Gaussian dataset   (b) Barley grain plant data

Fig. 2. Grid search for the optimal parameter based on the training accuracy for MRSLVQ incorporating Aczel-Alsina t-norm. The red line indicates the training accuracy for the non-parametric Product t-norm respectively standard MRSLVQ.

to 22.5% in the test accuracy (see Tab. 1). Again, the Hamacher t-norm is less effective. Setting $\gamma = 0.01$ yields an improved classification accuracy of only 3.9% (see Tab. 1).

6   Conclusions

In this work we proposed an extension of Multivariate Robust Soft LVQ incorporating t-norms in the learning dynamic. This is the first proposal of this type for prototype based learning to the authors best knowledge. Unsafe label information is very common for many real life data but not yet sufficiently addressed by appropriate learning methods and our method is a proposal to improve the current situation. The data can reflect the fuzziness in the labeling e.g. by similar scores for different class indices. This is a very similar setting to classical fuzzy-theory and a motivation for the use of t-norms to judge the similarity of label vectors. We considered different t-norms for MRSLVQ and observed that the used t-norms might lead to (slight) improvements in the model accuracy. Especially we found that the implicitly and unwittingly used Product t-norm may not be the best choice. The Aczel-Alsina t-norm performed best in our experiments but a wider study is necessary to get a sufficient support for generic statements. In future work we will address in more detail the theoretical links of the used label norm with respect to a large margin classifier and its generalization capabilities.

Acknowledgment.   Marie Curie Intra-European Fellowship (IEF): FP7-PEOPLE-2012-IEF (FP7-327791-ProMoS) is greatly acknowledged. We would like to thank Petra Schneider for providing the code of the MRSLVQ implementation.

References

[1] Schneider, P., Geweniger, T., Schleif, F.M., Biehl, M., Villmann, T.: Multivariate class labeling in Robust Soft LVQ. In: Verleysen, M. (ed.) 19th European Symposium on Artificial Neural Networks (ESANN 2011), pp. 17–22. d-side publishing (2011)

[2] Bhattacharya, S., Bhatnagar, V.: Fuzzy data mining: A literature survey and classification framework. International Journal of Networking and Virtual Organisations 11(3-4), 382–408 (2012)

[3] Bonilla, E.V., Robles-Kelly, A.: Discriminative probabilistic prototype learning. In: ICML. icml.cc / Omnipress (2012)

[4] Geweniger, T., Villmann, T.: Extending FSNPC to handle data points with fuzzy class assignments. In: Verleysen, M. (ed.) Proc. of European Symposium on Artificial Neural Networks (ESANN 2010), Brussels, Belgium. d-side publications (2010)

[5] Schleif, F.-M., Villmann, T., Hammer, B., Schneider, P.: Efficient kernelized prototype based classification. Int. J. Neural Syst. 21(6), 443–457 (2011)

[6] Schneider, P., Biehl, M., Hammer, B.: Distance learning in discriminative vector quantization. Neural Computation 21(10), 2942–2969 (2009)

[7] Hammer, B., Schleif, F.-M., Zhu, X.: Relational extensions of learning vector quantization. In: Lu, B.-L., Zhang, L., Kwok, J. (eds.) ICONIP 2011, Part II. LNCS, vol. 7063, pp. 481–489. Springer, Heidelberg (2011)

[8] Seo, S., Obermayer, K.: Soft learning vector quantization. Neural Computation 15, 1589–1604 (2003)

[9] Schneider, P.: Advanced methods for prototype-based classification. PhD thesis, Rijksuniveriteit Groningen (2010)

[10] Menger, K.: Statistical metrics. Proceedings of the National Academy of Sciences 28(12), 535–537 (1942)

[11] Gosztolya, G., Dombi, J., Kocsor, A.: Applying the generalized dombi operator family to the speech recognition task. CIT 17(3), 285–293 (2009)

[12] Senthil Kumar, A.V.: Diagnosis of heart disease using fuzzy resolution mechanism. Journal of Artificial Intelligence 5(1), 47–55 (2012)

[13] Ciaramella, A., Tagliaferri, R., Pedrycz, W., di Nola, A.: Fuzzy relational neural network. Int. J. Approx. Reasoning 41(2), 146–163 (2006)

[14] Quost, B., Masson, M.-H., Denaux, T.: Classifier fusion in the dempster-shafer framework using optimized t-norm based combination rules. International Journal of Approximate Reasoning 52(3), 353–374 (2011)

[15] Farahbod, F., Eftekhari, M.: Comparison of different t-norm operators in classification problems. International Journal of Fuzzy Logic Systems 2(3), 33–39 (2012)

[16] Lange, M., Villmann, T.: Derivatives of l_p-norms and their approximations. Machine Learning Reports 04/2013, pp. 43–59 (2013)

[17] Schneider, P., Biehl, M., Hammer, B.: Hyperparameter learning in probabilistic prototype-based models. Neurocomputing 73(7-9), 1117–1124 (2010)

Rejection Strategies for Learning Vector Quantization – A Comparison of Probabilistic and Deterministic Approaches

Lydia Fischer[1,2], David Nebel[3], Thomas Villmann[3],
Barbara Hammer[2], and Heiko Wersing[1]

[1] HONDA Research Institute Europe GmbH, Offenbach, Germany
[2] Bielefeld University, Germany
[3] University of Applied Sciences Mittweida, Germany

Abstract. In this contribution, we focus on reject options for prototype-based classifiers, and we present a comparison of reject options based on statistical models for prototype-based classification as compared to alternatives which are motivated by simple geometric principles. We compare the behavior of generative models such as Gaussian mixture models and discriminative ones to results from robust soft learning vector quantization. It turns out that (i) reject options based on simple geometric show a comparable quality as compared to reject options based on statistical approaches. This behavior of the simple options offers a nice alternative towards making a probabilistic modeling and allowing a more fine-grained control of the size of the remaining data in many settings. It is shown that (ii) discriminative models provide a better classification accuracy also when combined with reject strategies based on probabilistic models as compared to generative ones.

Keywords: prototype-based reject option, classification.

1 Introduction

Learning vector quantization (LVQ) [15] constitutes a powerful and efficient classification strategy particularly suited for multi-class classification or online scenarios. It can be substantiated by strong mathematical guarantees for generalization behavior as well as learning dynamics for modern cost function based versions such as generalized LVQ (GLVQ) [21] or robust soft LVQ (RSLVQ) [23]. In application scenarios, however, perfect classification can rarely be achieved due to inherent noise in the data, overlap of classes, missing sensors, etc. Essentially, a reject option relaxes the constraint of a classifier to provide a class label for a given input with a low confidence value, rather an explicit 'don't know' is accepted as a return in such cases.

Note that most classifiers actually do provide a continuous value rather than a crisp output only such as the distance of a given data point to the decision boundary. Together with an appropriate threshold, these numbers could be taken as a reject option. However, the real-valued outputs provided by the classifiers

T. Villmann et al. (eds.), *Advances in Self-Organizing Maps and Learning
Vector Quantization,* Advances in Intelligent Systems and Computing 295,
DOI: 10.1007/978-3-319-07695-9_10, © Springer International Publishing Switzerland 2014

can usually not be interpreted as a confidence measure because their scaling is unclear and can vary locally. A variety of approaches is concerned with techniques how to turn these values into a statistical confidence [20,27], or how to define appropriate, possibly local thresholds for a reject option which respects a different scaling of the values [9,25]. Interestingly, while a number of efficient strategies have been realized for popular classification schemes like support vector machines or k-nearest neighbor classifiers [4,20,27,7,12,9,6], relatively little approaches address prototype-based learning strategies such as LVQ [25,5,13]. Another idea is the distance-based two stage approach from [16] which separately addresses outliers and ambiguous regions. An approach, which combines a reject option with empirical risk minimization for a binary classifier, is proposed in [11] which could be a direction of further research.

In this approach we investigate reject options for prototype-based learning schemes such as LVQ. In particular, we investigate approaches which are inspired by the geometric nature of LVQ classifiers and we compare these reject options to reject options based on confidence values. We consider the key question: Are these geometric approaches comparable to reject strategies based on confidence values of probabilistic models which can be optimal as shown in [4], and if so under which conditions? Therefore, we systematically compare the behavior of the measures to rejection strategies for probabilistic models. We vary (i) the rejection strategy, ranging from deterministic, geometric measures to reject options based on confidence values, (ii) the data set, ranging from artificial data to typical benchmarks, and (iii) the nature of the prototype-based model for which the reject option is taken, considering purely discriminative models in comparison to generative ones. Albeit both classifiers are derived as explicit probabilistic models. Purely discriminative ones are tailored to the classification task rather than the data, such that it is not clear whether reject strategies can be based on their confidence values. Similarly, it is not clear whether efficient deterministic strategies based on simple geometric quantities can reach the performance of rejection strategies on confidence values, the latter is supposed to require valid probabilistic models of the data. We will show that this is indeed the case for real life settings: heuristic reject strategies based on geometric considerations offer an alternative to measures based on a confidence value, thus offering a way towards reject strategies for purely deterministic LVQ schemes.

2   Probabilistic Prototype-Based Classification

Assume a data set $\boldsymbol{X}$ with elements of the real vector space $\mathbb{R}^n$. A prototype-based classifier is characterized by a set of prototypes $\boldsymbol{W} = \{\boldsymbol{w}_i \in \mathbb{R}^n\}_{i=1}^k$, which are equipped with labels $c(\boldsymbol{w}_i) \in \{1, \ldots, C\}$, if a classification into C classes is considered. Classification of a data point $\boldsymbol{x} \in \mathbb{R}^n$ takes place by a winner takes all (WTA) scheme: $\boldsymbol{x}$ is mapped to the label $c(\boldsymbol{x}) = c(\boldsymbol{w}_i)$ of the prototype $\boldsymbol{w}_i$ which is closest to $\boldsymbol{x}$ as measured in some distance measure. Often, the standard squared euclidean distance $\|\boldsymbol{x} - \boldsymbol{w}_i\|^2$ or a generalized quadratic form

$(\boldsymbol{x} - \boldsymbol{w}_i)^T \Lambda (\boldsymbol{x} - \boldsymbol{w}_i)$ with positive semi-definite matrix Λ is considered; generalizations to more general dissimilarity measures such as divergences, functional metrics, or general dissimilarities have also been proposed [26,10].

Due to its simple classification scheme and the representation of the model in terms of few prototypes, prototype-based classification enjoys a wide popularity. Additional there are diverse learning techniques available to induce an appropriate model from a given data set. Popular learning techniques include the classical family of LVQ as proposed by Kohonen [15], generalizations of LVQ which establish the model by cost functions [21,23], or unsupervised learning schemes equipped with posterior labeling like neural gas or extensions thereof [17,2]. Here, we have a glimpse at two different strategies which play a role in the subsequent experiments. We only consider probabilistic LVQ models, because the results allow a direct use of a reject option on their confidence values.

RSLVQ: Robust soft learning vector quantization (RSLVQ) has been proposed as a probabilistic model which, in the limit of small bandwidth, yields update rules very similar to classical LVQ 2.1 [23]. The objective is given as

$$E = \sum_j \log p(y_j|\boldsymbol{x}_j, \boldsymbol{W}) = \sum_j \log \frac{p(\boldsymbol{x}_j, y_j|\boldsymbol{W})}{p(\boldsymbol{x}_j|\boldsymbol{W})} \tag{1}$$

where $p(\boldsymbol{x}_j|\boldsymbol{W}) = \sum_i p(\boldsymbol{w}_i)p(\boldsymbol{x}_j|\boldsymbol{w}_i)$ constitutes a mixture of Gaussians with prior probability $p(\boldsymbol{w}_i)$ usually taken uniformly over all prototypes. The probability $p(\boldsymbol{x}_j|\boldsymbol{w}_i)$ is usually taken as an isotropic Gaussian centered in $\boldsymbol{w}_i$ with fixed variance σ^2, or a generalization thereof with a more general covariance matrix. The probability $p(\boldsymbol{x}_j, y_j|\boldsymbol{W}) = \sum_i \delta_{c(\boldsymbol{x}_j)}^{c(\boldsymbol{w}_i)} p(\boldsymbol{w}_i)p(\boldsymbol{x}_j|\boldsymbol{w}_i)$ (δ_i^j being the Kronecker delta) restricts to the mixture components with the correct labeling. This likelihood ratio is optimized using a gradient technique. RSLVQ provides an explicit confidence value $p(y|\boldsymbol{x}, \boldsymbol{W})$ for every class y of a given data point $\boldsymbol{x}$.

GMM: Albeit RSLVQ is derived from a probabilistic model, its cost function is purely discriminative. This means model parameters do not necessarily yield to a good generative model for the observed data $\boldsymbol{x}$. As shown in [22], for example, this is not the case in general. In practice, generative data models are often trained in an unsupervised way, directly aiming at a representation of the data distribution $p(\boldsymbol{x})$, popular examples being Gaussian mixture models for density estimation. Here we consider a class-wise Gaussian mixture model (GMM) which aims at a representation of every class by optimizing the following data log-likelihood

$$E = \sum_j \log \left(\sum_i \delta_{c(\boldsymbol{x}_j)}^{c(\boldsymbol{w}_i)} p(\boldsymbol{w}_i)p(\boldsymbol{x}_j|\boldsymbol{w}_i) \right) \tag{2}$$

where $p(\boldsymbol{x}_j|\boldsymbol{w}_i)$ is a Gaussian distribution centered in $\boldsymbol{w}_i$, and $p(\boldsymbol{w}_i)$ is the class-wise prior of the prototype with $\sum_j \delta_{c(\boldsymbol{x}_j)}^{c(\boldsymbol{w}_j)} p(\boldsymbol{w}_j) = 1$. The model parameters can be optimized by means of a gradient technique or, alternatively, a classical EM scheme for every class, since the objective decomposes according to the class labels [3]. A GMM provides for each class y an explicit confidence measure

$p(y|\boldsymbol{x}, \boldsymbol{W}) = p(y)p(\boldsymbol{x}, c(\boldsymbol{x})|\boldsymbol{W})/\sum_{z\in\{1,...,C\}} p(z)p(\boldsymbol{x}, z|\boldsymbol{W})$ where, due to the training procedure, a generative data model representing the distribution on $\boldsymbol{x}$ is present. In this context $p(y)$ is the prior of the class with $\sum_{y\in\{1,...,C\}} p(y) = 1$.

Since GMM and RSLVQ offer probabilistic models, the classification of a data point $\boldsymbol{x}$ can be based on the most likely class $\mathrm{argmax}_y p(y|\boldsymbol{x}, \boldsymbol{W})$. In practice, the resulting maximum y often corresponds to the class of the closest prototype such that a close resemblance to a classical WTA scheme is obtained.

3   Reject Options

What are possible rejection measures of prototype-based models which correlate to the confidence of a classification and, together with a rejection strategy such as a simple threshold, lead to a reject option? In general, a rejection measure constitutes a function $r : \mathbb{R}^n \to \mathbb{R}$, $r(\boldsymbol{x})$ indicating the certainty of the classification of a data point $\boldsymbol{x}$, together with an ordering direction, which specifies whether low or high values of $r(\boldsymbol{x})$ correspond to a high certainty of the classification. We assume that a rejection measure is always scaled in such a way that smaller values correspond to a lower certainty. We consider the following rejection measures.

Conf: Chow proved for a Bayes classifier with known class densities that a reject option on $r_{\mathrm{Conf}}(\boldsymbol{x}) = \max_y p(y|\boldsymbol{x})$ reaches the optimum error-reject trade-off: for a certain error rate (error probability) it minimizes the reject rate (reject probability)[4]. This means to reject a data point if $r_{\mathrm{Conf}}(\boldsymbol{x}) < \theta$. This strategy relies on the assumption that a good probabilistic model of the data is given, otherwise guarantees as proved e. g. in [11] do not necessarily hold. Note that in regions with low class densities this measure can return high confidence values caused by normalization, thus it cannot exclude outliers. Our measure (Fig. 1) is inspired by the one of Bayes but the values are calculated by the mentioned models and not by a Bayes classifier.

Dist: This error measure is inspired by geometric considerations. It returns the distance of $\boldsymbol{x}$ to the closest decision boundary. Assume $\boldsymbol{w}^+$ and $\boldsymbol{w}^-$ correspond to prototypes with a different labeling and neighbored receptive fields with the belonging distances d^+ and d^- to $\boldsymbol{x}$. Then, the distance of a data point $\boldsymbol{x}$ to the decision boundary defined by these two prototypes is given as $r_{\mathrm{Dist}}(\boldsymbol{x}) = \frac{|d^+ - d^-|}{2\|\boldsymbol{w}^+ - \boldsymbol{w}^-\|^2}$ (Fig. 1). If only one prototype per class is present, the prototypes $\boldsymbol{w}^+$ and $\boldsymbol{w}^-$ are given by the two closest prototypes of the data point $\boldsymbol{x}$. Provided a class is represented by more prototypes than one, the underlying topology has to be estimated using e.g. the Hebbian learning strategy as proposed in [18].

d^+: This error measure is also geometrically inspired, treating points which are outliers with low confidence. This is measured by the squared distance to the closest prototype $r_{d+}(\boldsymbol{x}) = -d^+(\boldsymbol{x})$ (Fig. 1).

Note that these reject options differ in the following items:

- *Motivation of r:* There are essentially two different reasons to reject a data point, which are referred to in the literature as a rejection because of an

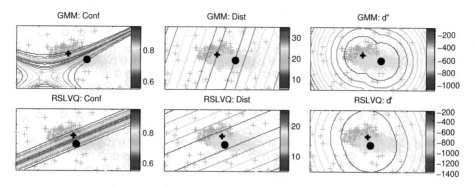

Fig. 1. Level curves of the considered reject options for a GMM and a RSLVQ model of an artificial 2D Gaussian data set. The black symbols are prototypes.

ambiguous classification, or a rejection because of the data point being an outlier [25]. The reject measures as given above follow different principles. **Conf** realizes a rejection because of ambiguity, since it requires that the maximum class probability reaches the threshold θ. Due to the normalization of probabilities, this results in a gap of the class probabilities. **Dist** explicitly realizes an ambiguous reject option by referring to the class boundary, while $\mathbf{d}^+$ realizes an outlier reject option.

- *Scaling of r:* For **Conf**, values are in the interval $[0, 1]$ allowing a direct interpretation as statistical confidence value. This fact offers a simple way to set an appropriate threshold due to external requirements regarding the confidence, for example. In contrast, the other measures take values in the real numbers, but their scaling is not clear. Since the scaling can even vary locally and it can depend nonlinearly on the confidence, a proper choice of a threshold is unclear. We will investigate global threshold strategies in experiments, yielding results comparable to reject options based on the confidence.
- *Requirements as regards the model:* The scaling of **Conf** as a confidence measures requires that a probabilistic model of the data is available. We investigate the effect of having a discriminative versus generative model in experiments, only the latter actually providing a valid representation of the input distribution in general.

These measures provide values indicating the confidence of a classification such that they give rise to a direct threshold-based rejection strategy: given $\theta \in \mathbb{R}$, points which fulfill $r(\boldsymbol{x}) < \theta$ are rejected. Since measures such as **Dist** and $\mathbf{d}^+$ aim at a rejection caused by different reasons. It can be worthwhile to combine several measures [25]. This leads to a more complex rejection strategy which depends on two thresholds. We refer to this measure as follows:

Comb: This measure combines the previous two reject options $r_{\mathrm{Comb}}(\boldsymbol{x}) = (r_{\mathrm{Dist}}(\boldsymbol{x}), r_{d+}(\boldsymbol{x}))$ leading to a reject strategy based on a threshold vector $\boldsymbol{\theta} = (\theta_1, \theta_2)$: $\boldsymbol{x}$ is rejected if $r_{\mathrm{Dist}}(\boldsymbol{x}) < \theta_1$ or $r_{d+}(\boldsymbol{x}) < \theta_2$.

4 Experiments

We test the behavior of the different rejection measures in experiments, focusing on the following questions: What is the behavior of the measures regarding different characteristics of the model ranging from a discriminative to a generative one? What is the behavior of simple deterministic heuristics in comparison to rejection strategies based on confidence measures and do the latter require valid probabilistic models? Since probabilistic models are needed for an evaluation of **Conf**, we use the two probabilistic models RSLVQ and GMM. For all settings, RSLVQ and GMM are trained using one prototype per class. For RSLVQ, a global parameter σ^2 is optimized via cross-validation. For GMM, correlations are set to zero and local scalings of the dimensions are adapted by means of diagonal matrices attached to the prototypes which are optimized in an EM scheme. Training takes place until convergence using random initialization and without leave-one-out method. Convergence is assumed if the training error changes less than 10^{-5} during two sequenced training steps. We use the following data sets:

- *Gaussian clusters*: This data set consists of two artificially generated Gaussian clusters in two dimensions with overlap. These are overlaid with uniform noise in the plane. Data are randomly divided into training and test set.
- *Image Segmentation*: The image segmentation data set consists of 2310 data points representing small patches from outdoor images with 7 different classes with equal distribution such as brickface, sky, ... [1]. Each data point consists of 19 real-valued image descriptors. The data set is decomposed into a training set of 210 data points and a test set of 2100 data points. Due to zero variance, dimensions 3 to 5 are deleted, and data are normalized by a z-transformation before training.
- *Tecator data*: The Tecator data set consists of 215 spectra with 100 spectral bands ranging from 850 nm to 1050 nm [24]. The task is to predict the fat content of the probes, which is turned into a two class classification problem to predict a high/low fat content by means of binning the real values into two classes of equal size. Data are randomly split into a training set with 144 samples and test set with 71 samples.
- *Haberman*: The Haberman survival data set contains 306 instances from two classes indicating the survival for more than 5 years after breast cancer surgery [1]. Data are represented by three attributes related to the age, the year, and the number of positive axillary nodes detected. Data are randomly split into training and test set of equal size.

For all data sets, two models are trained: a probabilistic generative model by means of class-wise GMM, and a probabilistic discriminative model by means of RSLVQ. For the resulting models, the effect of a reject option is compared for different possible strategies as introduced above. We vary the reject threshold θ in small steps from no reject (which corresponds to the original model) to full reject (i.e. no data point is classified). For **Comb**, a threshold vector is varied accordingly, and we report the result of the respective best combination. We denote the set of data points which are not rejected using θ as $\boldsymbol{X}_\theta$. The results are depicted as graphs plotting the relative size $|\boldsymbol{X}_\theta|/|\boldsymbol{X}|$ versus the classification accuracy on $\boldsymbol{X}_\theta$ normalized by its size.

Figure 2 shows the results obtained for the different rejection strategies and data sets. The resulting graphs [19] display a smooth transition from the accuracy of the model without reject options to the limit value 1 (in the case of Gaussian clusters it goes to 0) which results if $|\boldsymbol{X}_\theta|$ approaches 0 (we leave out the value for the empty set at $|\boldsymbol{X}_\theta| = 0$). The classification accuracy on $\boldsymbol{X}_\theta$ does not change with θ if the classification accuracy is already 100 % (as is the case for the Tecator data set for RSLVQ), or if the errors are uniformly distributed over the range of the rejection measure r which is the case for the Haberman data set, for example. In the latter case, classes are imbalanced with the second class accounting for roughly one third of the data only, and LVQ models tend to represent only class one properly, such that class two accounts for errors equally distributed according to r. Note that the graphs are subject of noise if the size $|\boldsymbol{X}_\theta|$ approaches 0 which can be attributed to the small sample size $\boldsymbol{X}_\theta$. Accordingly, the graphs are not reliable for $|\boldsymbol{X}_\theta|/|\boldsymbol{X}| < 0.1$, and the corresponding parts of the graphs should be seen as an indicator only. We choose the values of θ equidistant between the extremal values of each single measure.

Interestingly, the control of the number of points which are not rejected, $|\boldsymbol{X}_\theta|$, depending on the threshold θ partially has gaps, as indicated in Fig. 2 by the straight parts of the curves and the ending of the curves at some size of $|\boldsymbol{X}_\theta| \gg 0$. Such gaps can occur provided the size of $\boldsymbol{X}_\theta$ changes abruptly with the threshold, which seems to be the case in some settings where a further increase of the thresholds leads to a rejection of all remaining data points. This is the fact for **Conf** for Gaussian clusters, Image Segmentation and Tecator for the GMM model, indicating that no points with confidence larger than a maximum threshold value θ exist. Interestingly these gaps can be observed for **Conf** for the generative models only, not the discriminative ones. Further, this behavior is observed for $\mathbf{d}^+$ for the data sets Gaussian clusters and Image Segmentation (both models) and Tecator (generative model). In contrast, the graphs of **Dist** and **Comb** do not have large gaps.

We can draw a few general conclusions from the graphs displayed in Fig. 2: In all cases, the discriminative model RSLVQ yields the same or better results as compared to generative GMM models, albeit the latter have a higher degree of freedom because of an adaptive diagonal matrix per prototype unlike RSLVQ, which relies on a global bandwidth only. This also holds for the full range of certainty values taken for the reject strategies, regardless of whether deterministic of probabilistic rejection measures are used. Thus, it seems advisable to focus on the discriminative task, where confidence based measure or deterministic measures can be used. As expected, reject strategies based on the confidence yields the best behavior in most cases, but it does not allow a smooth variation of the size of $\boldsymbol{X}_\theta$ for a large range in two of the settings. As mentioned in Section 3 **Conf** cannot exclude outliers. This is apparently not a problem for the used data sets, highlighting the applicability of the optimality criterion of Chow [4].

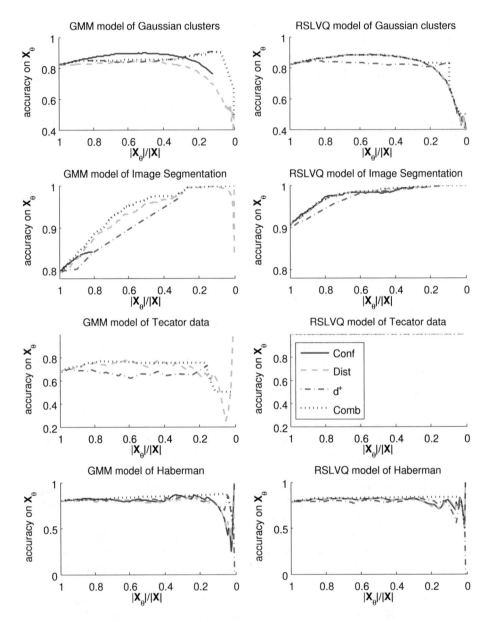

Fig. 2. Results of different rejection options when applied to generative or discriminative models trained for different data sets. We report the relative size of $\boldsymbol{X}_\theta$ as compared to the accuracy of the classifier on this set [19].

Dist seems to offer a reasonable strategy in all other settings, whereby the behavior is universally good for generative as well as discriminative models, and it relaxes the burden of computing an explicit confidence value. $\mathbf{d}^+$ gives better results than **Dist** in only one case (Gaussian clusters, GMM), and worse results than **Dist** in three cases (Gaussian cluster, RSLVQ; Image segmentation, both models; Tecator, GMM). Thus, in general, focusing on the discriminative nature seems advisable also as concerns the rejection strategy. As expected, **Comb** shows results comparable to the best of the two geometric reject options **Dist** and $\mathbf{d}^+$, but also requiring a more complex reject strategy by the combination of both values.

5   Conclusions

We have compared direct geometric reject options and their combination with Bayesian motivated reject options in a couple of benchmarks using models with different characteristics. The resulting observations are that geometric measures such as **Dist** behave equally good as probabilistic measures, while often allowing a more fine-grained control of the size of the rejected data set. In addition, they do not require explicit probabilistic models thus opening the way for an integration into powerful deterministic alternatives such as GLVQ [21]. The suitability of the approach to these settings is the topic of ongoing work [8].

While allowing for simple measures which are applicable for a wider range of models, the scaling of appropriate thresholds is not clear a priori and it depends on the data set at hand. In the literature, a few proposals how to automatically determine data-adapted values have been proposed [25], which can be transferred to our setting. They can even be extended to online scenarios, and LVQ classifiers offer intuitive life-long learning strategies [14].

Acknowledgement. BH gratefully acknowledges funding by the CITEC center of excellence. DN acknowledges funding by EFS. LF acknowledges funding by the CoR-Lab Research Institute for Cognition and Robotics and gratefully acknowledges the financial support from Honda Research Institute Europe.

References

1. Bache, K., Lichman, M.: UCI machine learning repository (2013)
2. Beyer, O., Cimiano, P.: Online Semi-Supervised Growing Neural Gas. Int. J. Neural Syst. 22(5) (2012)
3. Bishop, C.M.: Pattern Recognition and Machine Learning. Springer (2006)
4. Chow, C.K.: On Optimum Recognition Error and Reject Tradeoff. IEEE Transactions on Information Theory 16(1), 41–46 (1970)
5. De Stefano, C., Sansone, C., Vento, M.: To Reject or Not to Reject: That is the Question-An Answer in Case of Neural Classifiers. IEEE Transactions on Systems, Man, and Cybernetics, Part C 30(1), 84–94 (2000)
6. Delany, S.J., Cunningham, P., Doyle, D., Zamolotskikh, A.: Generating Estimates of Classification Confidence for a Case-Based Spam Filter. In: Muñoz-Ávila, H., Ricci, F. (eds.) ICCBR 2005. LNCS (LNAI), vol. 3620, pp. 177–190. Springer, Heidelberg (2005)

7. Devarakota, P.R., Mirbach, B., Ottersten, B.: Confidence Estimation in Classification Decision: A Method for Detecting Unseen Patterns. In: Int. Conf. on Advances in Pattern Recognition, ICAPR 2007 (2006)
8. Fischer, L., Hammer, B., Wersing, H.: Rejection Strategies for Learning Vector Quantization (submitted)
9. Fumera, G., Roli, F., Giacinto, G.: Reject option with multiple thresholds. Pattern Recognition 33(12), 2099–2101 (2000)
10. Hammer, B., Hofmann, D., Schleif, F.-M., Zhu, X.: Learning vector quantization for (dis-)similarities. Neurocomputing (accepted)
11. Herbei, R., Wegkamp, M.H.: Classification with reject option. Can. J. Statistics 34(4), 709–721 (2006)
12. Hu, R., Delany, S.J., Namee, B.M.: Sampling with Confidence: Using k-NN Confidence Measures in Active Learning. In: Proceedings of the UKDS Workshop at 8th Int. Conf. on Case-based Reasoning, ICCBR 2009, pp. 181–192 (2009)
13. Ishidera, E., Nishiwaki, D., Sato, A.: A confidence value estimation method for handwritten Kanji character recognition and its application to candidate reduction. Int. J. on Document Analysis and Recognition 6(4), 263–270 (2004)
14. Kirstein, S., Wersing, H., Gross, H.-M., Körner, E.: A Life-Long Learning Vector Quantization Approach for Interactive Learning of Multiple Categories. Neural Networks 28, 90–105 (2012)
15. Kohonen, T.: Self-Organization and Associative Memory, 3rd edn. Springer Series in Information Sciences. Springer (1989)
16. Landgrebe, T., Tax, D.M.J., Paclík, P., Duin, R.P.W.: The interaction between classification and reject performance for distance-based reject-option classifiers. Pattern Recognition Letters 27(8), 908–917 (2006)
17. Martinetz, T., Berkovich, S., Schulten, K.: Neural-gas Network for Vector Quantization and its Application to Time-Series Prediction. IEEE-Transactions on Neural Networks 4(4), 558–569 (1993)
18. Martinetz, T., Schulten, K.: Topology representing networks. Neural Networks 7(3), 507–522 (1994)
19. Nadeem, M.S.A., Zucker, J.-D., Hanczar, B.: Accuracy-Rejection Curves (ARCs) for Comparing Classification Methods with a Reject Option. In: MLSB, pp. 65–81 (2010)
20. Platt, J.C.: Probabilistic Outputs for Support Vector Machines and Comparisons to Regularized Likelihood Methods. In: Advances in Large Margin Classifiers, May 23. MIT Press (1999)
21. Sato, A., Yamada, K.: Generalized Learning Vector Quantization. In: Advances in Neural Information Processing Systems, vol. 7, pp. 423–429 (1995)
22. Schneider, P., Biehl, M., Hammer, B.: Distance learning in discriminative vector quantization. Neural Computation 21(10), 2942–2969 (2009)
23. Seo, S., Obermayer, K.: Soft Learning Lector Quantization. Neural Computation 15(7), 1589–1604 (2003)
24. Thodberg, H.H.: Tecator data set, contained in StatLib Datasets Archive (1995)
25. Vailaya, A., Jain, A.K.: Reject Option for VQ-Based Bayesian Classification. In: Int. Conf. on Pattern Recognition (ICPR), pp. 2048–2051 (2000)
26. Villmann, T., Haase, S.: Divergence-Based Vector Quantization. Neural Computation 23(5), 1343–1392 (2011)
27. Wu, T.-F., Lin, C.-J., Weng, R.C.: Probability Estimates for Multi-class Classification by Pairwise Coupling. J. of Machine Learning Research 5, 975–1005 (2004)

Part III

Classification and Non-standard Metrics

Prototype-Based Classifiers and Their Application in the Life Sciences

Michael Biehl

Univ. of Groningen - Johann Bernoulli Inst. for Math. and Computer Science
P.O. Box 407, 9700 AK Groningen - The Netherlands

Abstract. This talk reviews important aspects of prototype based systems in the context of supervised learning. Learning Vector Quantization (LVQ) serves as a particularly intuitive framework, in which to discuss the basic ideas of distance based classification. A key issue is that of chosing an appropriate distance or similarity measure for the task at hand. Different classes of distance measures, which can be incorporated into the LVQ framework, are introduced. The powerful framework of relevance learning will be discussed, in which parameterized distance measures are adapted together with the prototypes in the same training process. Recent developments and theoretical insights are discussed and example applications in the bio-medical domain are presented in order to illustrate the concepts.

T. Villmann et al. (eds.), *Advances in Self-Organizing Maps and Learning Vector Quantization,* Advances in Intelligent Systems and Computing 295,
DOI: 10.1007/978-3-319-07695-9_11, © Springer International Publishing Switzerland 2014

Generative versus Discriminative Prototype Based Classification

Barbara Hammer[1], David Nebel[2], Martin Riedel[2], and Thomas Villmann[2]

[1] Bielefeld University, Germany
bhammer@techfak.uni-bielefeld.de
[2] University of Applied Sciences Mittweida, Germany
{nebel,riedel,villmann}@hs-mittweida.de

Abstract. Prototype-based models such as learning vector quantization (LVQ) enjoy a wide popularity because they combine excellent classification and generalization ability with an intuitive learning paradigm: models are represented by few characteristic prototypes, the latter often being located at class typical positions in the data space. In this article we investigate inhowfar these expectations are actually met by modern LVQ schemes such as robust soft LVQ and generalized LVQ. We show that the mathematical models do not explicitly optimize the objective to find representative prototypes. We demonstrate this fact in a few benchmarks. Further, we investigate the behavior of the models if this objective is explicitly formalized in the mathematical costs. This way, a smooth transition of the two partially contradictory objectives, discriminative power versus model representativity, can be obtained.

1 Introduction

Since its invention by Kohonen [9], learning vector quantization (LVQ) enjoys a great popularity by practitioners for a number of reasons: the learning rule as well as the classification model are very intuitive and fast; the resulting classifier is interpretable since it represents the model in terms of typical prototypes which can be treated in the same way as data; unlike popular alternatives such as SVM the model can easily deal with an arbitrary number of classes; the representation of data in terms of prototypes lends itself to simple incremental learning strategies by referring to the prototypes as statistics for the already learned data. Due to these properties, LVQ has been successfully applied in diverse areas ranging from telecommunications and robotics to the biomedical domain [9,8].

Despite this success, LVQ has long been thought of as a mere heuristic [2] and some mathematical guarantees concerning its convergence properties or its generalization ability have been investigated more than ten years after its invention only [3,1,13]. Today, LVQ is usually no longer used in its basic form, rather variants which can be derived from mathematical cost functions are used such as generalized LVQ (GLVQ) [12], robust soft LVQ (RSLVQ) [16], or soft nearest prototype classification [15]. Further, one of the success stories of LVQ is linked to its combination with more powerful, possibly adaptive metrics instead

T. Villmann et al. (eds.), *Advances in Self-Organizing Maps and Learning Vector Quantization*, Advances in Intelligent Systems and Computing 295,
DOI: 10.1007/978-3-319-07695-9_12, © Springer International Publishing Switzerland 2014

of the standard Euclidean one, including, for example, an adaptive quadratic form [7,13], a general kernel [11,6], a functional metric [17], or extensions to discrete data structures [4].

Depending on the application domain, the objective of LVQ to find a highly discriminative classifier is accompanied by additional demands such as sparsity of the models or model interpretability. Modern LVQ techniques such as RSLVQ or GLVQ are explicitly derived from cost functions, such that it is possible to link the objectives of a practitioner to the mathematical objective as modeled in these cost functions. In this contribution, we argue that, while often used as an interpretable model, the objective of arriving at representative prototypes is usually not included in this mathematical objective. We propose an extension of LVQ schemes which explicitly takes this objective into account and which allows a weighting of the two partially contradictory objectives of discriminative power and representativity. We demonstrate the behavior of the resulting models in benchmark data sets where, depending on the setting, models with very different characteristics can be obtained this way.

2   LVQ Schemes

A LVQ classifier is given by a set of prototypes $w_i \in \mathbb{R}^n$, $i = 1, \ldots, k$ together with their labeling $c(w_i) \in \{1, \ldots, C\}$, assuming C classes. Classification of a point $x \in \mathbb{R}^n$ takes place by a winner takes all scheme: x is mapped to the label $c(x) = c(w_i)$ of the prototype w_i which is closest to x as measured in some distance measure, a probability in case of a RSLVQ classifier, respectively. For simplicity, we restrict to the Euclidean metric, even though general metrics could be used.

Given a training data set $x_j \in \mathbb{R}^n$, $j = 1, \ldots, m$, together with labels $y_j \in \{1, \ldots, C\}$, LVQ aims at finding prototypes such that the resulting classifier achieves a good classification accuracy, i.e. $y_j = c(x_j)$ for as many j as possible. Classical LVQ schemes such as LVQ 1 or LVQ 2.1 rely on Hebbian learning heuristics, but they do not relate to a valid underlying cost function in the case of a continuous data distribution [2]. A few alternative models have been proposed which are derived from explicit cost functions and which lead to learning rules resembling the update rules of classical LVQ schemes [12,16].

Generalized LVQ (GLVQ) [12] addresses the following cost function

$$E = \sum_j \Phi \left(\frac{d^+(x_j) - d^-(x_j)}{d^+(x_j) + d^-(x_j)} \right) \tag{1}$$

where $d^+(x_j)$ refers to the squared Euclidean distance of x_j to the closest prototype labeled with y_j, and $d^-(x_j)$ refers to the squared Euclidean distance of x_j to the closest prototype labeled with a label different from y_j. Φ refers to a monotonic function such as the identity or the sigmoidal function. Optimization typically takes place using a gradient technique. As argued in [13], the numerator of the summands can be linked to the so-called hypothesis margin of the classifier, such that a large margin and hence good generalization ability is aimed for while training. The denominator prevents divergence and numerical instabilities by normalizing the costs.

Robust soft LVQ (RSLVQ) [16] yields similar update rules based on the following probabilistic model

$$E = \sum_j \log \frac{p(\boldsymbol{x}_j, y_j | W)}{p(\boldsymbol{x}_j | W)} = \sum_j \log p(y_j | \boldsymbol{x}_j, W) \tag{2}$$

where $p(\boldsymbol{x}_j | W) = \sum_i p(\boldsymbol{w}_i) p(\boldsymbol{x}_j | \boldsymbol{w}_i)$ constitutes a mixture of Gaussians with prior probability $p(\boldsymbol{w}_i)$ (often taken uniformly over all prototypes) and probability $p(\boldsymbol{x}_j | \boldsymbol{w}_i)$ of the point $\boldsymbol{x}_j$ being generated from prototype $\boldsymbol{w}_i$, usually taken as an isotropic Gaussian centered in $\boldsymbol{w}_i$, or a slightly extended version described by a diagonal covariance matrix. The probability $p(\boldsymbol{x}_j, y_j | W) = \sum_i \delta_{y_j}^{c(\boldsymbol{w}_i)} p(\boldsymbol{w}_i) p(\boldsymbol{x}_j | \boldsymbol{w}_i)$ (δ - Kronecker delta) restricts to the mixture components with the correct labeling. This likelihood ratio is optimized using a gradient technique.

When inspecting these cost functions, the question occurs to what extend these LVQ schemes mirror the following objectives:

- **Discriminative Power:** the primary objective of LVQ schemes is to provide a classifier with small classification error on the underlying data distribution. Thus, its objective is to minimize the training error and, more importantly, classification error for new data points.
- **Representativity:** the resulting prototypes should represent the data in an accurate way such that it is possible to interpret the model by inspecting the learned prototypes.

Inhowfar are these objectives accounted for by the GLVQ or RSLVQ costs? Interestingly, RSLVQ aims at a direct optimization of the Bayesian error. Hence, its primary goal is the discriminative power of the model. RSLVQ has no incentive to find representative prototypes unless this fact directly contributes to a good discriminative model. This behavior has been observed in practice [14]: prototypes usually do not lie at class typical positions; they can be located outside the convex hull of the data, for example, provided a better classification accuracy. This behavior has also theoretically been investigated for the limit of small bandwidth in [1]: in the limit of small bandwidth, learning from mistakes takes place, i.e. prototype locations are adapted only if misclassifications are present. We will show one such example for original RSLVQ in the experiments.

What about the GLVQ costs? The numerator of GLVQ is negative if and only if the classification of the considered data point is correct. In addition, it resembles the hypothesis margin of the classifier. Due to this fact, one can expect a high correlation of the classification error and the cost function, making GLVQ suitable as a discriminative model. Nevertheless, since this correlation is not an exact equivalence, minima of this cost function do not necessarily correspond to good classifications in all situations: for highly imbalanced data, for example, the GLVQ costs prefer trivial solutions with all data being assigned to the majority class. This observation is also demonstrated by the fact that the classification accuracy of GLVQ can be inferior as compared to RSLVQ, the latter focussing on discrimination only, see e.g. [14] and our results in the experiments section.

Interestingly, the GLVQ costs have a mild tendency to find representative prototypes due to this form: The term $d^+(\boldsymbol{x})$ in the numerator aims at a small class-wise quantization error of the data. Further, solutions with small denominator are preferred, i.e. there is an emphasis to place all prototypes within the data set. We will see in experiments, that this compromise of representativity and discriminative behavior can yield to classification results inferior to RSLVQ for the sake of more representative models, but still an increase of model representativity is possible by adding a corresponding term to the costs.

3   Extending LVQ Schemes by Generative Modes

We are interested in a model-consistent extension of the RSLVQ and GLVQ costs which explicitly take the goal of representativity into account. Generally, we refer to the cost function of RSLVQ (2) or GLVQ (1) as $E_{\text{discr}}(W)$. The idea is to substitute these costs by the form

$$E - (1 - \alpha) \cdot E_{\text{discr}}(W) + \alpha \cdot E_{\text{repr}}(W) \tag{3}$$

where $E_{\text{repr}}(W)$ emphasizes the objective to find representative prototypes $\boldsymbol{w}_j$. The parameter $\alpha \in [0, 1]$ weights the influence of both parts for the optimization.

First, we have a look at how to choose $E_{\text{repr}}(W)$ for RSLVQ schemes. The idea is to add a term which maximizes the likelihood of the observed data being generated by the underlying model. Similar to RSLVQ, we can consider a class-wise Gaussian mixture model $p(\boldsymbol{x}_j, y_j|W) = \sum_i \delta_{y_j}^{c(\boldsymbol{w}_i)} p(\boldsymbol{w}_i)p(\boldsymbol{x}_j|\boldsymbol{w}_i)$ with prior probability $p(\boldsymbol{w}_i)$ and Gaussian $p(\boldsymbol{x}_j|\boldsymbol{w}_i)$. The costs aim at a generative model, i.e. we address the class-wise data log likelihood $\log \prod_j \delta_{y_j}^c p(\boldsymbol{x}_j|c, W) = \sum_j \delta_{y_j}^c \log \sum_i \delta_{y_j}^{c(\boldsymbol{w}_i)} p_c(\boldsymbol{w}_i)p(\boldsymbol{x}_j|\boldsymbol{w}_i)$ with prior $p_c(\boldsymbol{w}_i) = p(\boldsymbol{w}_i)/p(c)$ summing to one for every class c. Adding this generative term for all class-wise distributions, we arrive at the form

$$E_{\text{repr}}(W) = \sum_c \sum_j \delta_{y_j}^c \log \sum_i \delta_{y_j}^{c(\boldsymbol{w}_i)} p_c(\boldsymbol{w}_i)p(\boldsymbol{x}_j|\boldsymbol{w}_i) \tag{4}$$

We often assume equal prior for all classes c and prototypes $\boldsymbol{w}_i$ for simplicity. We choose Gaussians of the form

$$p(\boldsymbol{x}_j|\boldsymbol{w}_i) = \frac{1}{\sqrt{(2\pi)^n|\Sigma_i|}} \exp\left(-\frac{1}{2}(\boldsymbol{x}_j - \boldsymbol{w}_i)^T \Sigma_i^{-1}(\boldsymbol{x}_j - \boldsymbol{w}_i)\right) \tag{5}$$

where Σ_i is taken as diagonal matrix with entries $(\sigma_{i1}^2, \ldots, \sigma_{in}^2)$. Optimization takes place by means of a gradient ascent of these costs. The derivative of $E_{\text{discr}}(W)$ can be found in [16]. See [14] for update rules in case of an adaptive covariance matrix. For $E_{\text{repr}}(W)$ prototypes $\boldsymbol{w}_i$ are adapted according to

$$\frac{\partial E_{\text{repr}}(W)}{\partial \boldsymbol{w}_i} = \sum_j \delta_{y_j}^{c(\boldsymbol{w}_i)} \frac{p_{y_j}(\boldsymbol{w}_i) \cdot p(\boldsymbol{x}_j|\boldsymbol{w}_i)}{\sum_l \delta_{y_j}^{c(\boldsymbol{w}_l)} p_{y_j}(\boldsymbol{w}_l) \cdot p(\boldsymbol{x}_j|\boldsymbol{w}_l)} \cdot \Sigma_i^{-1} \cdot (\boldsymbol{x}_j - \boldsymbol{w}_i), \tag{6}$$

while the variances σ_{jk} are simultaneously updated referred to

$$\frac{\partial E_{\text{repr}}(W)}{\partial \sigma_{in}} = \sum_j \delta_{y_j}^{c(\boldsymbol{w}_i)} \frac{p(\boldsymbol{w}_i)p(\boldsymbol{x}_j|\boldsymbol{w}_i)}{p(\boldsymbol{x}_j,y_j|W)} \left(\frac{[\boldsymbol{x}_j - \boldsymbol{w}_i]_n^2}{\sigma_{in}^3} - \frac{1}{\sigma_{in}} \right). \tag{7}$$

In the limit of small bandwidth, this amounts to a class wise vector quantization scheme.

In a similar way, we enhance the GLVQ cost function by a term emphasizing the representativity of the prototypes in model consistent way. Here we choose the class-wise quantization error

$$E_{\text{repr}}(W) = \sum_j d^+(\boldsymbol{x}_j), \tag{8}$$

Taking the derivative overlays the update rules with a vector quantization step.

As we will see in experiments, depending on the data set, these two objectives can be contradictory, such that the choice of α can severely influence the outcome. Thereby, the scaling of the two objectives is not clear a priori: while a probabilistic modeling such as RSLVQ places the two objectives into the interval $(-\infty, 0]$ corresponding to a log likelihood, the discriminative part of GLVQ lies in $E_{\text{discr}}(W) \in (-1, 1)$, but $E_{\text{repr}}(W) \in [0, \infty)$ for GLVQ. Hence, without normalizing these terms, the scaling of the parameter α has different meanings in both settings. We will report results for the whole range $\alpha \in [0, 1]$ with step size 0.05 in case of RSLVQ, 0.001 for GLVQ, respectively.

4   Experiments

We test the behavior of the models for different values α in three benchmarks:

- **Gauss:** two two-dimensional Gaussian clusters with different covariance matrices and some degree of overlap are generated.
- **Tecator:** the data set consists of 215 spectra with 100 spectral bands ranging from 850 nm to 1050 nm [10]. The task is to predict the fat content of the probes.

To avoid local optima as much as possible, initial training takes place to distribute the prototypes in the data space, as proposed in [9]. In our experiments we simply start with an initial training phase where $\alpha = 1$ and we anneal the value α afterwards to the desired weighting parameter. For RSLVQ, diagonal entries of the covariance matrix are adapted individually for every mixture component. In all cases, we use one prototype/mixture component per class. Training takes place until convergence. To validate representativity we determine the following ratio for both models:

$$R = \frac{1}{C} \sum_{c_k} \sum_{j:\, c(\boldsymbol{x}_j)=c_k} \frac{d^+(\boldsymbol{x}_j)}{\sum\limits_{i:\, c(\boldsymbol{x}_i)=c_k} d(\boldsymbol{x}_i, \boldsymbol{\mu}_{c_k})}, \tag{9}$$

which is the class-wise quantization error according to the class mean $\boldsymbol{\mu}_{c_k}$.

Gauss: Due to the data generation, prototypes lying in the two class centers define a decision boundary which is close to the optimum decision boundary, albeit not being identical due to the non-isotropic Gaussians. This fact is mirrored in the dependency of the classification accuracy in respect to the parameter α as depicted in Fig. 1: the accuracy is widely constant for varying parameter α for both, RSLVQ and GLVQ schemes.

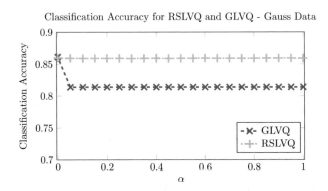

Fig. 1. Classification accuracy for RSLVQ and GLVQ for the Gauss data set varying parameter α

Interestingly, the classification accuracy for RSLVQ is higher than GLVQ which can be attributed to the fact that only the first model explicitly aims at an optimization of the Bayes error and an implicitly fitting of Gaussians, while the GLVQ costs are only correlated to a class discrimination.

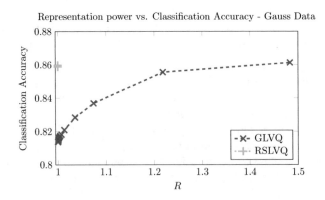

Fig. 2. Class-wise quantization error for the Gauss data set vs. accuracy for varying parameter α

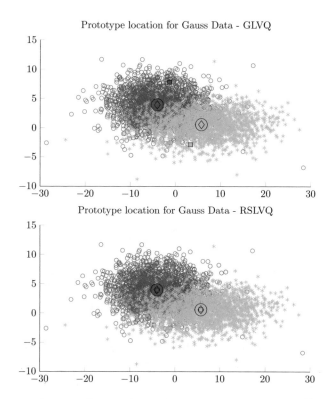

Fig. 3. Prototype location for the Gauss data set for extremal $\alpha \in \{0, 1\}$. squares $\,\hat{=}\,$ $\alpha = 1$; diamonds $\,\hat{=}\,$ $\alpha = 0$; filled circle $\,\hat{=}\,$ class mean.

For both approaches the prototype locations for extremal values $\alpha \in \{0, 1\}$ are depicted in Fig. 3. The prototypes which are obtained with RSLVQ do not change its position, as mirrored in the class-wise quantization error with increasing value α, see Fig. 2. These are at the class centers and obviously do not enormously differ from the respective class means. Unlike GLVQ, where for $\alpha = 1$ the prototypes do not coincide with the class means to better follow the optimum decision boundary for the given case. Contrary to RSLVQ, covariances are not used by standard GLVQ.

Tecator: For the tecator data set, there seems a clear difference between a good generative or good discriminative model as found by LVQ schemes. When varying the parameter α, the classification accuracy decreases (Fig. 4), while the representativity increases, see Fig. 5.

Interestingly, the prototypes lie at atypical positions for the purely discriminative models in this case, making their interpretability problematic: as depicted in Fig. 6, the spectral curves display a very characteristic shape which has no resemblance to spectra as observed in the data. These forms facilitate the class discrimination while interpretability is questionable. This setting also demonstrates

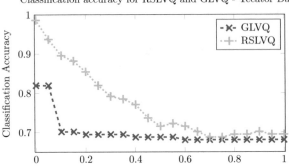

Fig. 4. Classification accuracy for RSLVQ and GLVQ for the Tecator data set varying parameter α

the partially problematic choice of an appropriate parameter α in particular for the GLVQ model. In this case, due to the inherent scaling, already small values of α have a dramatic effect on the classification accuracy of the result.

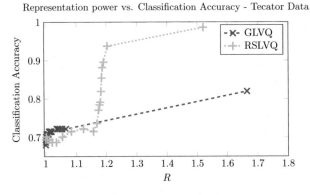

Fig. 5. Class-wise quantization error for RSLVQ and GLVQ for the Tecator data set vs. Classification accuracy for varying parameter α

5   Discussions

We have discussed the correlation of popular LVQ cost functions to the two aims, to obtain a small classification error and to obtain a representative model where prototypes are interpretable. By means of examples, we have seen that LVQ usually models the former objective, but the latter is only implicitly taken into account. An explicit integration of this objective enables enhanced models where the discriminative power versus the representativity of the prototypes can be controlled by the user, leading to better interpretable models in case the two

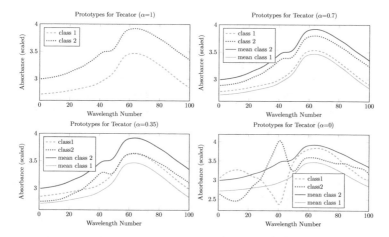

Fig. 6. Prototype locations for the Tecator data set and different choices of the parameter α. Interestingly, for the discriminative case $\alpha = 0$, atypical shapes with little resemblance of the class averages are obtained, while $\alpha = 1$ boosts class averages.

objectives are contradictory for the given data. We have shown the effect of such a control on the form of the prototypes in a few benchmarks.

So far, the two objectives are combined in one cost function and an appropriate balance parameter α has to be set. To make both algorithms comparable according to the used distance a localized relevance GLVQ approach [5] is mandatory. In this contribution our focus is on pointing out that both LVQ variants can be extended to make their results more interpretable. As an alternative, one can consider formulations which emphasize the primary aim of correct classification as a hard constraint, but integrate representativity as a soft constraint. This way, one can aim for the most representative solutions among a set of possible solutions which are invariant with respect to the classification error. Such an approach would result in formalizations of the form

$$\min \sum_j d^+(\boldsymbol{x}_j)$$
$$\text{such that } d^+(\boldsymbol{x}_j) \leq d^-(\boldsymbol{x}_j) + \epsilon \quad \forall j$$

for GLVQ, incorporating slack variables if no feasible solution exists, or

$$\max \sum_c \sum_j \delta^c_{y_j} \log \sum_i \delta^{c(\boldsymbol{w}_i)}_{y_j} p_c(\boldsymbol{w}_i) p(\boldsymbol{x}_j | \boldsymbol{w}_i)$$
$$\text{such that } p(y_j | \boldsymbol{x}_j, W) \geq p(c | \boldsymbol{x}_j, W) + \epsilon \quad \forall j \quad \forall c \neq y_j$$

for RSLVQ, again incorporating slack variables if necessary. The investigation of these alternatives will be the subject of future work.

Acknowledgement. BH has been supported by the CITEC center of excellence. DN and MR acknowledge funding by ESF.

References

1. Biehl, M., Ghosh, A., Hammer, B.: Dynamics and generalization ability of LVQ algorithms. Journal of Machine Learning Research 8, 323–360 (2007)
2. Biehl, M., Hammer, B., Schneider, P., Villmann, T.: Metric learning for prototype based classification. In: Bianchini, M., Maggini, M., Scarselli, F. (eds.) Innovations in Neural Information – Paradigms and Applications. SCI, vol. 247, pp. 183–199. Springer, Heidelberg (2009)
3. Crammer, K., Gilad-Bachrach, R., Navot, A., Tishby, A.: Margin analysis of the lvq algorithm. In: Advances in Neural Information Processing Systems, vol. 15, pp. 462–469. MIT Press, Cambridge (2003)
4. Hammer, B., Mokbel, B., Schleif, F.-M., Zhu, X.: White box classification of dissimilarity data. In: Corchado, E., Snávsel, V., Abraham, A., Woźniak, M., Graña, M., Cho, S.-B. (eds.) HAIS 2012, Part III. LNCS, vol. 7208, pp. 309–321. Springer, Heidelberg (2012)
5. Hammer, B., Schleif, F.-M., Villmann, T.: On the generalization ability of prototype-based classifiers with local relevance determination (2005)
6. Hammer, B., Strickert, M., Villmann, T.: Supervised neural gas with general similarity measure. Neural Processing Letters 21(1), 21–44 (2005)
7. Hammer, B., Villmann, T.: Generalized relevance learning vector quantization. Neural Networks 15(8-9), 1059–1068 (2002)
8. Kirstein, S., Wersing, H., Körner, E.: A biologically motivated visual memory architecture for online learning of objects. Neural Networks 21(1), 65–77 (2008)
9. Kohonen, T.: The self-organizing map. Proc. of the IEEE 78(9), 1464–1480 (1990)
10. D. of Statistics at Carnegie Mellon University,
    http://lib.stat.cmu.edu/datasets/
11. Qin, A.K., Suganthan, P.N.: A novel kernel prototype-based learning algorithm. In: ICPR (4), pp. 621–624 (2004)
12. Sato, A., Yamada, K.: Generalized learning vector quantization. In: Touretzky, M.C.M.D.S., Hasselmo, M.E. (eds.) Proceedings of the 1995 Conference on Advances in Neural Information Processing Systems 8, pp. 423–429. MIT Press, Cambridge (1996)
13. Schneider, P., Biehl, M., Hammer, B.: Adaptive relevance matrices in learning vector quantization. Neural Computation 21, 3532–3561 (2009)
14. Schneider, P., Biehl, M., Hammer, B.: Distance learning in discriminative vector quantization. Neural Computation 21, 2942–2969 (2009)
15. Seo, S., Bode, M., Obermayer, K.: Soft nearest prototype classification. IEEE Transactions on Neural Networks 14, 390–398 (2003)
16. Seo, S., Obermayer, K.: Soft learning vector quantization. Neural Computation 15(7), 1589–1604 (2003)
17. Villmann, T., Haase, S.: Divergence-based vector quantization. Neural Computation 23(5), 1343–1392 (2011)

Some Room for GLVQ:
Semantic Labeling of Occupancy Grid Maps

Sven Hellbach[1], Marian Himstedt[1], Frank Bahrmann[1], Martin Riedel[2], Thomas Villmann[2], and Hans-Joachim Böhme[1,*]

[1] University of Applied Sciences Dresden,
Artificial Intelligence and Cognitive Robotics Labs,
POB 12 07 01, 01008 Dresden, Germany
{hellbach,himstedt,bahrmann, boehme}@informatik.htw-dresden.de
[2] University of Applied Sciences Mittweida,
Computational Intelligence and Technomathematics,
POB 14 57, 09648 Mittweida, Germany
{riedel,thomas.villmann}@hs-mittweida.de

Abstract. This paper aims at an approach for labeling places within a grid cell environment. For that we propose a method that is based on non-negative matrix factorization (NMF) to extract environment specific features from a given occupancy grid map. NMF also computes a description about where on the map these features need to be applied. We use this description after certain pre-processing steps as an input for generalized learning vector quantization (GLVQ) to achieve the classification or labeling of the grid cells. Our approach is evaluated on a standard data set from University of Freiburg, showing very promising results.

Keywords: NMF, GLVQ, semantic labeling, occupancy grid maps.

1 Introduction

One of the major goals in cognitive robotics is to develop algorithms and hence robots that can be used intuitively. In particular, while thinking about a future practical application the user wants to use the robot out of the box without the need of a trained technician.

A possible scenario for such a situation would be that the robot is shown around by a human operator - as we would do with a new colleague. For this the robots needs to follow its operator. On the tour the operator should be able to input (e.g. verbalize) in which kind of room the robot currently is located. During the acquisition of the knowledge about room concepts the robot should already be able to predict the current room after a while of training.

This knowledge allows the robot to fulfill tasks, like finding a resting or parking position in an office instead in the hallway. Furthermore, the robot would be able

---

* This work was supported by ESF grand number 100076162.

T. Villmann et al. (eds.), *Advances in Self-Organizing Maps and Learning Vector Quantization,* Advances in Intelligent Systems and Computing 295,
DOI: 10.1007/978-3-319-07695-9_13, © Springer International Publishing Switzerland 2014

to fulfill place specific tasks or show adequate behavior, like silence in an office while welcoming arrivers in the hallway.

Achieving a semantic understanding of its environment of a mobile robot platform is an on-going topic in many research teams. A popular approach with training and testing datasets was presented by Mozos [1]. This algorithm mainly used geometrical features and an AdaBoost classifier to differentiate between three classes (room, corridor, doorway) within a metric map. The mildly noisy output was smoothed afterwards using probabilistic relaxation labeling. A variation of Mozos' solely laser-range-finder-based solution was published in [2, 3]. They used $\mathcal{L}_2$-regularized logistic regression on geometrical (area of polygonal approximation of the laser scan) as well as statistical laser scan features (standard deviations of lengths of consecutive scans and of ranges). Another approach using Support Vector Machines is proposed in [4].

The concept of teaching the mobile robot semantic labels at runtime, provided by a human guide, was followed by [5]. During a tour in its new surroundings, the robot obtains place labels where each spatial region is represented by one or more Gaussians. The complete map is later classified by region growing. A similar approach is described in [6].

Assuming laser range data is insufficient to fully understand our complex environment, more sensor cues are introduced. [7] used Mozos' laser range features and combined them with visual features obtained by SIFT (Scale-invariant feature transform) and CRFH (composed receptive field histogram). Each cue produces a scoring value which are then combined by SVM-DAS (SVM-based Discriminative Accumulation Scheme) to a final class label.

Due to new inexpensive RGB-D sensors, object recognition has gained much attention among research communities. For instance,[8, 9] build 3-dimensional maps via RGB-D-SLAM. Within the resulting maps, preconceived coarse (wall, ceiling, ground) and individual (printer, monitor, etc.) labels are recognized. Different kinds of rooms are inferred using an associative coupling of these lables. Conversely, it is possible to find out the most likely position of an individual label.

The remainder of this paper is organized as follows. Section 2 summarizes the proposed approach, as well as explains the details of non-negative matrix factorization (Sec. 2.1), Generalized Learning Vector Quantization (Sec. 2.3), and the data pre-processing (Sec. 2.2). The experimental results are discussed in Sec. 3, while the paper concludes in Sec. 4.

2   Approach

Similar as described in [10], our approach takes an occupancy grid map as input. From this occupancy grid map a set of basis primitives and corresponding activities is computed using Non-negative Matrix Factorization (NMF). The basis primitives in a practical application would be stored and used to derive the activity based description for the local maps as also already explained in [10].

Beyond this, in [10] Beyond this, in [10] we computed a representation of the environment with help of NMF. Based on the activities, we derived a histogram-like description, which describes the amount of environmental characteristics

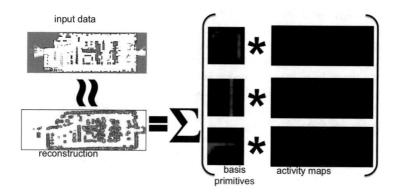

Fig. 1. Application of the NMF algorithm on the global occupancy grid map

(e.g. corners, walls) for a certain position. We could show that these histograms are already sufficient for solving the localization - more precisely the place recognition problem. With that in mind, in this work we go back to the original NMF representation to perform the clustering/classification of our environment. For that application it is important to be able to recognize a certain position in the environment in a unique way.

For the proposed method, we continue with the activity maps and apply a distance transform. This is necessary, since the activity maps only have distinct peaks (compare Fig. 2) and a slight translation within the map results in large dissimilarity. Using distance transform smoothens our representation and also encode the distance metric within the grid map into our vector space.

Finally, we use Generalized Learning Vector Quantization for prediction of the class labels for each grid cell. Combined with a segmentation approach (e. g. as proposed in [6]) this information can be used to determine the class of building structures, like rooms.

2.1   Non-negative Matrix Factorization

Like other approaches, non-negative matrix factorization (NMF) [11] is meant to solve the source separation problem. Hence, a set of training data is decomposed into basis primitives W and their respective activations H:

$$V \approx W \cdot H \tag{1}$$

Each training data sample is represented as a column vector V_i within the matrix V. Each column of the matrix W stands for one of the basis primitives. In matrix H the element H_i^j determines how the basis primitive W_j is activated to reconstruct training sample V_i.

Unlike PCA or ICA, NMF performs a decomposition, which only consists of non-negative elements. This means that the basis primitives can only be accumulated. There exists no primitive which is able to erase a 'wrong' superposition of other primitives. This usually results in primitives that are more interpretable.

Theoretically, the basis primitives are invariant to several transformations such as rotation, translation, and scale. This is achieved by adding a transformation matrix $\boldsymbol{T}$ to the decomposition formulation [12]. For each allowed transformation the corresponding activity has to be trained individually. To avoid trivial or redundant solutions a further sparsity constraint is necessary. Its influence can be controlled using the parameter λ [13].

For generating the decomposition, optimization-based methods are used. Hence, an energy function E has to be defined:

$$E(\boldsymbol{W}, \boldsymbol{H}) = \frac{1}{2} \|\boldsymbol{V} - \boldsymbol{T} \cdot \boldsymbol{W} \cdot \boldsymbol{H}\|^2 + \lambda \sum_{i,j} H_i^j \tag{2}$$

By minimizing this energy function (2), it is now possible to achieve a reconstruction using the matrices $\boldsymbol{W}$ and $\boldsymbol{H}$. This reconstruction is aimed to be as close as possible to the training data $\boldsymbol{V}$. The data has to be vectorized to be held as a column vector in $\boldsymbol{V}$. Hence a column in $\boldsymbol{H}$ is the vectorial form of a activity map

Enabling only translational invariance reduces the problem to a convolution over all translations. This is efficiently implemented by transforming the data into the frequency domain based on FFT. In order to reduce the complexity, rotation and scale are not taken into account for the presented approach.

The minimization of the energy function can be done by gradient descent. The factors $\boldsymbol{H}$ and $\boldsymbol{W}$ are updated alternately with a variant of exponentiated gradient descent or using NMFs multiplicative update rule until convergence (see [10] for algorithmic details).

The NMF formulation as it is written in this section allows an arbitrary transform of the basis primitives $\boldsymbol{W}$ with respect to their position on $\boldsymbol{H}$ and $\boldsymbol{V}$ respectively. If transformation invariance is limited to translational invariance all equations can be simplified to a convolution of $\boldsymbol{W}$ over $\boldsymbol{H}$. For details on the derivation and the exact implementation we would like to refer to [14]. However, intuitively it should be clear that if all possible shifts (translations) of $\boldsymbol{W}$ are coded with different $\boldsymbol{T}$ is equivalent with the mentioned convolution. Using convolution instead of iteration over all possible transformations immensely reduces computational effort. Furthermore, the well known trick of multiplying in frequency space can be applied.

2.2 Data Representation

In [10] a number of histogram based descriptors are evaluated. For this paper, we go back to a representation relying on the activities $\boldsymbol{H}$ computed by the NMF. As it is depicted in Fig. 2, the column vectors $\boldsymbol{H}^p$ of the activity matrix $\boldsymbol{H}$ with $\boldsymbol{H} = (H_k^p) = (\boldsymbol{H}^p)_i$ can be regarded as a map of activities with the same width and height $w \times h$ of the training map used in $\boldsymbol{V}$. We define this map for basis primitive $\boldsymbol{W}_p$ as:

$$\tilde{\boldsymbol{H}}(p) = \left(\tilde{H}(p)_j^i \right) \text{ with } \tilde{H}(p)_j^i = H_k^p \text{ with } k = j \cdot w + i \tag{3}$$

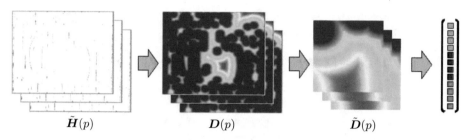

$\tilde{H}(p)$ $D(p)$ $\tilde{D}(p)$

Fig. 2. Preprocessing before GLVQ. The activities H computed by the NMF for the global map are thresholded and then distance transformed. After that, local patches are cut out representing the description for the local map. Finally, the local maps are vectorized.

These maps are then thresholded by θ to gain a binary activity map, we are only interested in cells that are definitely occupied. Defining $\mathcal{O}$ as a set as follows, also reduces data complexity, and computational costs:

$$\mathcal{O} = \left\{ (i,j) \in \Omega \, \middle| \, \tilde{H}(p)^i_j > \theta \right\}. \tag{4}$$

with $\Omega = \{1, \ldots, w\} \times \{1, \ldots, h\}$ being the set of grid cells in the map. Consequently, the grid cells where the corresponding basis primitive is to be placed belong to the set $\mathcal{O}$.

Subsequently, the binary map $\mathcal{O}$ undergoes a Euclidian distance transform [15]. This results in a map D where each grid cell contains the Euclidian distance to the nearest grid position, where a basis primitive is activated:

$$D(p) = \left(D^i_j \right) \text{ with } D^i_j = \min \left\{ d((i,j), q) \, | \, q \in \mathcal{O} \right\} \tag{5}$$

For this, $d(\cdot, \cdot)$ stands for a metric defined over Ω. We simply use the Euclidian distance here. The distance transform becomes necessary, since a minimal translation of the peaky activities leads to maximal dissimilarity. Distance transformation is applied to code the distance of each position to the occurrence of each primitive.

For the practical application these two steps need to be computed on the local maps during run time. To eliminate errors coming from the construction of these local maps, we decided to cut out patches of size $u \times v$ from the distance transformed activities $D(\mathrm{p})$ of the global map.

$$\tilde{D}(p) = \left(D^i_j \right) \text{ with } i = k - \frac{u}{2}, \ldots, k + \frac{u}{2}, j = l - \frac{v}{2}, \ldots, l + \frac{v}{2} \tag{6}$$

Otherwise, the experimental evaluation would also consider errors not caused by the proposed method.

The next step would be to transform the distance transformed activities into a vector space used by GLVQ. For this, each distance transformed activity is traversed row-wise. The resulting vectors are then simply concatenated to a single vector.

Before the vectorization takes place the patches or local maps respectively can be subsampled with step size s.

$$\tilde{D}'(p) = \left(D_j^i\right) \text{ with } i = k - \frac{u}{2}, k - \frac{u}{2} + s, \ldots, k + \frac{u}{2}, \tag{7}$$
$$j = l - \frac{v}{2}, l - \frac{v}{2} + s, \ldots, l + \frac{v}{2}$$

This results in a reduction of the number of input dimensions for the classifier. With the reduced number of dimensions it becomes possible to train the classifier with fewer training samples. For the practical application this brings an tremendous advantage since we aim to learn characteristics from the few sensor reading of a single room.

2.3   Generalized Learning Vector Quantization

Since learning vector quantization (LVQ) was introduced by Kohonen [16], this classification algorithm is frequently used by practitioners for several reasons: the learning rule as well as the classification model are very fast; due to that learning rule adjusting an existing model by new data could be done very efficiently; unlike popular alternatives such as SVM the model can easily deal with an arbitrary number of classes.

The basic LVQ scheme was introduced as a heuristic to minimize the classification error on a training set. In this contribution we use a very famous enhancement, known as generalized LVQ (GLVQ) [17]. Based on a mathematical cost function, it enables also the usage of different dissimilarity measures, e.g. adaptive quadratic forms, kernels or functional metrics. Extension of GLVQ using other metrics instead of the standard Euclidean can be found in [18–22].

However, a LVQ classifier is given by a set of prototypes $\boldsymbol{w}_i \in \mathbb{R}^n$, $i = 1, \ldots, k$ equipped with label $c(\boldsymbol{w}_i) \in \{1, \ldots, C\}$, assuming C classes. Classification of a datum $\boldsymbol{x} \in \mathbb{R}^n$ takes place by a winner takes all rule: $\boldsymbol{x}$ gets label $c(\boldsymbol{x}) = c(\boldsymbol{w}_i)$ of the prototype $\boldsymbol{w}_i$ which is closest to $\boldsymbol{x}$ measured in some distance measure. Given a training data set $\boldsymbol{x}_j \in \mathbb{R}^n$, $j = 1, \ldots, m$, together with labels $y_j \in \{1, \ldots, C\}$, the objective of a LVQ classifier is to place the prototypes $\boldsymbol{w}_i$ within the data space in such a way that the prototypes represent their corresponding classes as best as possible. Therefor, GLVQ addresses the following cost function

$$E = \sum_j \varPhi\left(\mu\left(\boldsymbol{x}_j\right)\right) \text{ with } \mu\left(\boldsymbol{x}_j\right) = \frac{d^+(\boldsymbol{x}_j) - d^-(\boldsymbol{x}_j)}{d^+(\boldsymbol{x}_j) + d^-(\boldsymbol{x}_j)}, \tag{8}$$

where $d^+(\boldsymbol{x}_j)$ denotes the distance of $\boldsymbol{x}_j$ to the closest prototype $\boldsymbol{w}^+$ with the same label $c(\boldsymbol{w}^+) = y_j$, and $d^-(\boldsymbol{x}_j)$ refers to the best matching prototype $\boldsymbol{w}^-$ with a class label $c(\boldsymbol{w}^-)$ different to y_j. $\varPhi$ refers to a monotonic function such as the identity or the sigmoidal function. Its argument becomes negative if a datum is classified correctly. If a sigmoidal transfer function is used, e.g.

$$\varPhi(z) = \frac{1}{1 + e^{-\alpha z}}, \quad \alpha > 0,$$

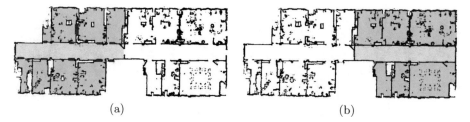

(a)   (b)

Fig. 3. Data set used for evaluation. The occupancy grid map of a building at the University of Freiburg. Three different classes are labeled: rooms (cyan), hallway (yellow) and passages (red). The map is split in the middle for training (a) and test (b) set.

E approximately counts the number of misclassifications. As exemplary mentioned in [23] increasing α further improves the classifier since the prototypes become border sensitive. However, minimizing this cost function leads to the maximization of the classification accuracy. Therefore, a stochastic gradient descent scheme is used. Given a sample $\boldsymbol{x}_i$ out of the training set we have to determine the derivatives according to the local error

$$\frac{\partial_S E}{\partial \boldsymbol{w}^+} = \xi^+ \cdot \frac{\partial d^+}{\partial \boldsymbol{w}^+} \text{ and } \frac{\partial_S E}{\partial \boldsymbol{w}^-} = \xi^- \cdot \frac{\partial d^-}{\partial \boldsymbol{w}^-} \tag{9}$$

with $\xi^+ = \pm f' \cdot \frac{2 \cdot d^\mp(\boldsymbol{x}_i)}{(d^+(\boldsymbol{x}_i)+d^-(\boldsymbol{x}_i))^2}$. For the squared Euclidean metric we simply have the derivative $\frac{\partial d^\pm(\boldsymbol{x}_i)}{\partial \boldsymbol{w}^\pm} = -2(\boldsymbol{x}_i - \boldsymbol{w}^\pm)$ realizing a vector shift of the prototypes.

3   Experiments

For the experimental evaluation we use an online available data set[1], which has already been used in [24]. Hence, our results can directly be compared to the other methods. The data set consists of already computed $700x289$ grid map, for which three classes of building structures have been labeled: room, hallway and passage. Furthermore, the set is already split up into training and test set, as it is shown in Fig. 3.

For computational reasons we subsample the given grid map by taking into account only each 4th pixel in horizontal and vertical direction. Hence, the classes room, hallway and passage consist of 2745, 1029, 99 in the training set and 3084, 937, 32 for the test set, respectively. As it can be seen, the data set offers a challenging problem since the classes are highly unbalanced - in particular the passage class.

Preparation: To gain first insides on the structure of the data, we started with an unsupervised clustering of the vector space described in Sec. 2.2. For this purpose, a k-means clustering was chosen, due to its simplicity and availability.

---

[1] http://webpages.lincoln.ac.uk/omozos/place_data_sets.html

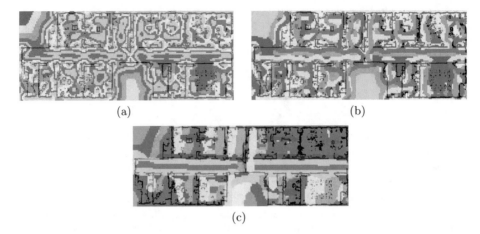

(a)   (b)

(c)

Fig. 4. Results of an unsupervised k-Means Clustering with different sets of parameters for patch size and subsampling of patches (resulting in a number of GLVQ input dimensions): (a) 10×10, 2 (360) (b) 20×20, 4 (360) (c) 50×50, 5 (1210)

Figure 4 shows the results for different patch sizes with different subsampling as described in Sec. 2.2 transferred back to the grid map for better visualization. A closer look reveals the distance transform, as the classes look a little like contour lines around structuring elements of the map, like walls. Depending on the patch size, the clusters vary in size from a coarse to fine coverage. Furthermore, the cluster also depends on which structuring element is close by. This can be explained by the different basis primitives which describe environment structures.

None of the test runs shows clear benefits above the other. Hence, we decided to use a 50×50 patch with a subsampling of 5 grid cells for further experiments.

Label Prediction: To understand how the supervised GLVQ works on our data, we used both test and training set for training for the preliminary tests depicted in Fig. 5 (a) and (b). Figure 5 (a) reveals that due to unbalanced classes the algorithm has problems with the passage class even on the training set. Hence, we decided to dilate the class regions. The improvement can be seen in 5 (b).

After the preliminary tests GLVQ was trained with the left part of the map, while the evaluation was performed on the right part (Fig. 5 (c)), gaining the following confusion matrices, confirming that passages are difficult to classify:

		Prediction (training)			Prediction (test)		
	in %	room	hallway	passage	room	hallway	passage
Actual class	room	99.82	0	0.18	92.99	2.15	4.86
	hallway	0	99.66	0.34	0.50	92.59	6.91
	passage	2.75	3.44	93.81	28.57	27.92	43.51

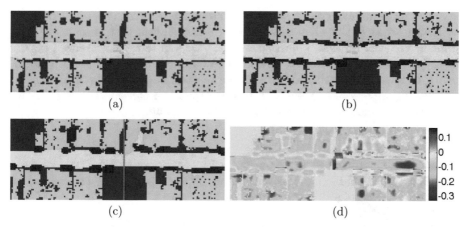

(a)   (b)

(c)   (d)

Fig. 5. (a)-(c) Experimental results using the same coloring as in Fig. 3 with a local map size of 50 × 50 (a) an additional dilation of class *passage* (b) and a separation in train and test set (left and right of the green line, respectively) (c). (d) Visulization of μ (Eqn. 8) for each grid cell as an indicator for the confidence of the classification.

In Fig. 5 (d) we show the values for μ visualized for each grid cell. Values close to zero (yellow) mean that the representation of the grid cell is close to the class border. Colors from yellow over green to blue show the confidence of a correct classification, while reddish colors show a misclassification.

For all three classes we reach accuracies of 99.05% for training and 88.64% for the test set with. Mozos [1] reached on the same data set an accuracies of 89.5% for the first step of their cascade. Shi [2] left out the doors and reached an overall accuracy of 99.78% An evaluation on a different data set in [4] reaches 86% classification accuracy.

4   Conclusion

Obviously, using only laser range data is not sufficient to derive a reliable conclusion for the classes of the different rooms. However, we understand our approach as an additional cue for a more complex system or as a system to determine a fast decision or something like a "first guess" for subsequent algorithms. Even though, we could show that our system produces results that compete with the results of other, even visual methods.

In addition to the promising prediction results, our approach offers the possibility for online adaptation. With that we can refine our training knowledge with each additional observed location. The proposed pre-processing step before GLVQ allows to subsample the data and hence reduces the number of input dimensions for GLVQ. With the reduced number of dimension it becomes possible to decrease the amount of training data. This, as well, supports the possibility to learn from the data of a single room.

Despite the already good results, it would be interesting to try different types of LVQ instead of the used GLVQ. Our results confirm that it seems to be a difficult problem to classify passages. Hence, we aim to try Generalized Matrix LVQ (GMLVQ) or the Kernel version of LVQ. Furthermore, it would be interesting to include the knowledge about the relation between different dimension of the LVQ input space. Some dimensions depend on the same basis primitive or share the same spatial location within the grid map.

References

1. Mozos, O.M., Triebel, R., Jensfelt, P., Rottmann, A., Burgard, W.: Supervised semantic labeling of places using information extracted from sensor data. RAS 55(5), 391–402 (2007)
2. Shi, L., Kodagoda, S., Dissanayake, G.: Laser range data based semantic labeling of places. In: IROS, pp. 5941–5946. IEEE (2010)
3. Shi, L., Kodagoda, S., Dissanayake, G.: Multi-class classification for semantic labeling of places. In: ICARCV, pp. 2307–2312. IEEE (2010)
4. Sousa, P., Araujo, R., Nunes, U.: Real-Time Labeling of Places using Support Vector Machines. In: ISIE, pp. 2022–2027 (2007)
5. Nieto-Granda, C., Rogers, J.G., Trevor, A.J., Christensen, H.I.: Semantic map partit. in indoor environments using regional analysis. In: IROS, pp. 1451–1456 (2010)
6. Bahrmann, F., Hellbach, S., Böhme, H.J.: Please tell me where I am: A fundament for a semantic labeling approach. In: KI, pp. 120–124 (2012)
7. Pronobis, A., Mozos, O.M., Caputo, B., Jensfelt, P.: Multi-modal semantic place classification. Int J. Robot. Res. 29(2-3), 298–320 (2010)
8. Koppula, H.S., Anand, A., Joachims, T., Saxena, A.: Semantic labeling of 3d point clouds for indoor scenes. In: NIPS, pp. 244–252 (2011)
9. Anand, A., Koppula, H.S., Joachims, T., Saxena, A.: Contextually guided semantic labeling and search for three-dimensional point clouds. Int. J. Robot. Res. 32(1), 19–34 (2013)
10. Hellbach, S., Himstedt, M., Boehme, H.J.: Towards Non-negative Matrix Factorization based Localization. In: ECMR (2013)
11. Lee, D.D., Seung, H.S.: Algorithms for non-negative matrix factorization. Adv. Neural Inf. Process. Syst 13, 556–562 (2001)
12. Eggert, J., Wersing, H., Körner, E.: Transformation-invariant representation and NMF. In: IJCNN, pp. 2535–2539 (2004)
13. Eggert, J., Körner, E.: Sparse Coding and NMF. In: IJCNN, pp. 2529–2533 (2004)
14. Vollmer, C., Hellbach, S., Eggert, J., Gross, H.M.: Sparse coding of human motion trajectories with non-negative matrix factorization. Neurocomp. (2013)
15. Paglieroni, D.W.: Distance transforms: properties and machine vision applications. CVGIP: Graph. Models Image Process. 54(1), 56–74 (1992)
16. Kohonen, T.: The self-organizing map. Proc. of the IEEE 78(9), 1464–1480 (1990)
17. Sato, A., Yamada, K.: Generalized learning vector quantization. In: NIPS, pp. 423–429. MIT Press, Cambridge (1996)
18. Hammer, B., Villmann, T.: Generalized relevance learning vector quantization. Neural Networks 15(8-9), 1059–1068 (2002)
19. Schneider, P., Biehl, M., Hammer, B.: Adaptive relevance matrices in learning vector quantization. Neural Computation 21, 3532–3561 (2009)

20. Qin, A.K., Suganthan, P.N.: A novel kernel prototype-based learning algorithm. In: ICPR (4), pp. 621–624 (2004)
21. Hammer, B., Strickert, M., Villmann, T.: Supervised neural gas with general similarity measure. Neural Processing Letters 21(1), 21–44 (2005)
22. Villmann, T., Haase, S.: Divergence-based vector quantization. Neural Computation 23(5), 1343–1392 (2011)
23. Kästner, M., Riedel, M., Strickert, M., Villmann, T.: Class border sensitive generalized learning vector quantization - an alternative to support vector machines. Machine Learning Reports 6(MLR-04-2012), 40–56 (2012)
24. Mozos, O.M.: Semantic Place Labeling with Mobile Robots. PhD thesis, Dept. of Computer Science, University of Freiburg (July 2008)

Anomaly Detection Based on Confidence Intervals Using SOM with an Application to Health Monitoring

Anastasios Bellas[1], Charles Bouveyron[2], Marie Cottrell[1], and Jerome Lacaille[3]

[1] SAMM, Université Paris 1 Panthé on-Sorbonne
90 rue de Tolbiac, 75013 Paris, France
{anastasios.bellas,marie.cottrell}@univ-paris1.fr
[2] Laboratoire MAP5, Université Paris Descartes & Sorbonne Paris Cité
45 rue des Saints-Pères, 75006 Paris, France
charles.bouveyron@parisdescartes.fr
[3] SNECMA, Rond-Point René Ravaud-Réau,
77550 Moissy-Cramayel CEDEX, France
jerome.lacaille@snecma.fr

Abstract. We develop an application of SOM for the task of anomaly detection and visualization. To remove the effect of exogenous independent variables, we use a correction model which is more accurate than the usual one, since we apply different linear models in each cluster of context. We do not assume any particular probability distribution of the data and the detection method is based on the distance of new data to the Kohonen map learned with corrected healthy data. We apply the proposed method to the detection of aircraft engine anomalies.

Keywords: Health Monitoring, aircraft, SOM, clustering, anomaly detection, confidence intervals.

1 Introduction, Health Monitoring and Related Works

In this paper, we develop SOM-based methods for the task of anomaly detection and visualization of aircraft engine anomalies.

The paper is organized as follows : Section 1 is an introduction to the subject, giving a small review of related articles. In Section 2, the different components of the system proposed are being described in detail. Section 3 presents the data that we used in this application, the experiments that we carried out and their results. Section 4 presents a short conclusion.

1.1 Health Monitoring

Health monitoring consists in a set of algorithms which monitor in real time the operational parameters of the system. The goal is to detect early signs of failure, to schedule maintenance and to identify the causes of anomalies.

T. Villmann et al. (eds.), *Advances in Self-Organizing Maps and Learning
Vector Quantization,* Advances in Intelligent Systems and Computing 295,
DOI: 10.1007/978-3-319-07695-9_14, © Springer International Publishing Switzerland 2014

Here we consider a domain where Health Monitoring is especially important: aircraft engine safety and reliability. Snecma, the french aircraft engine constructor, has developed well-established methodologies and innovative tools: to ensure the operational reliability of engines and the availability of aircraft, all flights are monitored. In this way, the availability of engines is improved: operational events, such as D&C (Delay and Cancellation) or IFSD (In-flight Shut Down) are avoided and maintenance operations planning and costs are optimized.

1.2 Related Work

This paper follows other related works. For example, [9] have proposed the *Continuous Empirical Score* (CES), an algorithm for Health Monitoring for a test cell environment based on three components: a clustering algorithm based on EM, a scoring component and a decision procedure.

In [8,3,7], a similar methodology is applied to detect change-points in Aircraft Communication, Addressing and Reporting System (ACARS) data, which are basically messages transmitted from the aircraft to the ground containing on-flight measurements of various quantities relative to the engine and the aircraft.

In [4], a novel *star* architecture for Kohonen maps is proposed. The idea here is that the center of the star will capture the normal state of an engine with some rays regrouping normal behaviors which have drifted away from the center state and other rays capturing possible engine defects.

In this paper, we propose a new anomaly detection method, using statistical methods such as projections on Kohonen maps and computation of confidence intervals. It is adapted to large sets of data samples, which are not necessarily issued from a single engine.

Note that typically, methods for Health Monitoring use an extensive amount of expert knowledge, whereas the proposed method is fully automatic and has not been designed for a specific dataset.

Finally, let us note that the reader can find a broad survey of methods for anomaly detection and their applications in [2] and [10,11].

2   Overview of the Methodology

Flight data consist of a series of measures acquired by sensors positioned on the engine or the body of the aircraft. Data may be issued from a single or multiple engines. We distinguish between *exogenous* or *environmental* measures related to the environment and *endogenous* or *operational* variables related to the engine itself. The reader can find the list of variables in Table 1. For the anomaly detection task, we are interested in operational measures. However, environmental influence on the operational measures needs to be removed to get reliable detection.

The entire procedure consists of two main phases.

1. The first phase is the *training* or *learning* phase where we learn based on healthy data.

Table 1. Description of the variables of the cruise phase data

Name	Description
Operational variables	
EXH	Exhaustion gas temperature
N2	Core speed
Temp1	Temperature at the entrance of the fan
Pres	Static pressure before combustion
Temp2	Temperature before combustion
FF	Fuel flow
Environmental variables	
ALT	Altitude
Temp3	Ambient temperature
SP	Aircraft speed
N1	Fan speed
Other variables	
ENG	Engine index
AGE	Engine age

- We cluster data into clusters of environmental conditions using only environmental variables.
- We correct operational measures variables from the influence of the environment using a linear model, and we get the residuals (corrected values).
- Next, a SOM is being learned based on the residuals.
- We calibrate the anomaly detection component by computing the confidence intervals of the distances of the corrected data to the SOM.

2. The learning phase is followed by the *test* phase, where novel data are taken into account.

   - Each novel data sample is being clustered in one of the environment clusters established in the training phase.
   - It is then being corrected of the environment influence using the linear model estimated earlier.
   - The test sample is projected to the Kohonen map constructed in the training phase and finally, the calibrated anomaly detection component determines if the sample is normal or not.

Clustering of the Environmental Contexts. An important point is the choice of the clustering method. Note that clustering is carried out on the *environmental* variables. The most popular clustering method is the Hierarchical Ascending Classification [5] algorithm, which allows us to choose the number of clusters based on the explained variance at different heights of the constructed tree.

However in this work our goal is to develop a more general methodology that could process even high-dimensional data and it is well-known that HAC is not adapted to this kind of data. Consequently, we are particularly interested

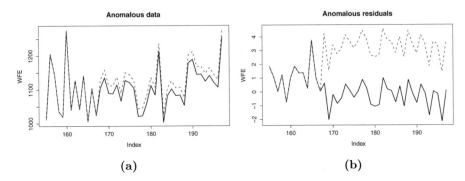

Fig. 1. An example of an anomaly of the FF variable of the cruise flight data (a) Superposition of the healthy data (solid black lines) and the data with anomalies (dashed red line) (b) Superposition of the corrected data obtained from the healthy data and corrected data obtained from corrupted data. The anomaly is visible only on corrected data.

in methods based on subspaces such as HDDC [1], since they can provide us with a parsimonious representation of high-dimensional data. Thus, we will use HDDC for the environment clustering, despite its less good performance for low-dimensional data.

Corrupting Data. In order to test the capacity of the proposed system to detect anomalies, we need data with anomalies. However, it is very difficult to get them due to the extraordinary reliability of the aircraft engines and we cannot fabricate them because deliberately damaging the engine or the test cell is clearly not an option. Therefore, we create artificial anomalies by corrupting some of the data based on expert specifications that have been established following well-known possible malfunctions of aircraft engines.

Corrupting the data with anomalies is carried out according to a *signature* describing the defect (malfunction). A signature is a vector $s \in \mathbb{R}^p$. Following s, a corruption term is added to the nominal value of the signal for a randomly chosen set of successive data samples.

Figure 1a gives an example of the corruption of the FF variable for one of the engines. Figure 1b shows the corrupted variable of the corrected data, that is, after having removed the influence of the environmental variables.

2.1 Clustering the Corrected Data Using a SOM

In order to build an anomaly detection component, we need a clustering method to define homogeneous subsets of corrected data. We choose to use the SOM algorithm [6] for its well-known properties of clustering organized with respect to each variable of the data as well as its visualization ability.

The output of the algorithm is a set of prototype vectors that define an "organized" map, that is, a map that respects the topology of the data in the input space. We can then color the map according to the distribution of the data for each variable. In this way, we can visually detect regions in the map where low or high values of a given variable are located. A smooth coloring shows that it is well organized. In the next section, we show how to use these properties for the anomaly detection task. ˙

2.2   Anomaly Detection

In this subsection, we present two anomaly detection methods that are based on confidence intervals. These intervals provide us with a "normality" interval of healthy data, which we can then use in the test phase to determine if a novel data sample is healthy or not.

We have already seen that the SOM algorithm associates each data sample with the nearest prototype vector, given a selected distance measure. Usually, the Euclidean distance is selected. Let L be the number of the units of the map, $\{m_l, \; l = 1, \ldots, L\}$ the prototypes. For each data sample, we calculate $\mathbf{x}_i$, its distance to the map, namely the distance to its nearest prototype vector:

$$d(\mathbf{x}_i) = \min_l \|\mathbf{x}_i - \mathbf{m}_l\|^2 \tag{1}$$

where $i = 1, \ldots, n$. Note that this way of calculating distance will give us a far more useful measure than if we had just utilized the distance to the global mean, i.e. $d(\mathbf{x}_i) = \|\mathbf{x}_i - \bar{\mathbf{x}}\|^2$.

The confidence intervals that we use here are calculated using distances of training data to the map. The main idea is that the distance of a data sample to its prototype vector has to be "small". So, a "large" distance could possibly indicate an anomaly. We propose a global and a local variant of this method.

Global Detection. During the training phase, we calculate the distances $d(\mathbf{x}_i)$, $\forall i$, according to Equation (1). We can thus construct a confidence interval by taking the 99-th percentile of the distances, $P_{99}(\{d(\mathbf{x}_i), \; \forall i\})$, as the upper limit. The lower limit is equal to 0 since a distance is strictly positive. We define thus the confidence interval $\mathcal{I}$

$$\mathcal{I} = [0, P_{99}(\{d(\mathbf{x}_i), \; \forall i\})] \tag{2}$$

For a novel data sample $\mathbf{x}$, we establish the following decision rule:

$$\begin{cases} \text{The novel data sample is healthy, if } d(\mathbf{x}) \in \mathcal{I} \\ \text{The novel data sample is an anomaly, if } d(\mathbf{x}) \notin \mathcal{I}. \end{cases} \tag{3}$$

The choice of the 99-th percentile is a compromise taking into account our double-sided objective of a high anomaly detection rate with the smallest possible false alarm rate. Moreover, since the true anomaly rate is typically very small in civil aircraft engines, the choice of such a high percentile, which also serves as an upper bound of the normal functioning interval, is reasonable.

Local Detection. In a similar manner, in the training phase, we can build a confidence interval for every cluster l. In this way, we obtain L confidence intervals $\mathcal{I}_l$, $l = 1, \ldots, L$ by taking the 99-th percentile of the *per* cluster distances as the upper limit

$$\mathcal{I}_l = [0, P_{99}\left(\{d(\mathbf{x}_i) : \mathbf{x}_i \text{ in SOM cluster } l\}\right)] \tag{4}$$

For a novel data sample $\mathbf{x}$ (in the test phase), we establish the following decision rule:

$$\begin{cases} \text{The novel data sample, affected to SOM cluster } l, \text{ is healthy, if } d(\mathbf{x}) \in \mathcal{I}_l \\ \text{The novel data sample, affected to SOM cluster } l, \text{ is an anomaly if } d(\mathbf{x}) \notin \mathcal{I}_l. \end{cases} \tag{5}$$

3   Application to Aircraft Flight Cruise Data

In this section, we present the data that we used for our experiments as well as the processing that we carried out on them.

Data samples in this dataset are snapshots taken from the cruise phase of a flight. Each data sample is a vector of endogenous and environmental variables, as well as categorical variables. Data are issued from 16 distinct engines of the same type. For each time instant, there are two snapshots, one for the engine on the left and another one for the engine on the right. Thus, engines appear always in pairs. Snapshots are issued from different flights. Typically, there is one pair of snapshots per flight. The reader can find the list of variables in Table 1. The dataset we used here contains 2472 data samples and 12 variables.

We have divided the dataset into a training set and a test set. For the training set, we randomly picked $n = 2000$ data samples among the 2472 that we dispose of in total. The test set is composed of the 472 remaining data samples. We have verified that all engines are represented in both sets. We have sorted data based on the engine ID (primary key of the sort) and for a given engine, based on the timestamp of the snapshot. We normalize the data (center and scale) because the scales of the variables were very different.

Selection of the Number of Clusters in Environment Clustering. Clustering is carried out on environmental variables to define clusters of contexts. Due to the large variability of the different contexts (extreme temperatures very high or very cold and so on), we have to do a compromise between a good variance explanation and a reasonable number of clusters (to keep a sufficient number of data in each cluster). If we compare HDDC to the Hierarchical Ascending Classification (HAC) algorithm in terms of explained variance, we observe that the explained variance is about 50 % for five clusters for both algorithms. And as mentioned before, we prefer to use HDDC [1] to present a methodology which can be easily adapted to high-dimensional data. Let $K = 5$ be the number of clusters.

Correcting the Endogenous Data from Environmental Influence. We correct the operational variables of environmental influence using the procedure we described in section 2. After the partition into 5 clusters based on environmental variables, we compute the residuals of the operational variables as follows: if we set $X^{(1)} = \text{N1}$, $X^{(2)} = \text{Temp3}$, $X^{(3)} = \text{SP}$, $X^{(4)} = \text{ALT}$ et $X^{(5)} = \text{AGE}$, we write

$$Y_{rkj} = \mu + \alpha_r + \beta_k + \gamma_{1k}X_{rkj}^{(1)} + \gamma_{2k}X_{rkj}^{(2)} + \gamma_{3k}X_{rkj}^{(3)} +$$
$$\gamma_{4k}X_{rkj}^{(4)} + \gamma_5 X_{rkj}^{(5)} + \varepsilon_{rkj} \tag{6}$$

where Y is one of the $d = 6$ operational variables, $r \in \{1, \ldots, 16\}$ is the engine index, $k \in \{1, \ldots, 5\}$ is the cluster number, $j \in \{1, \ldots, n_{rk}\}$ is the observation index. Moreover, μ is the intercept, α_r is the effect of the engine and β_k the effect of the cluster.

Learning a SOM with Residuals. By analyzing the residuals, one can observe that the model succeeds in capturing the influence of the environment on the endogenous measures, since the magnitude of the residuals is rather small (between -0.5 and + 0.5). The residuals therefore capture behaviors of the engine which are not due to environmental conditions. The residuals are expected to be centered, *i.e.* to have a mean equal to 0. However, they are not necessarily scaled, so we re-scale them.

Generally speaking, since residuals are not smooth, we carry out smoothing using a moving average of width $w = 7$ (central element plus 3 elements on the left plus 3 elements on the right). We note that by smoothing, we lose $\lfloor \frac{w}{2} \rfloor$ data samples from the beginning and the end. Therefore, we end up with a set of 1994 residual samples instead of the 2000 that we had initially. Next, we construct a Self-Organizing Map (SOM) based on the residuals (Figure 2). We have opted here for a map of 49 neurons (7×7) because we need a minimum of observations per SOM cluster in order to calculate the normal functioning intervals with precision.

The last step is the calibration of the detection component by determining the global and local confidence intervals based on the distances of the data to the map. For the global case, according to Equation 2, we have:

$$\mathcal{I} = [0, 4.1707]$$

In a similar manner, we derive the upper limits of the local confidence intervals, ranging from 1.48 to 6.03.

Test Phase. In the test phase, we assume that novel data samples are being made available. We first corrupt these data following the technique proposed in Section 2. Snecma experts provided us with signatures of 12 known defects (anomalies), that we added to the data. For data confidentiality reasons, we are obliged to anonymize the defects and we refer to them as "Defect 1", "Defect 2" etc.

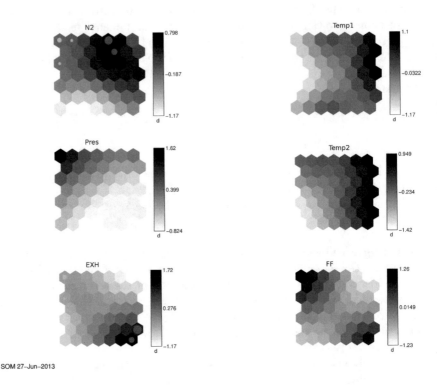

SOM 27–Jun–2013

Fig. 2. SOM built from the corrected training residuals for each of the $p = 6$ endogenous variables. Black cells contain high values of the variable while white ones contain low values. Red dots refer to anomalies and green dots to healthy data for two different types of defects bearing on the variables N2 and EXH. The proposed method clusters them in different regions of the map. The size of each dot is proportional to the number of points of the cluster.

We start by normalizing test data with the coefficients used to normalize training data earlier. We then cluster data into environment clusters using the model parameters we estimated on the training data earlier. Next, we correct data from environmental influence using the model we built on the training data. In this way, we obtain the test residuals, that we re-scale with the same scaling coefficients used to re-scale training residuals.

We apply a smoothing transformation using a moving average, exactly like we did for training residuals. We use the same window size, *i.e.* $w = 7$. Smoothing causes some of the data to be lost, so we end up with 466 test residuals instead of the 472 we had initially.

Finally, we project data onto the Kohonen map that we built in the training phase and we compute the distances $d(\mathbf{x})$ as in equation (1). We apply the decision rule, either the global decision rule of (3) or the local one of (5).

Table 2. Detection rate (*tpr*) and false alarm rate (*pfa*) for different types of defects and for both anomaly detection methods (global and local) for test data

	Global detection		Local detection	
Defect	tpr	pfa	tpr	pfa
Defect 1	100%	18,9%	100%	45,4%
Defect 2	100%	11,4%	100%	42,6%
Defect 3	100%	16,7%	100%	47,9%
Defect 4	100%	15,1%	100%	45,1%
Defect 5	96,7%	14,7%	100%	43,4%
Defect 6	100%	13,9%	100%	43,6%
Defect 7	96,7%	12,1%	96,7%	44,2%
Defect 8	100%	26,3%	100%	50%
Defect 9	100%	15,8%	100%	43,9%
Defect 10	100%	26,7%	100%	55,1%
Defect 11	100%	17,1%	100%	46,3%
Defect 12	100%	21%	100%	46,4%

In order to evaluate our system, we calculate the detection rate (*tpr*) and the false alarms rate (*pfa*):

$$tpr = \frac{\text{number of detections}}{\text{number of anomalies}}$$

$$pfa = \frac{\text{number of non-expected detections}}{\text{number of detections}}$$

In Table 2, we can see detection results for all 12 defects and for both detection methods (global and local). It is clear that both methods succeed in detecting the defects, almost without a single miss. The global method has a lower false alarm rate than the local one. This is because in our example, confidence intervals cannot be calculated reliably in the local case since we have few data per SOM cluster.

Figure 3 shows the distance d of each data sample (samples on the horizontal axis) to their nearest prototype vector (Equation 1). The light blue band shows the global confidence interval $\mathcal{I}$ that we calculated in the training phase. Red crosses show the false alarms and green stars the correct detections.

Due to limited space in this contribution, the figures related to the local detection can be found in the following URL: https://drive.google.com/folderview?id=0B0EJciu-PLatZzdqR25oVjNNaTg&usp=sharing

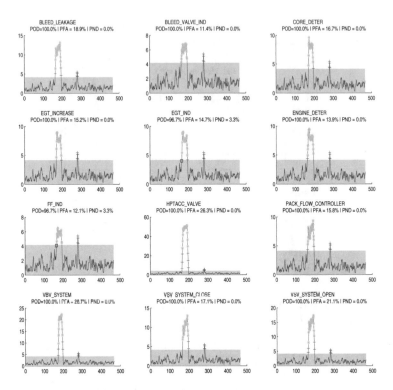

Fig. 3. Distances of the test data to their nearest prototype vector and the global confidence interval (in light blue). Red crosses show the false alarms and green stars show successful detection.

4 Conclusion and Future Work

We have developed an integrated methodology for the analysis, detection and visualization of anomalies of aircraft engines. We have developed a statistical technique that builds intervals of "normal" functioning of an engine based on distances of healthy data from the map with the aim of detecting anomalies. The system is first calibrated using healthy data. It is then fully operational and can process data that was not seen during training.

The proposed method has shown satisfying performance in anomaly detection, given that it is a general method which does not incorporate any expert knowledge and that it is, thus, a general tool that can be used to detect anomalies in any kind of data.

Another advantage of the proposed method is that the use of the dimension allows to carry out multi-dimensional anomaly detection in a problem of dimension 1. Moreover, the representation of the operational variables given by the use of the distance to the SOM is of a higher granularity than that of the distance from the global mean. Last but not least, the use of SOM allows us to give interesting visualizations of healthy and abnormal data, as seen in Figure 2.

An extension of our work would be to carry out anomaly detection for datastreams using this method. A naive solution would be to re-calibrate the components of the system with each novel data sample, but it would be very time-consuming. Instead, one can try to make each component of the system to operate on datastreams.

References

1. Bouveyron, C., Girard, S., Schmid, C.: High-dimensional data clustering. Computational Statistics & Data Analysis 52(1), 502–519 (2007a)
2. Chandola, V., Banerjee, A., Kumar, V.: Outlier detection: A survey. ACM Computing Surveys 41(3) (2009)
3. Côme, E., Cottrell, M., Verleysen, M., Lacaille, J.: Aircraft engine health monitoring using self-organizing maps. In: Perner, P. (ed.) ICDM 2010. LNCS, vol. 6171, pp. 405–417. Springer, Heidelberg (2010a)
4. Côme, E., Cottrell, M., Verleysen, M., Lacaille, J., et al.: Self organizing star (sos) for health monitoring. In: Proceedings of the European Conference on Artificial Neural Networks, pp. 99–104 (2010b)
5. Duda, R.O., Hart, P.E.: Pattern Classification and Scene Analysis. John Wiley & Sons, Inc., New York (1973)
6. Kohonen, T.: Self-organizing maps, vol. 30. Springer (2001)
7. Lacaille, J., Côme, E., et al.: Sudden change detection in turbofan engine behavior. In: Proceedings of the Eighth International Conference on Condition Monitoring and Machinery Failure Prevention Technologies, pp. 542–548 (2011)
8. Lacaille, J., Gerez, V.: Online abnormality diagnosis for real-time implementation on turbofan engines and test cells, pp. 579–587 (2011)
9. Lacaille, J., Gerez, V., Zouari, R.: An adaptive anomaly detector used in turbofan test cells. In: Proceedings of the Annual Conference of the Prognostics and Health Management Society (2010)
10. Markou, M.: Novelty detection: a review-part 1: statistical approaches. Signal Processing 83(12), 2481–2497 (2003a)
11. Markou, M.: Novelty detection. a review part 2: neural network based approaches. Signal Processing 83(12), 2499–2521 (2003b)

RFSOM – Extending Self-Organizing Feature Maps with Adaptive Metrics to Combine Spatial and Textural Features for Body Pose Estimation

Mathias Klingner[1], Sven Hellbach[1], Martin Riedel[2], Marika Kaden[2], Thomas Villmann[2], and Hans-Joachim Böhme[1,*]

[1] University of Applied Sciences Dresden,
Artificial Intelligence and Cognitive Robotics Labs,
POB 12 07 01, 01008 Dresden, Germany
{klingner,hellbach,boehme}@informatik.htw-dresden.de
[2] University of Applied Sciences Mittweida,
Computational Intelligence and Technomathematics,
POB 14 57, 09648 Mittweida, Germany
{riedel,kaestner,thomas.villmann}@hs-mittweida.de

Abstract. In this work we propose an online approach to compute a more precise assignment between parts of an upper human body model to RGBD image data. For this, a Self-Organizing Map (SOM) will be computed using a set of features where each feature is weighted by a relevance factor (RFSOM). These factors are computed using the generalized matrix learning vector quantization (GMLVQ) and allow to scale the input dimensions according to their relevance. With this scaling it is possible to distinguish between the different body parts of the upper body model. This method leads to a more precise positioning of the SOM in the 2.5D point cloud, a more stable behavior of the single neurons in their specific body region, and hence, to a more reliable pose model for further computation. The algorithm was evaluated on different data sets and compared to a Self-Organizing Map trained with the spatial dimensions only using the same data sets.

Keywords: Self-Organizing Maps, learning vector quantization, relevance learning, human machine interaction.

1 Introduction

A suitable human robot interface is of great importance for the practical usability of mobile assistance and service systems whenever such systems have to directly interact with persons. This interaction is often based on the learning and interpretation of the gestures and facial expressions of the dialog partner in order to avoid collision and to infer her intention. Therefore, it is necessary to track the motion of the human body or rather the movements of individual parts.

---

* This work was supported by ESF grand number 100130195.

T. Villmann et al. (eds.), *Advances in Self-Organizing Maps and Learning Vector Quantization*, Advances in Intelligent Systems and Computing 295,
DOI: 10.1007/978-3-319-07695-9_15, © Springer International Publishing Switzerland 2014

There exist similar works like the approach presented by Haker et.al. [1] and the one presented by Shotton et al. [2]. Shotton et al. utilize a precomputed classifier to assign image data to body regions for a final classification of body poses. The Shotton et al. approach is not usable for us because of the costly classifier computation. Our approach is an enhancement for the one presented by Haker et al. They use a SOM to compute an online estimation of body poses with a time-of-flight camera and without a precomputed classifier. We want to show that we are able to decrease the number of misplaced neurons of the SOM to ensure the accuracy of the subsequent body pose estimation without losing the ability to operate online.

The remainder of this paper is organized as follows. In the Sec. 2 our proposed method is outlined. Section 2.1 describes the algorithm used for the generation and training of the Self-Organizing Map. In Sec. 2.2 an introduction to our RF-SOM extension is given. Sec. 2.3 points out the used method for generating the adaptive metrics, being divided into two parts. The first part gives a short introduction to LVQ1 and the second one describes the used GMLVQ. In Sec. 3 the evaluation of the presented approach with different test cases will be described and Sec. 4 concludes this work.

2   Method

This work proposes an enhancement of the approach presented in [1] which uses a standard SOM approach with a body shaped topological map to model human upper body poses. We extend this approach by using an adaptive metric which is learned from data using GMLVQ presented in [3]. We use a Microsoft Kinect for the capture of the image data. The advantage of this camera is that in addition to the depth data a synchronized RGB image is delivered at the same time.

2.1   Self-Organizing Feature Map

For the original approach a Self-Organizing Map (SOM) is trained on the 2.5D point cloud data (Fig. 1(a)) of a depth camera to model the human upper body. We assume that only the foreground contains data of a person (Fig. 1(c)). The necessary separation of foreground and background in the captured scene is based on the Otsu threshold algorithm [4] using the 2.5D point cloud data as input. The segmentation is based on the spatial structure of the data. Furthermore, it doesn't matter how the background is designed in color or shape because of the dynamic character of the Otsu algorithm. In addition, we use a Viola Jones face detector [5] to find a face in the field of view of the camera to confirm this hypothesis. In contrast to the foreground segmentation, the face detection is computed on the RGB image data. Having successfully detected a face in the scene, the face detector will be discontinued, until the person leaves the field of view.

After the successful face detection and the extraction of the foreground, we initialize the pre-shaped SOM in the center of gravity of the resulting foreground point cloud [1]. Pre-shaped means that the SOM's topology is created in the form of a human upper body with horizontally outstretched arms.

| (a) 2.5D point Cloud | (b) Unsegmented RGB image. | (c) Segmented foreground. |

Fig. 1. Subfigure (a) shows a view of the 2.5D point cloud data recorded by a depth camera. In the foreground a person is standing in front of a monitor while the background contains the walls and the ceiling of the room. Subfigure (b) shows a RGB image from the camera. The corresponding foreground data after the foreground-background segmentation is shown in (c).

In [1] the best-matching neuron (BMN) for a presented stimulus is determined based on the Euclidean distance in the three spatial dimensions x, y and z. For this, the best-matching neuron $\mathbf{w}_B^S(t)$ for each data point $\mathbf{x}$ is computed and adapted to $\mathbf{w}_B^S(t+1)$ using the adaptation rate $\eta(t)$:

$$\mathbf{w}_B^S(t+1) = \mathbf{w}_B^S(t) + \eta(t) \cdot (\mathbf{x} - \mathbf{w}_B^S(t)) \tag{1}$$

Hence, the best-matching neuron is determined by the computation of the minimal Euclidean distance between the current data point $\mathbf{x}$ and all neuron w_k^S of the SOM.

$$\min_{\forall k} ||\mathbf{x} - \mathbf{w}_k^S|| \tag{2}$$

The adaptation rate $\eta(t)$ is set to $\eta_{w_B}(t)$ for the best matching neuron and is defined as

$$\eta_{\mathbf{w}_B^S}(t) = \eta_{in} \cdot \left(\frac{\eta_{fi}}{\eta_{in}}\right)^{\left(\frac{t}{t_{max}}\right)} \tag{3}$$

with η_{in} as the initial and η_{fi} as the final adaptation rate, t is the number of the current training step and t_{max} is the maximum number of training steps. Only direct neighbours of the best-matching neuron in the topological map are adapted. Hence, the adaptation rate for the neighborhood is set to $\eta_n(t)$ and defined as $\eta_n(t) = \eta_{\mathbf{w}_B^S}(t)/2$. A step t hereby contains the presentation of all data points of a single frame, which will be repeated t_{max} times per frame.

2.2 RFSOM Extension

As an extension of the standard Self-Organizing feature Map, we compute different textural features like Histograms of Oriented Gradients (HOG) [6], Local Binary Patterns (LBP) [7], Grayscale Co-ocurrence Matrix ($GLCM$) [8], and also standard color spaces, like RGB and HSV for each second voxel of the resulting

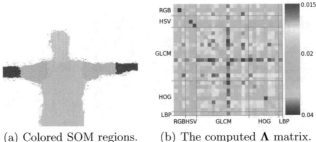

(a) Colored SOM regions.   (b) The computed Λ matrix.

Fig. 2. Subfigure (a) shows the colored regions of the SOM based on the SOM topology. Using the textural features of each voxel in the foreground a global relevance matrix (b) is computed. The right side of (b) shows the color scale from the minimal (top) to the maximal (bottom) relevance value.

foreground point cloud. Hence, the input space increases from $\mathbb{R}^3$ to $\mathbb{R}^n$ with n being much larger than 3. However, the main part of this extension is based on the approach presented in [3]. This approach describes the computation of a global relevance matrix Λ for a representation of the internal structure of some data.

We assume that with the use of textural features the regions (body parts) can be discriminated in a better way. However, our goal is to provide a number of additional texture features and let the approach select the relevant ones from the data. Hence, in this work Λ is computed using the textural features described above using GMLVQ as described in Sec. 2.3. For this, the matrix is trained in a supervised manner using regional information automatically gained from the SOM topology (Fig. 2(a)). Hence, a region label is assigned to each stimulus directed from the related best-matching neuron. For this, six regions were defined, based on the position of the neurons in the SOM topology: head, body, left arm, right arm, left hand, and right hand.

After the computation of matrix Λ, the Euclidian metric of the SOM (Sec. 2.1) will be replaced by an adaptive metrics. In this work, GMLVQ provides a Λ only for the textural information. From now on, each distance between a stimulus and a neuron will be computed using the adaptive metric in the $\mathbb{R}^n$ feature space, including the spatial dimensions. At the same time, the influence of the spatial dimensions in the adaptation process will be reduced by setting a general weight α for spatial and $1-\alpha$ for textural dimensions to give the textural features more influence. Hence, to combine spatial and textural information, we define a metric as follows:

$$d(\mathbf{x}, \mathbf{w}^S) = \alpha \cdot d_s(\mathbf{x}, \mathbf{w}^S) + (1 - \alpha) \cdot d_t(\mathbf{x}, \mathbf{w}^S) \tag{4}$$
$$= \alpha \cdot \|\mathbf{x}_s - \mathbf{w}_s^S\|_2 + (1 - \alpha) \cdot [\mathbf{x}_f - \mathbf{w}_f^S(t)]^T \Lambda [\mathbf{x}_f - \mathbf{w}_f^S(t)] \tag{5}$$

which is used to determine of the best-matching neuron. Because of this, Eq. 1 will be modified to

$$\mathbf{w}_B^S(t + 1) = \mathbf{w}_B^S(t) + \eta(t) \cdot [\mathbf{x} - \mathbf{w}_B^S(t)]^T \widetilde{\Lambda} [\mathbf{x} - \mathbf{w}_B^S(t)] \tag{6}$$

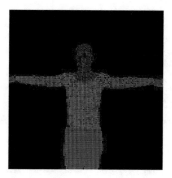

Fig. 3. The figure shows the colored mapping of the Voronoi regions into the image space for the trained lvq codebook vectors. The colorization is done in the R^3 but also represents all higher feature dimensions. This leads to different colors in the same region in the R^3.

as the adaptive metric including matrix $\widetilde{\Lambda}$ with

$$\widetilde{\Lambda} = \begin{bmatrix} \mathbb{I} & 0 \\ 0 & \Lambda \end{bmatrix} \tag{7}$$

which contains matrix Λ as the weight matrix for the textural features (Fig. 2(b)). We introduce the name Relevance-Feature-Self-Organizing Map (RFSOM) for this idea.

2.3   Learning Adaptive Metrics

Learning the adaptive metric is usually done with the GMLVQ. In addition a LVQ1 is used for the prototype initialisation and adaptation to accelerate convergence of the GMLVQ. Finally, a GMLVQ without adaptation of the lvq prototypes is used for the computation of the relevance metrics.

LVQ1 in a Nutshell: The target of the LVQ is to approximitate a number of different prototypes in the data space to represent each part of data as good as possible. The training data for the LVQ1 is given as $(\mathbf{x}_i, c_j) \in \mathbb{R}^N \times \{1, ..., C\}$. With N being the dimensionality of the data (number of features) and C the number of existing classes (body parts). Each prototype is characterized by the combination of its location in the feature space $\mathbf{w}_i^L \in \mathbb{R}^N$ and its class label $c(\mathbf{w}_i^L) \in \{1, ..., C\}$. For a data point $\mathbf{x}_i$ of class c_j the best-matching prototype is then determined. This prototype is then adapted towards the data point if it is from the same class, otherwise it will be adapted in the counter direction. Hence, to evolve distinctive class borders a best-matching prototype from a different class than the data point is pushed away from this part of the data space. However, this sometimes leads to an incorrect representation of the data space. To some extent, this is due to the fact that some dimensions provide only few contributions to the classification process.

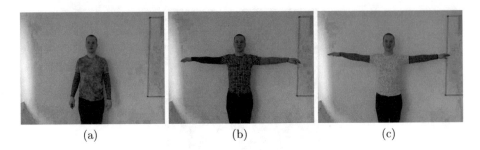

(a)   (b)   (c)

Fig. 4. The three different color and pattern styles for the data sets are shown in (a), (b) and (c)

GMLVQ: The GMLVQ algorithm adresses this problem with the introduction of a matrix $\mathbf{\Lambda}$ of similarity measures in the prototype adaptation. The matrix $\mathbf{\Lambda}$ then serves as a weight matrix for the features during training. Dimensions with a large absolute value of λ_{ij} can be interpreted as being more important for the classification than dimensions with a small or zero one. On top of this, also the negative correlation is important as a weight for a dimension.

To compute $\mathbf{\Lambda}$, a set of prototypes for the different classes is necessary. In contrast to the approach presented in [3] the initial adaptation of the prototypes to the characteristics of the specific classes will be computed with a standard LVQ1. In order to minimize known instabilities during the initialization of the LVQ1, the prototypes of a class are generated in the region of the point cloud they should represent. The necessary feature vectors for the prototype generation are taken randomly from the respective region. After the LVQ1 has converged, the matrix computation is processed without a further adaptation of the prototypes which is part of the standard GMLVQ. Hence, the prototypes will be unaffected during the matrix adaptation. We accept the possible error since our approach still improves the results in contrast to the classical SOM.

The GMLVQ introduces a new kind of generalized distance in the form

$$d^{\mathbf{\Lambda}}(\mathbf{w}^L, \mathbf{x}) = (\mathbf{x} - \mathbf{w}^L)^T \mathbf{\Lambda} (\mathbf{x} - \mathbf{w}^L) \tag{8}$$

where the matrix $\mathbf{\Lambda}$ itself is a full $N \times N$ matrix. The content of each cell λ_{ij} can be regarded as the relevance of the combined appearance of feature i and j. To ensure that $\mathbf{\Lambda}$ is positive semidefinite and symmetric we can substitute the matrix with $\mathbf{\Lambda} = \mathbf{\Omega}^T \mathbf{\Omega}$.

The adaptation of the matrix elements Ω_{lm} can be computed by

$$\begin{aligned}
\Delta\Omega_{lm} = -\epsilon \cdot 2 \cdot \Phi'(\mu(\mathbf{x})) \cdot \Big(& \mu^+(\mathbf{x}) \cdot \Big((x_m - w^L_{J,m})[\mathbf{\Omega}(\mathbf{x} - \mathbf{w}^L_J)]_l\Big) \\
& -\mu^-(\mathbf{x}) \cdot \Big((x_m - w^L_{K,m})[\mathbf{\Omega}(\mathbf{x} - \mathbf{w}^L_K)]_l\Big)\Big)
\end{aligned} \tag{9}$$

where $\mathbf{x}$ is the presented data, $\mathbf{w}^L_J$ the nearest class prototype, $\mathbf{w}^L_K$ the nearest non class prototype, l and m the feature index, and ϵ the learning rate [3].

The computation of Λ is a time consuming problem. Hence, the adaptation is carried out in parallel. During that time, the standard SOM is applied. As soon as the GMLVQ computation is completed, the matrix Λ is fed to the RFSOM which takes over. This means, we don't compute a new Λ for each time frame.

3 Evaluation

The experiments should show a noticeable increase of correct positioned neurons in the corresponding region for the RFSOM training in contrast to the standard euclidian SOM with only the three spatial dimensions. Therefore, the number of correctly positioned neurons for each region over a number of frames has been counted for both algorithm. Correctly positioned means the neuron stays in its corresponding region after the training of a single frame is completed.

Previous experiments have shown that the algorithm works with normal clothing. However, to get significant different classes for the body parts and to gain insight in the meaning of Λ in the GMLVQ algorithm, special clothes were used for the test data. Each of the clothing has a specific color and shape pattern (Fig. 4).

We recorded a test set, which contains 10 frames with different arm movements. Also, the movements carried out in the different sets are not the same. In set one and two the arm movements of the test person are smooth and the body pose of the first frame is close to the initial pose of the untrained SOM. In contrast, the third set contains a movement with a big step between two arm poses and an initial body pose which differs strongly from the initial pose of the SOM. For this, the third set can be understood as a test example for what happens when images are lost in the processing pipeline or as an example for a not perfect initialisation situation of the SOM. To gain ground truth data of the body regions in the test sets, all of them were previously labeled by hand.

To get a closer look on the effect of the various features, the features where combined in different feature sets and separately evaluated. Each feature set contains the three spatial dimensions x, y, z and also RGB and HSV. These features were then combined with $GLCM$, HOG, LBP, $HOG + LBP$ and $GLCM + HOG + LBP$. Furthermore, the grayscale levels for the image used for the $GLCM$ computation was changed between 9, 32, and 64 possible gray values for the $GLCM + HOG + LBP$ combination. This is due to the time-consuming calculation of the $GLCM$ if all 256 gray values are used and shows how the minimized gray level affects the accuracy of the approach.

The final and most significant question was, how the GMLVQ algorithm itself and simple modifications of it affect the accuracy of the whole approach. Therefore, each frame was processed with (i) the standard SOM training regime (spatials dimension only), (ii) using RFSOM with initial $LVQ1$, (iii) RFSOM without initial $LVQ1$ and (iv) RFSOM using the identity matrix as Λ matrix.

Results: As described in Sec. 3, the evaluation should show a noticeably increase of the correct positioned neurons in their region for the RFSOM approach. Therefore, a comparison between the SOM and the RFSOM for all feature set

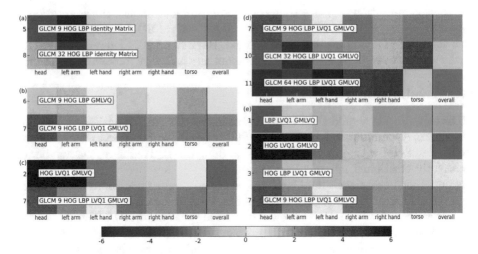

Fig. 5. The patches in this figure are visualizations of the percentage wise difference between the RFSOM and SOM for a region and feature set over a fixed number of 10 frames. Each feature set contains x,y,z and also RGB and HSV. This basic sets were then combined with features from HOG, LBP and $GLCM$ (e). The gray levels for the $GLCM$ where changed between 9, 32 and 64 values (d). Finally, each frame was computed with the standard SOM training regime, using RFSOM with and without initial $LVQ1$ (b) and RFSOM using the identity matrix as Λ matrix (a). The color scaling is set to a minimum of -6%. For this, dark red means an increase of 6% what is a really good result and dark blue a decrease by at least 6%. Hence, set number 4 seems constantly bad but with a wider color range the differences between the body parts will be more clearly represented.

combinations was computed. In this comparison, the correct positioned neurons for a region were counted over a number of frames. To get an average position for each neuron in a frame, every frame was computed 10 times. Finally, for each set the number of correct positioned neurons per region over all frames was divided by the number of all neurons computed for this region over all frames. The results of this comparison are shown in the accuracy plot in Fig. 5. Hence, each patch of each subfigure in Fig. 5 is the visualization of the percentage wise difference between the RFSOM and the SOM for a specific region and feature set over a fixed number of 10 frames.

The rows of the plots summarize the results for the different feature sets. In the horizontal direction from the left to the right the results for the body parts are shown. The last column is an overall computation. Hence, each correct positioned neuron of each region was counted and divided by the number of all computed neurons for all regions over all frames. Finally, the difference between the accuracy of the RFSOM and the SOM was computed and visuallized. A temperature scale between -6% and 6% is shown on the bottom of the figure. Referring to this, dark red means an increase of 6% what is a really good result and dark blue a decrease by at least 6%.

There is a number of different information which can be extracted from this matrix, basically it can be said that the RFSOM approach is functional. The feature sets that seem to be adequate for our problem show an increase of accuracy in comparison to the standard approach. This is a very good result for an online approach without a precomputed classifier and probably ensures a higher precision of the subsequent pose estimation.

The comparison between the RFSOM with the identity matrix as Λ and the standard SOM shows that the results for the RFSOM in this case are worse. In particular, using the identity matrix in set number 5 reduces the accuracy by 2% in contrast to the standard SOM and in set number 8 by 0.4% (compare Fig. 5(a)). This can be explained by the tremendous increase in the number of dimensions, while not all dimensions contribute to the classification problem. This assumption can be confirmed by the comparison to their counterpart (Fig. 5(d)) with a computed matrix based on the extracted features. In this figure the results are better than in the standard approach. Hence, the identity matrix is definitely not the correct choice as an alternative to a computed Λ.

On top of this, a comparison of set number 6 and 7 (in Fig. 5(b)) which differs in the use of the *LVQ1* revealed another information. Both of them are computed with the same feature combination but set number 7 made use of the *LVQ1*, whereas 6 did not. The difference for set 7 amounts approximately 2%. For set 6 it is at least 0.4% better in comparison to the original approach. This means that the integration in the processing pipeline ensures a better result for the Λ computation.

A closer look to the overall computation in Fig. 5(b) shows that set number 2 and 7 are the best candidates for a further examination. Both of them increase the accuracy of the neuron positioning by approximately 3%. In set number 2 an increase of the accuracy for the head and the left arm by 6% can be seen. Even for the left hand, the right hand and the right arm, but just up to maximal 3%. The difference between the left side and the right side in set 2 is due to the contrast diversity in the image. The lighting condition cause a higher contrast in the right part of the image resulting in more distinct gradients

Furthermore, Fig. 5(d) shows the accuracy for the combination of all features using different gray levels in the *GLCM* computation. It can be seen that a larger number of gray levels decreases the overal accuracy in contrast to the original approach despite the fact that the accuracy of some body parts are increasing. Besides, taking the computational effort into account, using a larger number of gray levels does not make sense.

Figure 5(e) visualizes the difference between some possible feature combinations increasing in complexity from the top to the bottom. It can be seen that *LBP* alone can not handle the problem and decreases the accuracy by up to 1.6%. It seems that the single *HOG* in 2 is much more capable to develop significant features. In contrast, the combination of both off them in 3 increases the accuracy of the head and the torso but decreases the accuracy of all other body parts. Hence, the overall computation of this set remains at the zero border and thus the combination is also not useful.

To conclude the set evaluation, we can say that set number 7 is the most promising candidate. This set is included in Fig. 5(b), (c), (d) and (e) because it is suitable for all of this test cases. It shows a homogeneous increasing of the accuracy for all body parts except the left hand. Even though, the increase is not as large as for set number 2 in Fig. 5(c) and (e) the homogeneous effect for all body parts is here more from interest.

Since the GMLVQ is processed in parallel, we only need to measure for the SOM and RFSOM thread which runs on a single core. For that we gain computation times between 300 ms up to 3 seconds on an i7 Q840 with 1.87 Ghz and 3GB of RAM. Hence, with some improvements our approach aims towards real-time performance.

4   Conclusion

The RFSOM is a robust approach to get a more precise development of the used Self-Organizinig Map under real-time condition. We reached promising results due to the used feature combination and the quality of the image data. Hence, the use of *GLCM* with 9 gray values, *HOG* and *LBP* in combination with the GMLVQ algorithm seems to be the best choice for an increase of accuracy for all body parts. The small number of gray values for the *GLCM* also ensures the online capability of the approach. Further examinations will be done to get a faster implementation and a further increase of the accuracy in the positioning of the neurons. After reliably knowing the body pose for each frame it makes sense to extract the trajectory of the person as described in [9].

References

1. Haker, M., Böhme, M., Martinetz, T., Barth, E.: Self-Organizing Maps for Pose Estimation with a Time-of-Flight Camera. In: Kolb, A., Koch, R. (eds.) Dyn3D 2009. LNCS, vol. 5742, pp. 142–153. Springer, Heidelberg (2009)
2. Shotton, J., Fitzgibbon, A.W., Cook, M., Sharp, T., Finocchio, M., Moore, R., Kipman, A., Blake, A.: Real-Time Human Pose Recognition in Parts from Single Depth Images. In: IEEE CVPR 2011, pp. 1297–1304 (2011)
3. Schneider, P., Biehl, M., Hammer, B.: Adaptive relevance matrices in learning vector quantization. Neural Computation 21, 3532–3561 (2009)
4. Otsu, N.: A Threshold Selection Method from Gray-level Histograms. IEEE Transactions on Systems, Man and Cybernetics 9(1), 62–66 (1979)
5. Viola, P., Jones, M.: Robust Real-Time Face Detection. International Journal of Computer Vision 57(2), 137–154 (2004)
6. Dalal, N., Triggs, B.: Histograms of oriented gradients for human detection. In: CVPR 2005, vol. 1, pp. 886–893 (2005)
7. Ojala, T., Pietikainen, M., Maenpaa, T.: Multiresolution gray-scale and rotation invariant texture classification with local binary patterns. IEEE Transactions on PAMI 24(7), 971–987 (2002)
8. Haralick, R., Shanmugam, K., Dinstein, I.: Textural features for image classification. IEEE Transactions on Systems, Man and Cybernetics SMC-3(6), 610–621 (1973)
9. Klingner, M., Hellbach, S., Kästner, M., Villmann, T., Böhme, H.J.: Modeling Human Movements with Self-Organizing Maps using Adaptive Metrics. In: NC2 2012, pp. 14–19 (2012)

Beyond Standard Metrics – On the Selection and Combination of Distance Metrics for an Improved Classification of Hyperspectral Data

Uwe Knauer, Andreas Backhaus, and Udo Seiffert

Fraunhofer IFF
Joseph-v.-Fraunhofer-Str. 1, 39106 Magdeburg, Germany
{uwe.knauer,andreas.backhaus,udo.seiffert}@iff.fraunhofer.de
http://www.iff.fraunhofer.de

Abstract. Training and application of prototype based learning approaches such as Learning Vector Quantization, Radial Basis Function networks, and Supervised Neural Gas require the use of distance metrics to measure the similarities between feature vectors as well as class prototypes. While the Euclidean distance is used in many cases, the highly correlated features within the hyperspectral representation and the high dimensionality itself favor the use of more sophisticated distance metrics. In this paper we first investigate the role of different metrics for successful classification of hyperspectral data sets from real-world classification tasks. Second, it is shown that considerable performance gains can be achieved by a classification system that combines a number of prototype based models trained on differently parametrized divergence measures. Data sets are tested using a number of different combination strategies.

Keywords: Divergence, Metrics, Hyperspectral, SNG, GLVQ, RBF.

1   Introduction

The optical characterization of organic and inorganic materials with hyperspectral imaging is becoming a widespread application within plant breeding, smart farming, material sorting, or quality control in food production. The generic behavior of the material to reflect, absorb, or transmit light is used to characterize its identity and even molecular composition. A hyperspectral camera records a narrowly sampled spectrum of reflected or transmitted light in a certain wavelength range and produces a high-dimensional pattern of highly correlated spectral channels per image pixel. Often, the direct relationship between this pattern and the target value, for example a material category is unknown. In the simple case exact spectral bands are known that correlate with the presence of certain chemical compounds. If such direct knowledge is unavailable, machine learning is used to learn a classification or regression task from available labeled reference data.

Prototype based models like the Learning Vector Quantization [12], Supervised Neural Gas [11], or Radial Basis Function Networks [21] provide a set of

T. Villmann et al. (eds.), *Advances in Self-Organizing Maps and Learning Vector Quantization,* Advances in Intelligent Systems and Computing 295,
DOI: 10.1007/978-3-319-07695-9_16, © Springer International Publishing Switzerland 2014

tools to learn a classification task from high dimensional data. These methods utilize a certain similarity measure to compare an input pattern to a number of stored prototypes in order to predict the pattern's category. Commonly the Euclidean distance is used to calculate the similarity of the input and proto- type pattern. Each feature is compared separately irrespective of its position in the high dimensional feature vector. In contrast, spectral pattern are data samples that describe a function or distribution of energy across a well ordered wavelength range. Therefore this type of data is also called 'functional data'.

An approach to calculate the similarity of statistical distributions are di- vergences which offer an alternative way to characterize dissimilarity between spectral patterns. Additionally, more general divergences like the γ-divergence include parameters that can be potentially tuned to adapt the dissimilarity mea- sure to the learning task at hand. Divergence dissimilarity measures have been successfully integrated into prototype based machine learning models but perfor- mance gains have been minimal so far on models using just a single dissimilarity measure [22,26,15,27].

This paper shows that considerable performance gains can be achieved by a classification system that combines a number of prototype based models trained on differently parametrised divergence measures. A number of hyperspectral data sets from real-world classification tasks are tested using a number of different combination strategies.

2   Related Work

The idea to include task-adaptive non-standard metrics and dissimilarity func- tion into a pattern recognition system has been widely researched. In [24] the Ma- halanobis distance replaces the standard Euclidean distance. The Mahalanobis matrix is not calculated as the co-variance matrix but a distance metric learning method is used to calculate a transformation which assures small distance be- tween nearest neighboring points from the same class and separation of points belonging to different classes by large margin. Likewise in [23], the label infor- mation of the data is used to calculate a task-specific distance function based on the Kullback-Leibler divergence. The distance is based on the conditional distribution of label information in dependence to the input data which is es- timated on a validation set. In [1] the behaviour of the Minkowski distance to measure proximity especially in high-dimensional feature spaces was investi- gated. The methods highlighted have in common that they treat the process to find an adaptive metric separately from the actually learning of the classifica- tion model. In contrast, parameterized metrics and dissimilarity functions can be directly integrated into the learning process of models like GLVQ, SNG, or RBF. One parameterization is the use of relevance weights or matrices in the Euclidean distance [12,25,20]. Another possibility is the use of the generalized metric, in the case of the Euclidean norm the Minkowski norm as well as the use of divergences [26,15], for example the γ-divergence and its special case the Cauchy-Schwarz divergence [22]. Parameters are either systematically explored or learned directly with other model parameters through minimizing the models object/energy function. So far the utilization of a distance function tuned to a single parameter setting has not shown significant performance improvements. The approach explored in this paper is the combination of a number of models, tuned to different γ parameters in order to create a classifier system whose global performance is significantly better then the performance of each model tuned to a single parameter.

In the field of multiple classifier fusion, several approaches have been proposed to create classifier ensembles with superior classification performance as well as to combine sets of existing classifiers to overcome the limitations of individual classifiers [13,4,17,2,6,14,16].

As we study the impact of parameters such as model size and distance metric, a large number of classifiers is trained for evaluation purposes by systematically varying these factors. However, this approach creates an ensemble of classifiers which may provide diverse as well as correlated decisions on the training and testing data.

While correlation and diversity between classifiers can be simply measured, it remains an open question which level of diversity and correlation provides the best results in classifier fusion [7]. In common approaches such as Bagging and Boosting, diversity is fostered by random sampling or by iteratively generating complementary classifiers for falsely classified feature vectors. However, these approaches also require a high level of correlation of the individual classifiers because final decisions are obtained by majority voting.

The existing approaches for classifier fusion can be roughly divided into trained and non-trained combiners [8]. Also early and late fusion can be easily discriminated. The topology of fusion methods is another important aspect to categorize the different approaches.

While the application of non-standard distance metrics is motivated by previous work on classification of functional data, its impact on the generation of classifier ensembles for the same problem is unknown. As trained combiners have shown superior performance in a previous study with non-functional data [18], we focus on ensemble learning with decision tree based learners. The advantages of using tree based learners are sketched in the next section.

3   Methods

3.1   Training and Evaluation of RBF, GLVQ, and SNG Classifiers

Classification models were implemented as published in [12,11,3]. For the GLVQ and SNG no non-linearity in the energy function was used. The distance function between a data vector $\mathbf{v}$ and a prototype vector $\mathbf{w}$ (respectively the hidden neurons in the RBF) was either the squared Euclidean distance defined as

$$d\left(\mathbf{v}, \mathbf{w}\right) = \sum_i \left(v_i - w_i\right)^2,\tag{1}$$

or the γ-divergence defined as

$$d\left(\mathbf{v},\mathbf{w},\gamma\right) = \log\left(\frac{\left(\sum_i v_i^{\gamma+1}\right)^{\frac{1}{\gamma(\gamma+1)}}\left(\sum_i w_i^{\gamma+1}\right)^{\frac{1}{\gamma+1}}}{\left(\sum_i v_i w_i^{\gamma}\right)^{\frac{1}{\gamma}}}\right). \tag{2}$$

The γ-divergence with $\gamma = 2$ is widely known as the Cauchy-Schwarz distance. The model training in all three classifier systems (RBF, SNG, GLVQ) is essentially an energy minimization problem. In the standard learning scheme, stochastic gradient descent with step-sizes manually set for different parameters are used. In order to avoid a manually chosen step-size, we used the non-linear conjugate gradient approach with automatic step size from the optimization toolbox 'minFunc' available for Matlab. For this we provided the energy function as well as the first derivatives according to all model parameters. The parameter γ was set varying from 1 to 10 in steps of one. Additionally, the generalized Kullback-Leiber divergence [10] was used to investigate the behavior for convergence of γ to zero. Prototype vectors and network weights were initialized randomly. The RBF used a 1-of-N coding scheme at its output to represent discrete class information. In the RBF, SNG, and GLVQ the prototypes were pre-trained using a Neural Gas with the Euclidean distance or γ-divergence as similarity function with an identical setup compared to the later classification model. In the GLVQ and SNG model, separate pre-learning runs for prototypes from identical classes were performed. The dataset was divided into training and test data according to a 5-fold cross validation scheme with stratified random sampling. After training, the predicted labels for the test data with the respective model were collected as well as scalar model outputs. In case of the RBF, the scalar output was the output of the linear output layer. For the GLVQ and SNG we used the distances to the closest prototype of the same class as well as the smallest distance to a prototype of any other class as scalar output. We set 20, 30, or 40 as total number of prototypes/hidden neurons in all three models. In the GLVQ and SNG an identical number of prototypes per class was used. In addition to the Euclidean distance we also used weighted Euclidean distance as an alternate distance metric where the weights are automatically adapted in the training phase.

3.2   Fusion of RBF, LVQ, and SNG Results

The real-valued scalar outputs of the different classifiers make a feature vector which is used as the input for learning a combining rule. In this study, we focus on the application of decision tree based learners. Algorithms such as C4.5 or its variant J4.8 use local optimization of a threshold value and selection of a single input feature to maximize the separation into given target classes. As the input features are the output values of classifiers itself this is similar to the selection of operating points as known from receiver operating characteristic (ROC) and precision recall (PR) analysis. Hence, any decision of the resulting trees can be easily interpreted as a sequence of operating point selections for the different input classifiers. To overcome known limitation of decision tree learning, ensembles of trees are used instead. Hence, for the combination of the different classifiers these methods are used:

1. AdaBoost with decision trees [9],
2. Random Forests [5], and
3. CRAGORS (cascaded reduction and growing of result sets) [19].

The used implementations of AdaBoost, decision trees, and Random Forests are part of the Spider toolbox and WEKA. For AdaBoost pruned decision trees are used for better generalization performance. For Random Forest classifiers unpruned trees are used. Boosting was set to 10 iterations and all Random Forest classifiers consist of 10 trees as well. CRAGORS is included to address the tradeoff between ensemble size and ensemble accuracy. This combining algorithm is expected to provide less accurate results, but to select small subset of relevant input classifiers which already provide a significant improvement in classification accuracy. The dataset for testing combination performance has been generated from the outputs of 5-fold cross-validation of the individual classifiers. For every spectrum the outputs of all the different classifiers have been collected. 10-fold cross-validation was used to obtain average accuracy values for the 3 combining methods.

4   Datasets

The hyperspectral datasets have been selected from several industrial applications where hyperspectral imaging can be used for the detection of a desired target material or defective objects for a subsequent material sorting. We deliberately chose classification tasks that showed mediocre classification accuracy on single prototype based models. Five binary classification problems were chosen for this publication:

1. Detection of aluminium within waste material,
2. Classification of mature vs. immature coffee beans,
3. Detection of putrid hazelnuts among healthy hazelnuts,
4. Detection of fungi infested hazelnuts among healthy hazelnuts, and
5. Anomality detection on the surface of fluffed pulp.

We limited our study to two-class problems for two major reasons. First, the detection of a single important class is a typical scenario in industrial applications. Hence, the above datasets have been collected separately. Especially the hazelnut datasets belong to different studies. Second, the current implementation of CRAGORS which we wanted to test on hyperspectral datasets is so far limited to two-class problems.

For the hyperspectral image acquisition, material samples of one class were positioned with a standard optical PTFE (polytetrafluoroethylene) calibration pad on a translation table. Hyperspectral images were recorded using a HySpex

Table 1. Average accuracy of base classifiers (5-fold cross-validation)

Datasets	RBF			SNG			GLVQ		
	L_2	γ	complete	L_2	γ	complete	L_2	γ	complete
D1	0.8150	**0.8265**	0.8265	0.7168	**0.7325**	0.7402	0.6853	**0.7110**	0.7110
D2	**0.8560**	0.8430	0.8560	0.6937	**0.7027**	0.7295	**0.6820**	0.6755	0.6820
D3	0.6680	**0.6813**	0.6813	0.5710	**0.6005**	0.6005	0.5370	**0.5493**	0.5530
D4	**0.9618**	0.9496	0.9618	0.7724	**0.7894**	0.7894	0.7626	0.7626	0.7626
D5	0.7452	**0.7635**	0.7635	0.6597	**0.6963**	0.6963	0.6312	**0.6575**	0.6575

SWIR-320m-e line camera (Norsk Elektro Optikk A/S). Spectra are from the short-wave infra-red range (SWIR) of 970 nm to 2,500 nm at 6 nm resolution yielding a 256 dimensional spectral vector per pixel. The camera line has a spatial resolution of 320 px and can be recorded with a maximum frame rate of 100 fps. Radiometric calibration was performed using the vendors software package and the PTFE reflectance measure. Material was segmented from background via Neural Gas clustering. From each material class, 2,000 labeled spectral samples for each class were chosen randomly and combined to the datasets representing the two-class problems listed above. Spectral vectors were normalized to unit length.

5   Results and Discussion

As a baseline we measured the accuracy gain of using γ-divergence instead of the Euclidean distance for the datasets D1 to D5. Tab. 1 lists the accuracies of GLVQ, RBF, and SNG classifiers with respect to the used metric. The comparison shows a minor improvement in the accuracy when using the γ-divergence only for a few datasets. This is in accordance to previously reported results on using alternative distance measures [22,26,15,27]. For SNG classifiers the γ-divergence yields better results on all datasets. However, RBF classifiers outperform SNG and GLVQ on all datasets. The column *complete* lists the best results obtained from a slightly extended set of base classifiers including the Kullback-Leibler divergence and Cauchy-Schwarz divergence measures.

As we set 20, 30, or 40 as total number of prototypes/hidden neurons in all three models and also trained models for different values of γ, only the average accuracy of the best performing classifier is shown. Tab. 2 lists the parameter settings for these classifiers. We found, that the best results are obtained by different settings of the number of prototypes/hidden neurons as well as different similarity measures. Especially, γ differs significantly for the datasets for which application of γ-diversity is beneficial.

Tab. 3 shows that considerable performance gains can be achieved by the proposed classification system that combines a number of prototype based models trained on differently parametrized divergence measures. We combined the results of different subsets of the GLVQ, RBF, and SNG classifiers. By considering only variants of Euclidean based classifiers a significant increase in accuracy

Table 2. Parameter settings of the best classifiers, including Kullback Leibler divergence (KLD)

RBF	RBF		
Datasets	L_2	γ	complete
D1	40 neurons	20 neurons, $\gamma = 5$	same as γ
D2	30 neurons	40 neurons, $\gamma = 1$	same as L_2
D3	20 neurons	40 neurons, $\gamma = 5$	same as γ
D4	40 neurons	30 neurons, $\gamma = 2$	same as L_2
D5	40 neurons	40 neurons, $\gamma = 2$	same as γ
	SNG		
Datasets	L_2	γ	complete
D1	30 neurons	30 neurons, $\gamma = 1$	40 neurons, KLD
D2	20 neurons	40 neurons, $\gamma = 6, \gamma = 2$	same as γ
D3	40 neurons	40 neurons, $\gamma = 1$	same as γ
D4	30 neurons	30 neurons, $\gamma = 10$	same as γ
D5	40 neurons	40 neurons, $\gamma = 10$	same as γ
	GLVQ		
Datasets	L_2	γ	complete
D1	40 neurons	40 neurons, $\gamma = 6$	same as γ
D2	20 neurons	40 neurons, $\gamma = 1$	same as L_2
D3	20 neurons, weighted Euclidean	20 neurons, $\gamma = 2$	same as γ
D4	40 neurons	40 neurons, $\gamma = 8$	same as γ and L_2
D5	40 neurons	30 neurons, $\gamma = 2$	same as γ

is found. However, γ-divergence based classifier ensembles perform better on all datasets. Especially, for datasets D3 and D5 a large difference between Euclidean and γ based ensembles exists for all groups of combined classifiers (RBF, GLVQ, SNG). Additional improvements are possible if Euclidean and γ-diversity based classifiers are merged. Adding the Kullback-Leibler divergence based classifiers does not further improve the results significantly. As before, only the results of the best combination algorithm is shown in Tab. 3. For all tested datasets Random Forest and Boosted Decision Trees are competitive and there is no clear winner among these two methods. Additionally, the algorithm CRAGORS was used to find a subset of classifiers which provide a trade-off between the number of considered input classifiers and the gain in classification performance. The column *complete set* lists the results of a combination without discriminating beetween the pools of L_2-based and γ-based classifiers.

The comparison of the accuracies of different combining methods is shown in Tab. 4. The presentation is limited to the results of combining RBF network classifiers trained with γ-divergence based distance measures. The number in brackets reports the number of the used input features to indicate the trade-off between accuracy gain and the number of required RBF classifiers. The difference between AdaBoost and Random Forests is not significant. Also, the chosen limitation to 10 decision trees leaves room for additional improvements of the accuracy.

Table 3. Average accuracy of combined classifiers, 10-fold cross-validation

RBF	Ensemble pool			
Datasets	L_2	γ	$L_2 + \gamma$	complete set
D1	0.9737	**0.9965**	0.9982	0.9972
D2	0.9775	**0.995**	0.994	0.9960
D3	0.8010	**0.8985**	0.9155	0.9133
D4	0.9959	**0.9984**	0.9984	0.9992
D5	0.9205	**0.9790**	0.9825	0.9822
SNG	Ensemble pool			
Datasets	L_2	γ	$L_2 + \gamma$	complete set
D1	0.8465	**0.9698**	0.9660	0.9723
D2	0.8550	**0.9517**	0.9540	0.9550
D3	0.6190	**0.7463**	0.7412	0.7410
D4	0.9276	**0.9951**	0.9976	0.9976
D5	0.7798	**0.9377**	0.9383	0.9390
LVQ	Ensemble pool			
Datasets	L_2	γ	$L_2 + \gamma$	complete set
D1	0.7773	**0.9412**	0.9417	0.9463
D2	0.82	**0.8605**	0.8865	0.9012
D3	0.5493	**0.5877**	0.5867	0.5962
D4	0.9024	**0.9911**	0.9919	0.9878
D5	0.7505	**0.8830**	0.8953	0.8920

It should be noted that the used comparison method also contributes to classifier diversity. 5-fold cross-validation was used to train base classifiers and to collect realistic classifier outputs for unseen samples. Hence, all the classifiers have been trained on different subsets representing 80 percent of all samples. However, the Euclidean and γ-diversity based classifiers have been obtained under the same conditions. Therefore, the observed difference in performance is clearly related to the used metric. To study the different contributions to classifier diversity and performance in more detail, hold-out testing with independently sampled data should be used.

In contrast to other combining methods such as boosting the proposed approach is built on top of a set of independently tuned classifiers. The main advantage of a separation between tuning of the base classifiers and their combination into an ensemble is that it can be easily adapted to existing classification frameworks. Additionally, using a supervised classification algorithm for the combination instead of a simple combining rule such as majority voting is beneficial.

The results indicate that optimizing γ in the training of a single classifier may not yield the significant gain in accuracy as reported for the ensembles. Because we variied γ systematically over a small but meaningful range we expect such an approach to achieve a result competetive to our baseline condition.

Table 4. Average accuracy of different combined classifiers (RBF networks with γ-distances only, baseline includes all distance measures), 10-fold cross-validation, numbers in brackets denote required RBF networks

RBF	Fusion method			
Datasets	Baseline	CRAGORS	AdaBoost	Random Forest
D1	0.8265 (1)	0.9355 (12)	**0.9965 (48)**	0.9953 (55)
D2	0.856 (1)	0.9435 (13)	**0.9950 (53)**	0.9925 (59)
D3	0.6813 (1)	0.7445 (20)	**0.8985 (59)**	0.8855 (59)
D4	0.9618 (1)	0.9846 (7)	**0.9984 (25)**	**0.9984 (59)**
D5	0.7635 (1)	0.8840 (18)	**0.9790 (55)**	0.9683 (57)

6   Summary

The results show that choosing another metric or modifying model size may slightly improve classification accuracy. However, the tuning of parameters is required. The question remains whether other classification algorithms, other parameter setting than the tested ones, different model sizes, or a different topology of neural networks may yield better results or not. The major contribution of this paper with respect to γ-metrics and multiple classifier fusion is that it was possible to demonstrate for all tested datasets, that systematically varying the γ value of the distance metric is an extraordinarily effective way to create a diverse ensemble of classifiers. Especially, the trade-off between diversity and correlation seems to be near optimal for classification of hyperspectral data. Hence, the major contribution from an engineering perspective is to provide an easy-to-use framework for the analysis of hyperspectral data. By the fusion of classifier results, great improvements in classification accuracy have been made for several real-world applications. Moreover, for the first time the improvements reached a level which meets application specific lower boundaries on the precision and the detection rate. However, a lot of future work has to be done to get a deep theoretic understanding of the role of the γ-metric with respect to ensemble diversity. We also limited our study to the analysis of hyperspectral data as a representative of functional data. For practical applications, the selection of classifier subsets or a parallel computation of classifier results may be required to meet application specific time constraints.

References

1. Aggarwal, C.C., Hinneburg, A., Keim, D.A.: On the surprising behavior of distance metrics in high dimensional space. In: Van den Bussche, J., Vianu, V. (eds.) ICDT 2001. LNCS, vol. 1973, pp. 420–434. Springer, Heidelberg (2000)
2. Al-Ani, A., Deriche, M.: A new technique for combining multiple classifiers using the dempster-shafer theory of evidence. Journal of Artificial Intelligence Research 17, 333–361 (2002)
3. Backhaus, A., Bollenbeck, F., Seiffert, U.: Robust classification of the nutrition state in crop plants by hyperspectral imaging and artificial neural networks. In:

Proc. 3rd Workshop on Hyperspectral Image and Signal Processing: Evolution in Remote Sensing, Lisboa, Portugal (2011)

4. Bishop, C.M., Svensén, M.: Hierarchical Mixtures of Experts. In: 19th Conference on Uncertainty in Artificial Intelligence (2003)
5. Breiman, L.: Random forests. Machine Learning 45, 5–32 (2001)
6. Chen, X., Li, Y., Harrison, R., Zhang, Y.-Q.: Type-2 fuzzy logic-based classifier fusion for support vector machines. Applied Soft Computing 8(3), 1222–1231 (2008)
7. Didaci, L., Fumera, G., Roli, F.: Diversity in Classifier Ensembles: Fertile Concept or Dead End? In: Zhou, Z.-H., Roli, F., Kittler, J. (eds.) MCS 2013. LNCS, vol. 7872, pp. 37–48. Springer, Heidelberg (2013)
8. Duin, R.P.W.: The combining classifier: to train or not to train. In: ICPR (2002)
9. Freund, Y., Schapire, R.E.: A decision-theoretic generalization of on-line learning and an application to boosting. In: European Conference on Computational Learning Theory, pp. 23–37 (1995)
10. Geweniger, T., Kästner, M., Villmann, T.: Optimization of parametrized divergences in fuzzy c-means. In: ESANN (2011)
11. Hammer, B., Strickert, M., Villmann, T.: Supervised Neural Gas with general similarity measure. Neural Processing Letters 21, 21–44 (2005)
12. Hammer, B., Villmann, T.: Generalized relevance learning vector quantization. Neural Networks 15, 1059–1068 (2002)
13. Jordan, M.I., Jacobs, R.A.: Hierarchical Mixtures of experts and the EM-algorithm. Neural Computation 6(2), 181–214 (1994)
14. Kang, S., Park, S.: A fusion neural network classifier for image classification. Pattern Recogn. Lett. 30(9), 789–793 (2009)
15. Kästner, M., Backhaus, A., Geweniger, T., Haase, S., Seiffert, U., Villmann, T.: Relevance learning in unsupervised vector quantization based on divergences. In: Laaksonen, J., Honkela, T. (eds.) WSOM 2011. LNCS, vol. 6731, pp. 90–100. Springer, Heidelberg (2011)
16. Khreich, W., Granger, E., Miri, A., Sabourin, R.: Iterative Boolean combination of classifiers in the ROC space: An application to anomaly detection with HMMs. Pattern Recognition 43(8), 2732–2752 (2010)
17. Kittler, J., Hatef, M., Duin, R.P.W., Matas, J.: On combining classifiers. IEEE Transactions on Pattern Analysis and Machine Intelligence 20(3), 226–239 (1998)
18. Knauer, U., Seiffert, U.: A Comparison of Late Fusion Methods for Object Detection. In: IEEE International Conference on Image Processing, pp. 1–8 (2013)
19. Knauer, U., Seiffert, U.: Cascaded Reduction and Growing of Result Sets for Combining Object Detectors. In: Zhou, Z.-H., Roli, F., Kittler, J. (eds.) MCS 2013. LNCS, vol. 7872, pp. 121–133. Springer, Heidelberg (2013)
20. Mendenhall, M.J., Merényi, E.: Relevance-based feature extraction for hyperspectral images. IEEE Transactions on Neural Networks 19(4), 658–672 (2008)
21. Moody, J., Darken, C.J.: Fast learning in networks of locally tuned processing units. Neural Computation 1, 281–294 (1989)
22. Mwebaze, E., Schneider, P., Schleif, F.-M., Haase, S., Villmann, T., Biehl, M.: Divergence based Learning Vector Quantization. In: Verleysen, M. (ed.) 18th European Symposium on Artificial Neural Networks (ESANN 2010), pp. 247–252. d-side publishing (2010)
23. Peltonen, J., Klami, A., Kaski, S.: Learning more accurate metrics for self-organizing maps. In: Dorronsoro, J.R. (ed.) ICANN 2002. LNCS, vol. 2415, pp. 999–1004. Springer, Heidelberg (2002)

24. Płoński, P., Zaremba, K.: Improving performance of self-organising maps with distance metric learning method. In: Rutkowski, L., Korytkowski, M., Scherer, R., Tadeusiewicz, R., Zadeh, L.A., Zurada, J.M. (eds.) ICAISC 2012, Part I. LNCS, vol. 7267, pp. 169–177. Springer, Heidelberg (2012)
25. Schneider, P., Schleif, F.-M., Villmann, T., Biehl, M.: Generalized matrix learning vector quantizer for the analysis of spectral data. In: ESANN, pp. 451–456 (2008)
26. Villmann, T., Haase, S.: Divergence based vector quantization of spectral data. In: 2010 2nd Workshop on Hyperspectral Image and Signal Processing: Evolution in Remote Sensing (WHISPERS), pp. 1–4 (2010)
27. Villmann, T., Haase, S.: Divergence-based vector quantization. Neural Comput. 23(5), 1343–1392 (2011)

Part IV

Advanced Applications of SOM and LVQ

The Sky Is Not the Limit*

Erzsébet Merényi

Rice University, Department of Statistics and Department of Electrical & Computer Engineering, 6100 Main Street MS-138, Houston, TX 77005, USA

We live in the era of Big Data, or at least our awareness of Big Data's presence and impact has sharpened in the past ten years. Compared to data characteristics decades ago, Big Data not only means a deluge of unfiltered bytes, but even more importantly it represents a dramatic increase in data dimensionality (the number of variables) and complexity (the relationships among the often interdependent variables, intricacy of cluster structure). Along with the opportunities for nuanced understanding of processes and for decision making, these data created new demands for information extraction methods in terms of the detail that is expected to be identified in analysis tasks such as clustering, classification, regression, and parameter inference. Many traditionally favored techniques do not meet these challenges if one's aim is to fully exploit the rich information captured by sophisticated sensors and other automated data collection techniques, to ensure discovery of surprising small anomalies, discriminate important, subtle differences, and more. A flurry of technique developments has been spawned, many augmenting existing algorithms with increasingly complex features.

Self-Organizing Maps [1] have shown their staying power in the face of these changes and stood out with their simplicity and elegance in capturing detailed knowledge of manifold structures. In our research we have not yet encountered a limit in terms of data complexity. SOMs learn astonishingly well. They are extremely good "listeners" to what the data has to say. An outstanding challenge rather seems to be in equally sharp interpretation of what an SOM has learned.

I will present methods and tools we have developed for deciphering SOMs and for using their knowledge in various ways [2–7]. They are aimed at "precision mining" of large and high-dimensional, complex data, separating important from unimportant details of data characteristics in the presence of noise and some quantifiable degree of topology violation. Components of these tools build on seminal works by several colleagues in the SOM community (e.g., [8–13]), further developing or engineering the original ideas.

I will highlight applications and effectiveness through three types of Big Data: remote sensing hyperspectral imagery for characterizing planetary surface

* This paper uses ALMA data ADS/JAO.ALMA#2011.0.00465.S. ALMA is a partnership of ESO (representing its member states), NSF (USA) and NINS (Japan), together with NRC (Canada) and NSC and ASIAA (Taiwan), in cooperation with the Republic of Chile. The Joint ALMA Observatory is operated by ESO, AUI/NRAO and NAOJ. The National Radio Astronomy Observatory is a facility of the National Science Foundation operated under cooperative agreement by Associated Universities, Inc.

182   E. Merényi

materials, functional Magnetic Resonance Images for brain mapping, and astronomical imagery obtained with the world's most advanced radiointerferometric array, ALMA (the Atacama Large Millimeter / Submillimeter Array, in Chile) for answering astrophysical questions ranging from star and planet formation to the formation of the universe.

In this abstract I briefly describe the challenges associated with these representative Big Data and I give a preview of some results we obtained with SOMs and related knowledge extraction approaches.

Hyperspectral images (spectral signatures acquired in hundreds of narrow, contiguous band passes on a regular spatial grid over a target area) have long been utilized for remote geochemical analyses of terrestrial and planetary surfaces. Typical hyperspectral imagery spans the visible to near- and thermal-infrared wavelengths with 5-20 nm band width, sufficient to resolve the discriminating spectral features of (near-)surface compounds. For example, hyperspectral imagery affords identification of individual plant species, soil constituents, the paints of specific cars, and a large variety of roof and building materials, creating a need to extract as many as a hundred different clusters from a single image. These clusters can be extremely variable in size, shape, density, proximities and other properties. Another demand arising from such sophisticated data is to differentiate among clusters that have subtle differences, as the ability to do so can enable important discoveries or increased customization in decision making.

For example, landslide risk models can be greatly improved by including (in addition to the traditional factors of mountain slope and rain fall) the types and

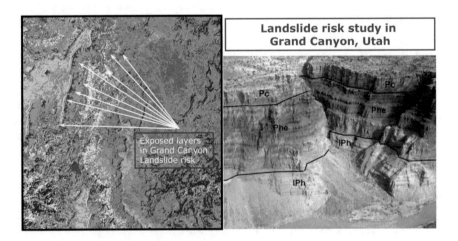

Fig. 1. Mapping clay distribution in soils for landslide risk assessment in Cataract Canyon, Grand Canyon, Utah, U.S.A. **Left:** Classification map produced from a remote sensing hyperspectral image. 15 of the 28 classes (each indicated by a different color) are exposed soil layers, several of which are indicated by the white arrows. **Right:** Photograph of some of the soil layers, in part of the imaged site. Figure adapted from [14].

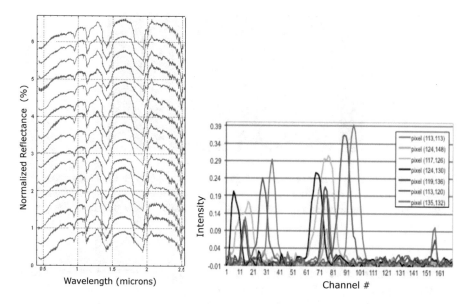

Fig. 2. Left: Visible-Near-Infrared hyperspectral signatures of clay-bearing soil classes, mapped in Fig. 1. Spectra are vertically offset for viewing convenience. The variations in the spectral window most discriminating for clay species (0.5 – 0.7 and 2.0 – 2.3 microns) are very subtle. Blue and red curves are the means of training and test samples for the classification experiment in [14], with standard deviations shown by the vertical bars. **Right:** Sample emission spectra, from combined C18O, 13CO, CS lines of ALMA receiver band 7, showing differences in composition, Doppler shift, depth and temperature. 170 channels were stacked from the C18O, 13CO, CS lines. Data credit: JVO, project 2011.0.00318.5.

amounts of clay minerals contained in the exposed soil layers of mountains. For assessment of large areas remote sensing is used, which detects the clay minerals highly diluted in the soil matrix, resulting in weakened signatures. To distinguish and map the 15 or so layers of different soils around a landslide area in the Grand Canyon (Fig. 1) hyperspectral signatures with such slight variations as in Fig. 2 must be discriminated precisely by a classifier and produce maps showing the spatial distribution of the various soils, as in Fig. 1. An SOM was instrumental in accomplishing the delicate task [14].

In stellar astronomy, where Ångström resolution is typical, the data complexity can grow even higher. 21st century observatories such as ALMA achieve, for the first time, data sets that begin to approach, and in some dimensions exceed, the richness of data from terrestrial and planetary remote sensing. High spatial and spectral resolution image cubes with thousands of frequency channels are extending into new and wider wavelength domains, and at the same time capturing several different physical quantities that characterize 3-dimensional plasma structures. The "spectra" are no longer vectors of homogeneous variables. Effects of spatial depth, Doppler shift, temperature and densities are influencing the signatures in addition to chemical composition. Fig. 2, right, gives an

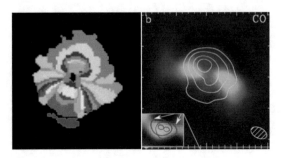

Fig. 3. Structure found in protostar HD 142527 from ALMA data. **Left:** SOM clustering from hyperspectral ALMA data cube by the author. **Right:** From single doppler line, by [16]. Figure reproduced with permission. Data credit: JVO, project 2011.0.00318.5.

illustrative sample of ALMA data, combined from three different emission lines. The left image in Fig. 3 shows structural details of a protostar produced (to my knowledge) by the first SOM clustering of a complex ALMA image cube [15], in comparison to details extracted from a single doppler line by [16]. This protostar has stirred great interest recently because of a planet formation process that has been detected deep in its interior.

Functional Magnetic Resonance Imagery (fMRI) poses many similar challenges as hyperspectral data, with typically higher-dimensional data vectors and potentially more clusters. The time courses — vectors of measurements of blood-oxygen-level dependece (BOLD) signals at hundreds of time points recorded during the observation of a subject at each of several hundred thousands of voxels in the brain volume — can be clustered to find brain areas with similar activation patterns. Correlation analysis of the characteristic time courses of the identified clusters can further reveal temporal relationships of various sub-networks in the brain. With SOM tools we can glean detailed maps of the entire brain with more complete coverage than seen in many published results.

Fig. 4, left, shows a pair of representative brain slices with our recent SOM clustering [17], for which all available voxels were used from the entire brain volume. The clusters coincide well with several known functional areas throughout all slices (not shown here). In comparison, clustering with statistical methods in [18] (center) was applied only to two selected slices, and the clusters identified are highly segmented, with very sparse coverage. The brain maps on the right, from [19] were obtained by SOM clustering, and have good coverage of selected functional areas. An important property in the face of Big Data, SOMs are not nearly as limited by large data volumes as many other methods (for example, graph-based clustering, where the number of vertices grows quadratically with the number of data vectors). The ability of learning well from large volumes of data allows precise identification of a large variety of functional regions, which in turn enables more nuanced investigation of such fundamental questions as — in our study — the generation of the conscious movement in healthy and impaired brains.

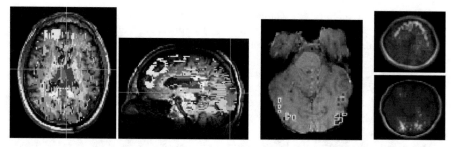

Fig. 4. Clusterings of fMRI images, based on the BOLD time courses. **Left pair of images:** SOM clusters obtained with our tools, in two selected brain slices, showing good coincidence with and coverage of known functional regions such as the thalamus (dark blue), insula (mauve symmetrically placed spots on either side of the thalamus in the left image), visual cortex (light blue, dark green and orange), and superior frontal gyrus (light yellow, at top of the left image, at front left in the right image) [17]. Data credit: The Methodist Research Institute, Houston, Texas. **Center:** Clusters generated by statistical hypothesis testing, from [18]. **Right:** Clusters found in the motor cortex (top) and visual cortex (bottom) by [19] using SOMs. Figures from [18] and [19] reproduced with permission.

SOMs arguably provide a key to accurate learning of diverse types of highly structured data. However, with this power come new puzzles. While sharpening knowledge extraction methods to match the richness of the data, we must also recognize that interpretation of the increasing detail emerging from data like these may be the next challenge in the Big Data picture. Sophisticated tools that allow penetration of previously unidentified relationships in the data may return "distilled" results that look complicated, hard-to-digest, and not straighforward to interpret or verify.

ALMA, for example, represents such advanced observational capability that fully exploiting the information content will require — as much as anything else — new capabilities to synthetize, visualize, and interpret the extracted knowledge, the already summarized information! The cluster map on the left in Fig. 3, for example, can only show part of the protostar structure detected by the SOM. We yet have to devise a visualization to layer on and meaningfully convey the full information. In closing, I will illustrate some cases of this interesting problem.

Acknowledgments. Special thanks to ALMA project scientist Al Wootten, for sharing and helping with ALMA data. Collaboration with Drs. Robert Grossman and Christof Karmonik at the Methodist Research Institute, Houston, Texas, on the analysis of their fMRI data is gratefully acknowledged.

References

1. Kohonen, T.: Self-Organizing Maps, 2nd edn. Springer, Heidelberg (1997)
2. Zhang, L., Merényi, E., Grundy, W.M., Young, E.Y.: Inference of surface parameters from near-infrared spectra of crystaline h2o ice with neural learning. Publications of the Astronomical Society of the Pacific 122(893), 839–852 (2010), doi:10.1086/655115

3. Merényi, E., Tasdemir, K., Zhang, L.: Learning highly structured manifolds: harnessing the power of SOMs. In: Biehl, M., Hammer, B., Verleysen, M., Villmann, T. (eds.) Similarity-Based Clustering. LNCS, vol. 5400, pp. 138–168. Springer, Heidelberg (2009)
4. Tasdemir, K., Merényi, E.: Exploiting data topology in visualization and clustering of Self-Organizing Maps. IEEE Trans. on Neural Networks 20(4), 549–562 (2009)
5. Mendenhall, M., Merényi, E.: Relevance-based feature extraction for hyperspectral images. IEEE Trans. on Neural Networks 19(4), 658–672 (2008)
6. Merényi, E., Jain, A., Villmann, T.: Explicit magnification control of self-organizing maps for "forbidden" data. IEEE Trans. on Neural Networks 18(3), 786–797 (2007)
7. Zhang, L., Merényi, E.: Weighted Differential Topographic Function: A Refinement of the Topographic Function. In: Proc. 14th European Symposium on Artificial Neural Networks (ESANN 2006), Brussels, Belgium, pp. 13–18. D facto publications (2006)
8. Hammer, B., Villmann, T.: Generalized relevance learning vector quantization. Neural Networks 15, 1059–1068 (2002)
9. Villmann, T., Der, R., Herrmann, M., Martinetz, T.: Topology Preservation in Self–Organizing Feature Maps: Exact Definition and Measurement. IEEE Transactions on Neural Networks 8(2), 256–266 (1997)
10. Bauer, H.U., Der, R., Herrmann, M.: Controlling the magnification factor of self–organizing feature maps. Neural Computation 8(4), 757–771 (1996)
11. Martinetz, T., Schulten, K.: Topology representing networks. Neural Networks 7(3), 507–522 (1994)
12. Ultsch, A., Simeon, H.P.: Kohonen's self organizing feature map for exploratory data analysis. In: Proc. INNC-1990-PARIS, Paris, vol. I, pp. 305–308 (1990)
13. DeSieno, D.: Adding a conscience to competitive learning. In: IEEE International Conference on Neural Networks, pp. 117–124. IEEE (1988)
14. Rudd, L., Merényi, E.: Assessing debris-flow potential by using AVIRIS imagery to map surface materials and stratigraphy in cataract canyon, Utah. In: Green, R. (ed.) Proc. 14th AVIRIS Earth Science and Applications Workshop, Pasadena, CA, May 24-27 (2005)
15. Merényi, E.: Hyperspectral image analysis in planetary science and astronomy. abstract. Presentation in Special Session "Building the Astronomical Information Sciences: From NASA's AISR Program to the New AAS Working Group on Astroinformatics and Astrostatistics" (January 7, 2014)
16. Casassus, S., van der Pas, G., Perez, M.S., et al.: Flowes of gas through a proto-planetary gap. Nature 493, 191 (2013)
17. O'Driscoll, P.: Using Self-Organizing Maps to discover functinal relationships of brain areas from fMRI images. Master's thesis, Rice University (June 2014)
18. Heller, R., Stanley, D., Yekutieli, D., Rubin, N., Benjamini, Y.: Cluster-based analysis of FMRI data. NeuroImage 33(2), 599–608 (2006) PMID: 16952467
19. Liao, W., Chen, H., Yang, Q., Lei, X.: Analysis of fMRI data using improved self-organizing mapping and spatio-temporal metric hierarchical clustering. IEEE Transactions on Medical Imaging 27(10), 1472–1483 (2008)

Development of Target Reaching Gesture Map in the Cortex and Its Relation to the Motor Map: A Simulation Study

Jaewook Yoo, Jinho Choi, and Yoonsuck Choe

Department of Computer Science and Engineering, The Texas A&M University,
College Station, Texas, USA
{jwookyoo,jhchoi84}@neo.tamu.edu, choe@tamu.edu

Abstract. The motor maps in the cortex are topologically organized, just like the sensory cortical maps, where nearby locations in the map represent behaviors of similar kinds. However, there is not much research on how such motor maps are formed. In this paper, as a first step in this direction, we developed a target reaching gesture map using a self-organizing map model of cortical development (the GCAL model, a simplified yet enhanced version of the LISSOM model). The inputs were target reaching behavior of a two-joint arm on a 2D plane (2 DOF), encoded as a time-lapse image where time was encoded as the pixel intensity. For training, 20,000 random arm movements were generated where each arm movement started at a random initial location and moved toward one of 24 predefined target locations. The resulting gesture map showed global topological order where nearby neurons represented gestures toward nearby target locations, comparable to the motor map reported in the experimental literature. Although our simulations were based on a sensory cortical map development framework, the results suggest that it could be easily adapted to transition into motor map development. Our work is an important first step toward a fully motor-based motor map development (e.g. using proprioceptive input), and we expect the findings reported in this paper to help us better understand the general nature of cortical map development, not just for the sensory but also for the motor modality.

Keywords: Motor map development, Self-organizing maps, Cortical development, Target reaching gesture.

1 Introduction

In the recent studies, Graziano et al. found that the motor cortex in the macaque brain forms a topographical map of complex behaviors [1], where the final posture of the movements form an organized map. As we can see in the Fig.1(a), the monkey's the hand target location of reaching behavior evoked from extended electrical microstimulation on a certain location of the mortor cortex was always the same, regardless of the initial hand position. Furthermore, the target location

T. Villmann et al. (eds.), *Advances in Self-Organizing Maps and Learning
Vector Quantization*, Advances in Intelligent Systems and Computing 295,
DOI: 10.1007/978-3-319-07695-9_18, © Springer International Publishing Switzerland 2014

188   J. Yoo, J. Choi, and Y. Choe

forms an organized map on the motor cortex, where ventral and anterior areas corresponded to the target locations in the upper space of the body, whereas dorsal and posterior areas in the motor cortex corresponded to target locations in the lower space of the body as in Fig.1(b). Based on theses findings, our question is how such motor maps are formed in the cortex through the development period.

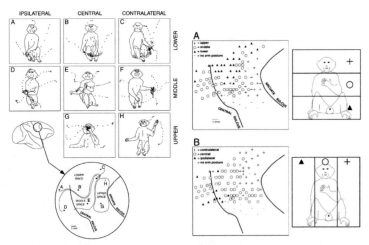

(a) Eight example posture illustrating the topographic map in the precentral cortex of monkey.

(b) Topography of hand and arm found postures in the precentral gyrus based on 201 stimulation sites in monkey.

Fig. 1. The topographic map found in precentral cortex of Monkey. (a) The enlarged view at the bottom shows the sites of the electrical microstimulation. The movements shown in the rectangles A - H were evoked by stimulating the sites A - H in the enlarged circle at the bottom. The stimulation of the right side of the brain caused mainly the left side of the body (left arm to move). (b) A shows the distribution of hand positions along the vertical axis, which are upper, middle, and lower space. After each stimulation, the evoked final target positions were used to categorize the site. B shows the distribution of hand positions along the horizontal axis, which are contralateral (right when using left hand), central, and ipsilateral (left when using left hand) space. Adapted from [1].

In existing works, simulation studies were conducted to mimic the development of visual and tactile maps in the cortex. The visual cortical neuron's receptive fields (RFs) and their map were computationally developed using a self-organizing map model of the cortex (the LISSOM model) by Miikkulainen et al. [2]. Park et al. showed that both visual RFs and tactile RFs can be derived by training the same self-organizing map model of the cortex with different types of inputs (natural-scene images and texture images, respectively) [3].

In this paper, we investigate the possibility that a motor map in the cortex can be developed based on the same cortical learning process as the visual and tactile maps in the cortex. A self-organizing map model of the cortex (the GCAL model[4][5]: a simplified yet enhanced version of the LISSOM model[2]) was trained with two-joint arm movements (2 DOF) on a 2D plane, which were subsequently encoded as a time-lapse image (cf. Motion History Images [6]) where time was encoded as the pixel intensity. We investigated if the experiment can give rise to a motor map organization similar to Fig. 1.

This paper is organized as follows. The related work is reviewed in Section 2. Next, the GCAL model will be explained in Section 3. Section 4 will explains the platforms and the procedure for the experiments. The results are presented in Section 5. The discussion and conclusion are in Section 6 and 7, respectively.

2   Related Work

There is not much work on how motor map is formed in the cortex. Recently, several related studies were conducted, where a simulation study for the motor map clustering of monkey using a standard self-organized map (SOM) learning with encoded movements, and a multi-modal reinforcement learning algorithm to form a map according to behavioral similarity.

Aflalo and Graziano [7] showed a computational topographic map organization with three constraints. The three constraints are the body parts that were being moved for movements, the position reached in Cartesian space, and the ethological (behavioral) category to which the movement belonged. Encoded movements (body parts, hand coordinates, and behavioral category) were used as inputs for training. The initial somatotopic body map from the literature was used to initialize the model. A standard Self-Organized Map (SOM) learning [8] was used for the motor map clustering. However, their map configuration through the SOM learning is purely computational since their experiment with the learning algorithm did not consider the neural connectivity or plasticity in the cortex. Also, the initial somatotopic body map which already represent a rough motor map of the adult brain significantly affected the final configuration of the map.

Ring et al. introduced a new approach to address the problem of continual learning [9][10], which was inspired by the recent research on the motor cortex [1]. Their system modules, called *mot*, are self-organized in a two-dimensional map according to behavioral similarity. However, their method was based on a multi-modal reinforcement learning algorithm, and did not consider the neural underpinnings. Their aim in the study was to improve learning performance through their new approach, not to understand how the motor map in the cortex is developed in the cortex.

While several related studies have been conducted, there is a lack of studies for fine-grained, biologically plausible motor map development in the cortex.

3   The GCAL Model

We trained the GCAL (Gain Control, Adaptation, Laterally connected) model of cortical development to investigate motor map development in the cortex with 2 DOF arm movements in 2D plane. The GCAL model is a simplified, but more robust and more realistic version of the LISSOM model, which has been developed recently by Bednar et al. [4][5]. The GCAL model was designed to remove some of the artificial limitations and biologically unrealistic features of the LISSOM model.

GCAL is a self-organizing map model of the visual cortex [4][5]. Even though GCAL was originally developed to model the visual cortex, it is actually a more general model of how the cortex organizes to represent correlations in the sensory input. Therefore, sensory modalities other than vision should work with GCAL. For example, earlier work with LISSOM, a precursor of GCAL, was used to model somatosensory cortex development [3] [11].

In our GCAL experiments, we decreased the retina size to 2.0 to fit the input image size (80×80 pixels) and enlarged the projection area (radius: 1.5) to project all parts of the arm movements. The sizes of LGN ON and LGN OFF maps in the thalamus (lateral geniculate nucleus) and their projection size were the same as that of the retina. The radius of the projection area for LGN ON (and LGN OFF) was calculated as $r = \frac{v+l}{2}$, where r, v, and l indicate the radius of the projection area for LGN ON (or LGN OFF), the V1 area, and the LGN ON (or LGN OFF) sheet size, respectively. This way, the arm movements can be projected to the V1 level without cropping. Also, the other parameters were adjusted according to the sheets and the projection sizes. *Note that in the following we will use the GCAL terminology of retina, LGN, and V1 to refer to the sensory surface, thalamus, and cortex, respectively.*

The following description of the GCAL model closely follows [4][5]. The basic GCAL model is composed of four two-dimensional sheets (three levels) of neural units, including the retinal photoreceptor (input) and the ON and OFF channel of RGC/LGN (retinal ganglion cells and the lateral geniculate nucleus), the pathway from the photoreceptors to V1 area. Fig. 2 shows the architecture of GCAL we used.

The GCAL training consists of four steps overall as below.

1. At each iteration (input), the retina (sheet) is activated by the time-lapse image of the 2-DOF arm movement.
2. LGN ON and LGN OFF sheets are activated according to the connection weights between the retina and the LGN ON and LGN OFF sheets. Also, the lateral connections from other neurons in the LGN ON and LGN OFF sheets affect the activations. The activation level η for a unit at position j in an RGC/LGN sheet L at time $t + \delta t$ is defined as:

$$\eta_{j,L}(t + \delta t) = f\left(\gamma L \sum_{i \in F_j} \psi_{i,P}(t)\,\omega_{i,j}\right) \tag{1}$$

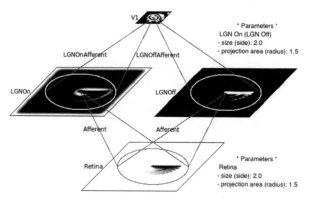

Fig. 2. The GCAL architecture. In the model, the retina size was increased to 2.0 (side) to fit the input image size (80×80 pixels) and the projection area enlarged to 1.5 (radius) to project all parts of the arm movements. *Note that in the text we will use the GCAL terminology of retina, LGN, and V1 to refer to the sensory surface, thalamus, and cortex in our gesture map model, respectively.*

where the activation function f is a half-wave rectifying function. The terms γL, $\psi_{i,P}$, and ω_{ij} are defined as follows:

- γL is an arbitrary multiplier for the overall strength of connections from the retina sheet to the LGN sheet.
- $\psi_{i,P}$ is the activation of unit i in the two-dimensional array of neurons on the retina sheet from which LGN unit j receives input (its connection field F_j).
- ω_{ij} is the connection weight from photoreceptor weight from the retina i to LGN unit j.

3. V1 sheet is activated by three different types of connections: 1) the afferent connection from the LGN ON and LGN OFF sheets ($p = A$), 2) the recurrent lateral excitatory connection ($p = E$), and 3) the recurrent lateral inhibitory connection from other neurons in V1 sheet ($p = I$). The V1 activation is settled through the lateral interactions. The contribution, X_{jp}, to the activation of unit j from each lateral projection type ($p = E, I$) is then updated for the settling steps as:

$$X_{jp}(t + \delta t) = \sum_{i \in F_{jp}} \eta_{i,V}(t)\omega_{ij,p} \qquad (2)$$

where $\eta_{i,V}$ indicates the activation of unit i taken from the set of neurons in V1 that connect to unit j. F_j is its connection field. The weight $\omega_{ij,p}$ is for the connections from unit i in V1 to unit j in V1 for the projection p. The afferent activity ($p = A$) remains constant during this setting of the lateral activity.

4. V1 neuron's activation level is calculated over time by a running average (smoothed exponential average), and the threshold automatically adjusted through a homeostatic mechanism.

5. LGN to V1 and V1 lateral connections are adjusted using a normalized Hebbian learning rule.

$$\omega_{ij,p}(t) = \frac{\omega_{ij,p}(t-1) + \alpha\eta_j\eta_i}{\sum_k \left(\omega_{kj,p}(t-1) + \alpha\eta_j\eta_k\right)} \tag{3}$$

where for unit j, α is the Hebbian learning rate for the afferent connection field F_j.

4   Experiment

We trained the GCAL model in Fig.2 with target reaching behavior of the two-joint arm on a 2D plane. We generated 20,000 movements in which each started from a random location (posture) and moved towards one of the 24 predefined target locations (postures). These input movements simulate the monkey's arm movement in Fig 1(a). Our main question was if the model can learn a target reaching gesture map such that the map has the characteristics of the motor map in Fig. 1(a) and Fig. 1(b). Details about the experiment platform, generating movements, and experiment procedures are as follows.

4.1   Experiment Platform

We ran the experiments on a Desktop PC (CPU: Intel Core 2 Duo 3.16GHz, Memory: 16GB) and Laptop (CPU: Intel Core i7 2 GHz, Memory: 8GB). Both machines ran on Ubuntu 10.04 (32bit). The installed Topographica version was 0.9.7. The python version was 2.6.5, and the gcc/g++ version 4.4.3.

For training the GCAL model, we mainly used the Topographica neural map simulator package, developed by Bednar et al. [2]. Topographica is a simulator for topographic maps in any two-dimensional cortical or subcortical region, such as visual, auditory, somatosensory, proprioceptive, and motor maps plus the relevant parts of the external environment [2]. The simulator is mainly written in Python, which makes it easily extendable and customizable according to the users' needs. The simulator is freely available including the full source code at http://topographica.org.

4.2   Generating Movements

Two-joint arm movements were generated on a 2D plane and used as inputs for the GCAL model training. Each of the 20,000 different arm movements started at a random location (posture) and moved towards one of the 24 predefined target locations (postures). These generated movements represent the arm movements of the monkey described in Fig. 1(a) and Fig. 1(b).

The arm consisted of two joints J1 ($\theta1$) and J2 ($\theta2$) in which the length ratio of the arm L1 : L2 is 1.6 : 1 (Fig. 3(a)). For each movement, first randomly pick initial angles for $\theta1$ and $\theta2$ (between -180 $\sim$ 180 degrees) and the 24 target

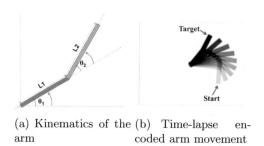

(a) Kinematics of the
arm

(b) Time-lapse en-
coded arm movement

Fig. 3. The kinematics and movement of the two-joint arm. (a) The arm consists of two joint J1 (θ1) : J2 (θ2), and the arm L1 : L2 (with the length ratio 1.6 : 1). The θ1 and θ2 are randomly picked initially and change toward the target. (b) The arm movement is encoded as time-lapse image where time is encoded as the pixel intensity. The darkest one is the target (most recent) posture.

locations as shown in Fig. 4. Then, J1 (θ1) and J2 (θ2) are changed toward the target locations from the initial angles either by 5 or 10 degrees each step, until they reach to the target posture. After each step of the angle update, the posture of the arm was plotted on the same sheet but with different opacity. The intensity was increased over time by 20% (Fig. 3(b)). Fig. 4 shows the examples of the generated arm movements. The 24 predefined target locations consisted of 16 distal locations (Fig. 4(a)) and 8 proximal locations (Fig. 4(b)). Note that in generating these motion patterns, we did not consider the natural movement statistics of the monkey's arm, largely due to the lack of such data.

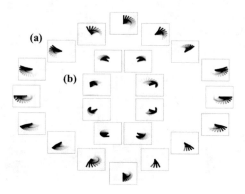

Fig. 4. Examples of movements with 24 target locations. Starting from a random posture, move toward to one of 24 target locations (postures). The movement over time is expressed using different pixel intensity (darker = more recent). (a) Example movements with 16 distal target locations. (b) Example movements with 8 proximal target locations. These movements simulate the arm movements in the experimental literature (Fig.1(a)). Note: For the same target location, many different time-lapse images were generated by varying the initial posture.

4.3   Experiment Procedure

After generating 20,000 random reaching movements encoded as a time-lapse image, we trained the GCAL model with these inputs. In each iteration, one of the 20,000 inputs was randomly picked for training. The density of the two LGN sheets and the single V1 sheet were 24 × 24 and 48 × 48, respectively. The parameters of the model for training were based on the default parameters in the Topographica package except some parameters such as retina sizes as described in Section 3. Once the training is done, we analyzed the resulting map, with a focus on the LGN to V1 afferent connection patterns.

5   Results

5.1   Local Topography Based on Target Location Similarity

The resulting gesture map is shown in Fig. 5. As we can see, the map is topologically ordered according to the target locations, where nearby locations (neurons) of the map represent nearby target locations (end-effector locations of final postures). For example, we can see that the similar target locations are clustered in the areas of top-left, top-right, bottom-left, bottom-right, center-left, center-right, and so on. In Fig. 6(a), each arrow of the grids shows the orientation and the distance of the target locations from the center. The vectors (arrows) with similar lengths and orientations represent similar target distance and angle.

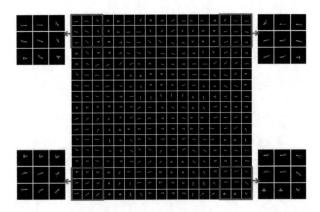

Fig. 5. The resulting gesture maps of LGN OFF to V1 projection. 17 × 17 RFs are plotted from 48 × 48 cortex density to see the details of them. The enlarged views show the zoomed in views of 3 × 3 RFs at each corner. Note: LGN OFF and LGN ON patterns are exact inverses of each other and thus contain the same information. The learned projections to V1 from these two sheets were similar as a result, so here we only showed the LGN OFF to V1 projections which is easier to visually inspect.

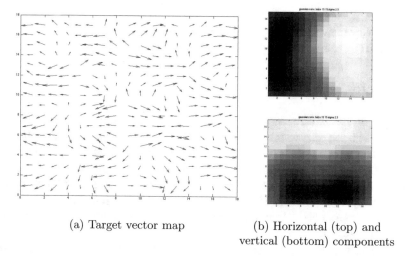

(a) Target vector map   (b) Horizontal (top) and
vertical (bottom) components

Fig. 6. (a) Target vectors estimated from the resulting gesture map receptive fields (Fig. 5). The direction and the length of each arrow show the target location's direction and the distance from the center. (b) The color maps of the horizontal and the vertical components of the vectors in (a): bright=high (right, up), dark=low (left, down). The color maps were convolved with a Gaussian filter of size 15×15 pixels and sigma 2.5 to show more clearly the global order.

5.2 Global Topographic Order

The resulting gesture map show global topographic order. The color maps of the horizontal and the vertical components of the vector field in Fig. 6(a) are shown in Fig. 6(b). As we can see, the vectors (the target locations) show horizontal order (Fig. 6(b), top) and vertical order (Fig. 6(b), bottom), which is comparable to the findings reported in the experimental literature (Fig. 1). Some neighboring vectors show opposite directions in Fig. 6(a), but this is coherent with the biological observations. As we can see in Fig. 1(b), some adjacent stimulation sites in the precentral gyrus in the monkey show targets in the opposite directions such as upper vs. lower, or left vs. right arm postures. Look for + and ▲ located right next to each other in Fig. 1(b).

6   Discussion

The main contribution of this paper is the use of a general cortical development model (GCAL) to show how fine-grained target reaching gesture maps can be learned, based on realistic arm reaching behavior. An immediate limitation is that the input itself was not a dynamic pattern of movement (i.e., it was just a static time-lapse image). However, as shown by Miikkulainen et al. [2], addition of multiple thalamus sheets with varying delay can address this kind of issue. Miikkulainen et al. [2] used such a configuration to learn visual (motion) direction sensitivity in V1. We intend to extend our model to include such a dynamic component in the input. Also, we are working on gesture map development using proprioceptive input from a simulated joint with muscle spindle afferent, departing from the visually oriented simulation framework used in this paper.

7   Conclusion

In this paper, we developed a target reaching gesture map using a biologically motivated self-organizing map model of the cortex (GCAL model, a simplified yet enhanced version of the LISSOM model) with two-joint arm movements as inputs. The resulting gesture map showed a golbal topographic order based on the target locations. The map is comparable to the motor map reported in the experimental study [1] (Fig. 1(a) and Fig. 1(b)). Although our simulations were based on a sensory cortical map development framework, the results suggest that it could be easily adapted to transition in to motor map development. Our work is an important first step toward a fully motor-based motor map development, and we expect the findings reported in this paper to help us better understand the general nature of cortical map development, not just for the sensory but also for the motor modality.

Acknowledgments. All simulations were done using Topographica (GCAL), available at http://topographica.org.

References

1. Graziano, M.S., Taylor, C.S., Moore, T.: Complex movements evoked by microstimulation of precentral cortex. Neuron 34, 841–851 (2002)
2. Miikkulainen, R., Bednar, J.A., Choe, Y., Sirosh, J.: Computational Maps in the Visual Cortex. Springer (2005)
3. Park, C., Bai, Y.H., Choe, Y.: Tactile or visual?: Stimulus characteristics determine receptive field type in a self-organizing map model of cortical development. In: Proceedings of the 2009 IEEE Symposium on Computational Intelligence for Multimedia Signal and Vision Processing, pp. 6–13 (2009)
4. Bednar, J.A.: Building a mechanistic model of the development and function of the primary visual cortex. Journal of Physiology-Paris 106(5-6), 194–211 (2012)
5. Law, J.S., Antolik, J., Bednar, J.A.: Mechanisms for stable and robust development of orientation maps and receptive fields. Technical report, School of Informatics, The University of Edinburgh, EDI-INF-RR-1404 (2011)
6. Bobick, A.F., Davis, J.W.: The recognition of human movement using temporal templates. IEEE Transactions on Pattern Analysis and Machine Intelligence 23(3), 257–267 (2001)
7. Aflalo, T.N., Graziano, M.S.A.: Possible origins of the complex topographic organization of motor cortex: Reduction of a multidimensional space onto a two-dimensional array. The Journal of Neuroscience 26, 6288–6297 (2006)
8. Kohonen, T.: Self-organizing maps, 3rd edn. Springer (2001)

9. Ring, M., Schaul, T., Schmidhuber, J.: The two-dimensional organization of behavior. In: 2011 IEEE International Conference on Development and Learning (ICDL), pp. 1–8 (2011)
10. Ring, M., Schaul, T.: The organization of behavior into temporal and spatial neighborhoods. In: 2012 IEEE International Conference on Development and Learning and Epigenetic Robotics (ICDL), pp. 1–6 (2012)
11. Wilson, S.P., Law, J.S., Mitchinson, B., Prescott, T.J., Bednar, J.A.: Modeling the emergence of whisker direction maps in rat barrel cortex. PLoS One 5(1) (2010)

A Concurrent SOM-Based Chan-Vese Model for Image Segmentation

Mohammed M. Abdelsamea[1], Giorgio Gnecco[1], and Mohamed Medhat Gaber[2]

[1] IMT Institute for Advanced Studies, Lucca, Italy
[2] Robert Gordon University, Aberdeen, UK
{mohammed.abdelsamea,giorgio.gnecco}@imtlucca.it,
m.gaber1@rgu.ac.uk

Abstract. Concurrent Self Organizing Maps (*CSOM*s) deal with the pattern classification problem in a parallel processing way, aiming to minimize a suitable objective function. Similarly, Active Contour Models (*ACM*s) (e.g., the Chan-Vese (*CV*) model) deal with the image segmentation problem as an optimization problem by minimizing a suitable energy functional. The effectiveness of *ACM*s is a real challenge in many computer vision applications. In this paper, we propose a novel regional *ACM*, which relies on a *CSOM* to approximate the foreground and background image intensity distributions in a supervised way, and to drive the active-contour evolution accordingly. We term our model Concurrent Self Organizing Map-based Chan-Vese (*CSOM-CV*) model. Its main idea is to concurrently integrate the global information extracted by a *CSOM* from a few supervised pixels into the level-set framework of the *CV* model to build an effective *ACM*. Experimental results show the effectiveness of *CSOM-CV* in segmenting synthetic and real images, when compared with the stand-alone *CV* and *CSOM* models.

Keywords: Image segmentation, Chan-Vese model, Concurrent Self Organizing Maps, global active contours, neural networks.

1 Introduction

Concurrent Self Organizing Maps (*CSOM*s) [1] combine several Self Organizing Maps (*SOM*s) to deal with the pattern classification problem (hence, the image segmentation problem as a particular case) in a parallel processing way, with the aim of minimizing a suitable objective function, usually the quantization error of the maps. In a *CSOM*, each *SOM* is constructed and trained individually on a subset of samples coming only from its associated class. The aim of this training is to increase the discriminative capability of the system. So, the training of the *CSOM* is supervised for what concerns the assigment of the training samples to the various *SOM*s, but each individual *SOM* is trained with the *SOM* specific self-organizing learning rule.

Similarly, Active Contour Models (*ACM*s) deal with the image segmentation problem as an(other) optimization problem, as they try to divide an image into

T. Villmann et al. (eds.), *Advances in Self-Organizing Maps and Learning
Vector Quantization,* Advances in Intelligent Systems and Computing 295,
DOI: 10.1007/978-3-319-07695-9_19, © Springer International Publishing Switzerland 2014

several regions by minimizing an "energy" functional, whose argument is the separating contour of the regions. Starting from an initial contour, the optimization is performed iteratively, evolving the current contour with the aim of gradually improving the approximation of the actual boundary of the regions (hence the denomination "active contour" models, which is used also for models that evolve the contour but are not based on the explicit minimization of a functional [2]). A particularly useful subclass of ACMs are the so-called global regional ACMs, which use statistical information about the regions (e.g., intensity, texture, colour distribution, etc.) to determine when to terminate the contour evolution [3,4]. Most of the existing global regional ACMs rely explicitly on a particular probability model (e.g., Gaussian, Laplacian, etc.) for the image intensity distribution, which results in restricting their scope in handling images in a global way, and affects negatively their performance when processing noisy images. On the other hand, SOM-based models have the advantage with respect to other neural-network models (e.g., multilayer perceptrons [5]) of being able to predict the underlying image intensity distributions relying on their topology preservation property [6], which is typical of SOMs. Another positive feature is that - likewise other neural network models - they can be also implemented in a parallel processing way. However, the application of existing SOM-based models in segmentation usually results in disconnected boundaries. Moreover, they are often quite sensitive to the noise. Motivated by the issues above, in the paper we propose a novel ACM, named Concurrent Self Organizing Map based Chan-Vese ($CSOM$-CV) model, which combines SOMs and global ACMs in order to deal with the image segmentation problem reducing the disadvantages of both approaches, while preserving the aforementioned advantages.

The paper is organized as follows. In Sections 2 and 3, we review briefly, resp., the general architecture of $CSOM$ as a classification tool, and the formulation of the CV model. Section 4 presents the formulation of the proposed $CSOM$-CV model. Section 5 presents experimental results comparing the segmentation accuracy of the proposed model and the ones of the stand-alone CV and $CSOM$ models, on the basis of a number of synthetic and real images. Finally, Section 6 provides some conclusions.

2   The $CSOM$ Model

In this section, we review the Concurrent Self Organizing Map ($CSOM$) model [1] as a pattern classification tool, hence also an image segmentation technique.

The classification process of a $CSOM$ starts by training a series of SOMs (one for each class) in a parallel way, using for each SOM a subset of samples coming from its associated class. During the training process, the neurons of each SOM are topologically arranged in the corresponding map on the basis of their prototypes (weights) and of the ones of the neurons within a certain geometric distance from them, and are moved toward the current input using the classical self-organization learning rule of a SOM, which is expressed by

$$w_n(t+1) := w_n(t) + \eta(t)h_{bn}(t)[x(t) - w_n(t)], \tag{1}$$

where $t = 0, 1, 2, 3, \ldots$ is a time index, $w_n(t)$ is the prototype of the neuron n at time t, $x(t)$ is the input at time t, $\eta(t)$ is a learning rate, and $h_{bn}(t)$ is the neighborhood kernel at time t of the neuron n around a specific neuron b, called best-matching unit (BMU). More precisely, in each SOM and at the time t, an input vector $x(t) \in \mathbb{R}^D$ is presented to feed the network, then the neurons in the map compete one with the other to be the winner neuron b, which is the chosen as the one whose weight $w_b(t)$ is the closest to the input vector $x(t)$ in terms of a similarity measure, which is usually the Euclidean distance $\| \cdot \|_2$. In this case, $\|x(t) - w_b(t)\|_2 := \min_n \|x(t) - w_n(t)\|_2$, where n varies in the set of neurons of the map. Once the learning of all the SOMs has been accomplished, the class label of a previously-unseen input test pattern is determined by the criterion of the minimum quantization error. More precisely, the BMU neuron associated with the input test pattern is determined for each SOM, and the winning SOM is the one for which the prototype of the associated BMU neuron has the smallest distance from the input test pattern, which is consequently assigned to the class associated with that SOM. We conclude mentioning that, when it is used as a supervised image segmentation technique, $CSOM$ usually results in disconnected boundaries, and is often sensitive to the noise.

3   The CV Model

In this section, we briefly review the formulation of the Chan-Vese (CV) model [3], which is a well-known representative state-of-the-art global regional ACM. The importance of the CV model among ACMs is emphasized by the very large number of citations of [3] (more than 3600 on Scopus at the present time).

   In the so-called "level set" formulation of the CV model, the current active contour C is represented as the zero level set of an auxiliary function $\phi : \Omega \to \mathbb{R}$, where Ω is the image domain: $C := \{x \in \Omega : \phi(x) = 0\}$. In the following, we denote by $\text{in}(C)$ and $\text{out}(C)$, resp., the approximations of the foreground and the background that are associated with the contour C, i.e.:

$$\text{in}(C) := \{x \in \Omega : \phi(x) > 0\}, \quad \text{out}(C) := \{x \in \Omega : \phi(x) < 0\}.$$

When dealing with contours evolving in time, the function $\phi(x)$ is replaced by a function $\phi(x, t)$. Then, the time evolution of the level set function ϕ in the CV model is described by the following Partial Differential Equation (PDE) (in the variables x and t, omitted from the next formula to shorten the notation):

$$\frac{\partial \phi}{\partial t} = \delta(\phi) \left[\mu \nabla \left(\nabla \phi / \|\nabla \phi\|_2 \right) - \nu - \lambda^+ \left(I - c^+ \right)^2 + \lambda^- \left(I - c^- \right)^2 \right], \qquad (2)$$

where $I(x)$ denotes the image intensity at the pixel location x, $\mu \geq 0$ and $\nu \geq 0$ are regularization parameters, and $\lambda^+ \geq 0$, and $\lambda^- \geq 0$ are parameters that control the influence of the last two terms. The first term in μ keeps the level set function smooth, the second one in ν influences the propagation speed of the contour, while the third and fourth terms can be interpreted, resp., as internal and external "forces" acting on the contour. Finally, $\delta\left(\cdot\right)$ denotes the Dirac generalized function, and c^+ and c^- are, resp., the mean intensity inside and outside the contour, defined as follows:

$$c^+ := \int_{\mathrm{in}(C)} I\left(x\right) dx \Big/ \int_{\mathrm{in}(C)} dx\,, \quad c^- := \int_{\mathrm{out}(C)} I\left(x\right) dx \Big/ \int_{\mathrm{out}(C)} dx\,. \quad (3)$$

The CV model can also be derived, in a Maximum Likelihood setting, by making the assumption that the foreground and background follow Gaussian intensity distributions with the same variance [7]. Then, the model approximates globally their distributions by the two scalars c^+ and c^-, resp., which are their mean intensities. As a global regional ACM, the CV model drives the contour to match regions that maximize the difference in their mean intensity. As a consequence, the evolution process is controlled completely by how accurate the mean intensities (3) are in representing the foreground/background distributions in each evolving step, starting from the initial contour. Furthermore, the implementation of this model often requires to re-initialize the evolution curve to be a signed distance function, which is a computationally expensive operation.

4   The *CSOM-CV* Model

In this section, we describe our Concurrent Self Organizing Map based Chan-Vese Model (*CSOM-CV*). Such model is composed of an off-line and on-line sessions, which are described, resp., in Subsections 4.1 and 4.2.

4.1   Training Session

The *CSOM-CV* model we propose makes use of two *SOM*s, one associated with the foreground, the other to the background. We make a distinction between the two *SOM*s by using, resp., the superscripts $+$ and $-$ for the associated weights. We assume that two sets of training samples belonging to the true foreground Ω^+ and the true background Ω^- of a training image $I^{(tr)}$ are available. They are defined as: $L^+ := \{x_1^+, \ldots, x_{|L^+|}^+ \in \Omega^+\}$ and $L^- := \{x_1^-, \ldots, x_{|L^-|}^- \in \Omega^-\}$, where $|L^+|$ and $|L^-|$ are their cardinalities.

In the following, we describe first the learning procedure of the SOM trained with the set of foreground training pixels L^+. In the training session, after choosing a suitable topology of the SOM associated with the foreground, the intensity $I^{(tr)}(x_t^+)$ of a randomly-extracted pixel $x_t^+ \in L^+$ of the foreground of the training image is applied as input to the neural map at time $t = 0, 1, \ldots, t_{\max}^{(tr)} - 1$, where $t_{\max}^{(tr)}$ is the number of iterations in the training of the neural map. Then, the neurons are self-organized in order to preserve - at the end of training - the topological structure of the image intensity distribution of the foreground. Each neuron n of the SOM is connected to the input by a weight vector w_n^+ of the same dimension D as the input (which - in the case of gray-level images considered in the paper - has dimension 1). After their random initialization, the weights w_n^+ of the neurons are updated by the self-organization learning rule (1), which we re-write in the form specific for the case considered here:

$$w_n^+(t+1) := w_n^+(t) + \eta(t)h_{bn}(t)[I^{(tr)}(x_t^+) - w_n^+(t)], \tag{4}$$

In this case, the *BMU* neuron b is the one whose weight vector is the closest to the input $I^{(tr)}(x_t)$ at time t. Both the learning rate $\eta(t)$ and the neighborhood kernel $h_{bn}(t)$ are designed to be time-decreasing in order to stabilize the weights $w_n^+(t)$ for t sufficiently large. In this way - due to the well-known properties [6] of the self-organization learning rule (4) - when the training session is completed, one can accurately model and approximate the input intensity distribution of the foreground by associating the intensity of each input to the weight of the corresponding *BMU* neuron. In particular, in the following we make the choice $\eta(t) := \eta_0 \exp(-t/(\tau_\eta))$, where $\eta_0 > 0$ is the initial learning rate and $\tau_\eta > 0$ is a time constant, whereas $h_{bn}(t)$ is selected as a Gaussian function centered on the *BMU* neuron, i.e., it has the form $h_{bn}(t) := \exp(-\|r_b - r_n\|_2^2/(2r^2(t)))$, where $r_b, r_n \in \mathbb{R}^2$ are the location vectors in the output neural map of neurons b and n, resp., and $r(t) > 0$ is a time-decreasing neighborhood radius (this choice of the function $h_{bn}(t)$ guarantees that, for fixed t, when $\|r_b - r_n\|_2$ increases, $h_{bn}(t)$ decreases to zero gradually to smooth out the effect of the *BMU* neuron on the weights of the neurons far from the *BMU* neuron itself, and when t increases, the influence of the *BMU* neuron becomes more and more localized). In particular, in the following we choose $r(t) := r_0 \exp(-t/\tau_r)$, where $r_0 > 0$ is the initial neighborhood radius of the map, and $\tau_r > 0$ is another time constant.

Finally, the learning procedure of the other *SOM* differs only in the random choice of the training pixel (which is now denoted by x_t^-, and belongs to the set L^-), and in the weights of the network, which are denoted by w_n^-.

4.2 Testing Session

Once the training of the two *SOM*s has been accomplished, the two trained networks are applied on-line in the testing session, during the evolution of the contour C, to approximate and describe globally the foreground and background intensity distributions of a similar test image $I(x)$. Indeed, during the contour evolution, the two mean intensities mean$(I(x)|x \in \text{in}(C))$ and mean$(I(x)|x \in \text{out}(C))$ in the current approximations of the foreground and background are presented as inputs to the two trained networks. We now define the quantities

$$w_b^+(C) := \operatorname{argmin}_n |w_n - \text{mean}(I(x)|x \in \text{in}(C))| , \tag{5}$$
$$w_b^-(C) := \operatorname{argmin}_n |w_n - \text{mean}(I(x)|x \in \text{out}(C))| ,$$

where $w_b^+(C)$ is the prototype of the *BMU* neuron to the mean intensity inside the current contour, while $w_b^-(C)$ is the prototype of the *BMU* neuron to the mean intensity outside it. Then, we define the functional of the *CSOM-CV* model as

$$E_{CSOM-CV}(C) := \lambda^+ \int_{\text{in}(C)} e^+(x,C)dx + \lambda^- \int_{\text{out}(C)} e^-(x,C)dx , \tag{6}$$

$$e^+(x, C) := \left(I(x) - w_b^+(C)\right)^2, e^-(x, C) := \left(I(x) - w_b^-(C)\right)^2,$$

where the parameters $\lambda^+, \lambda^- \geq 0$ are, resp., the weights of the two image energy terms $\int_{\text{in}(C)} e^+(x, C)dx$ and $\int_{\text{out}(C)} e^-(x, C)dx$, inside and outside the contour.

Now, as in [3], we replace the curve C with the level set function ϕ, obtaining

$$E_{CSOM-CV}(\phi) = \lambda^+ \int_{\phi>0} e^+(x, \phi)dx + \lambda^- \int_{\phi<0} e^-(x, \phi)dx,$$

where we have made explicit the dependence of e^+ and e^- on ϕ. In terms of the Heaviside step function $H(\cdot)$, the *CSOM-CV* functional can be also written as:

$$E_{CSOM-CV}(\phi) = \lambda^+ \int_\Omega e^+(x, \phi)H(\phi(x))dx + \lambda^- \int_\Omega e^-(x, \phi)(1 - H(\phi(x)))dx.$$

Finally, applying the gradient-descent technique in an infinite-dimensional setting likewise in [3], the contour evolution is described by the *PDE*

$$\frac{\partial \phi}{\partial t} = \delta(\phi)\left[-\lambda^+ e^+ + \lambda^- e^-\right], \tag{7}$$

which shows how the learned neurons of the two *SOM*s are used to determine the internal and external "forces" acting on the contour. Apart from this difference, Eq. (7) has a similar form as Eq. (2), and can be solved iteratively using the same smoothing and discretization techniques described in [3]. Moreover, in a similar way to [8], we perform - at each iteration of a finite-difference approximation of (7) - the regularization of the current level set function by replacing it with its convolution with a Gaussian filter of suitable width. Finally, the contour evolution is performed for $t_{\max}^{(evol)}$ iterations (unless convergence is obtained before). We conclude mentioning that the *CSOM-CV* model can also be extended to *RGB* images (this extension is not presented here, due to space limits).

5   Experimental Study

In this section, we demonstrate the effectiveness of the *CSOM-CV* model, compared to the stand-alone *CSOM* and *CV* models, in handling synthetic and real images. For a fair comparison, the *CSOM-CV*, *CV* and the *CSOM* models used in this experiment are all implemented in Matlab R2012a on a PC with the following configuration: 1.8 GHz Intel(R) Core(TM) i3-3217U, and 4.00 GB RAM. In each experiment, the *CSOM-CV* parameters are fixed as follows: $\eta_0 = 0.9$, $\sigma = 1.5$, and the weight parameters (i.e., λ^+, λ^-) are fixed to 1. Also, $r_0 = \max(M, N)/2$, where M and N are the numbers of rows and columns of the neural map, $t_{\max}^{(tr)} = 10000$, $t_{\max}^{(evol)} = 1000$, $\tau_\eta = t^{(tr)\max}$, $\tau_r = t_{\max}^{(tr)}/\ln(r_0)$, $\rho = 1$. The *SOM*s are composed of 3×3 neurons in most experiments (i.e., $M = N = 3$). In the *CV* model, λ^+, λ^- are also fixed to 1, μ is chosen such that the final contour is smooth enough and $\nu = 0$ (as made in [3, p. 268]). All the

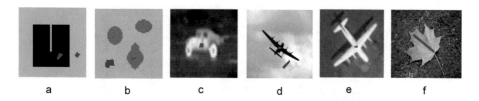

Fig. 1. The training images used in this paper together with the supervised foreground pixels (red) and the supervised background pixels (blue) used in training sessions of the *CSOM-CV* and the *CSOM* models. By its definition, no supervised pixels are used by the *CV* model

gray-level images considered in this section are 8-bit images, so the range of the values assumed by the intensity is 0-255.

To demonstrate the robustness of *CSOM-CV* to the presence of the noise, in the experiment described in Fig. 2 we have used the noise-free images of Fig. 1(a) and (b) in the training sessions of *CSOM-CV* and *CSOM*, then the so-learned *SOM*s have been applied on-line by the two models to their noisy versions as test images. As shown in Fig. 2, for this case *CSOM-CV* is less sensitive to the noise than *CV* (which does not make use of supervised training examples) and *CSOM*, since the regions of the foreground are detected more accurately by *CSOM-CV*. Fig. 3 illustrates the effectiveness of *CSOM-CV* in handling other images. The segmentation results of the *CSOM-CV* model shown in the first row demonstrate its ability to segment objects with blurred edges and background, while on the same images the *CSOM* and *CV* models incur, resp., in over- and under- segmentation problems. Similarly, as shown, resp., in the second and third rows, *CSOM-CV* outperforms *CSOM* and *CV* also in handling images characterized by nonhomogeneous background intensity distribution, and in the presence of a shadow. Simular results are obtained for the fourth image.

To demonstrate the computational efficiency of *CSOM-CV* when compared to *CSOM* and *CV*, Table 1 shows, for the images shown in Fig. 2 and 3, the CPU time (in seconds) required to perform the active-contour evolution by the *CSOM CV* and the *CV* models, and the CPU time (in seconds) of the *CSOM* model (which is nearly the same as the training time of the *CSOM-CV* model, as they share the same concurrent *SOM*). For the *CSOM-CV* and *CV* models, the number of iterations performed before convergence of the active-contour evolution is also reported in the table. As illustrated by Table 1, we can observe that, during the active-contour evolution, the *CSOM-CV* model has shown to be much faster than the *CV* model in all the listed cases, thus confirming its efficiency after training, especially when dealing with noisy versions of the same image (since in this case the training phase is performed only one time). Indeed, during the active-contour evolution, the *CSOM-CV* has required in general less iterations than the *CV* model to practically reach convergence, which compensates its larger computational time required by each iteration of the active-contour evolution. Moreover, as illustrated in Table 2, we have also used the Precision, Recall,

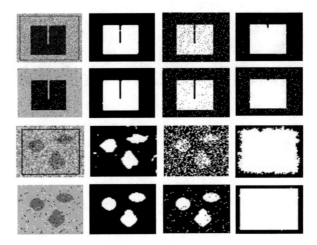

Fig. 2. The robustness of the *CSOM-CV* model to two different kinds of noise: the first column shows, from top to down, two noisy versions of the image shown in Fig. 1(a), and two noisy versions of the image shown in Fig. 1(b), resp., with the addition of Gaussian noise with standard deviation $SD = 50$ (first and third row) and salt and pepper noise (second and fourth row). The initial contours used by the *CSOM-CV* and *CV* models are also shown (first and third row); finally, the second, third, and fourth columns show, resp., the corresponding binary segmentation obtained by the *CSOM-CV*, *CSOM*, and *CV* models.

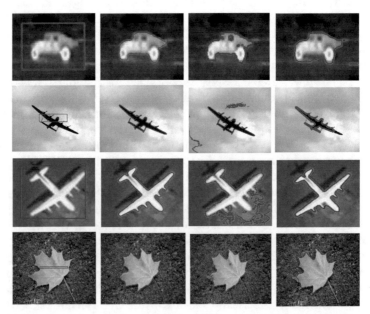

Fig. 3. The segmentation results obtained on real and synthetic gray-level images. The first column shows the original images with the initial contours, while the second, third, and fourth columns show, resp., the corresponding segmentation results obtained by the *CSOM-CV*, *CSOM*, and *CV* models.

Table 1. The contour evolution time and number of iterations required by the *CSOM-CV* and *CV* models to segment the foreground for the images shown in this paper. The CPU time of *CSOM* and the percentage of labeled pixels (lp) are also included.

Image in	Image size	lp (%)	CSOM-CV model		CSOM model	CV model	
			CPU t. (s)	# Iter.	CPU t. (s)	CPU t. (s)	# Iter.
Fig. 2 row 1	114×101	3.01	0.73	20	16.7	3.2	158
Fig. 2 row 2	114×101	3.01	0.62	18	14.7	3.64	219
Fig. 2 row 3	64×61	5.3	0.078	10	4.9	0.04	4
Fig. 2 row 4	64×61	5.3	0.07	10	5	0.98	30
Fig. 3 row 1	118×93	3.02	0.04	10	6.12	2.12	137
Fig. 3 row 2	300×225	1.38	0.62	37	42.03	6.68	205
Fig. 3 row 3	135×125	2.38	0.15	17	10.1	4.18	266
Fig. 3 row 4	300×203	4.32	0.81	33	33.79	9.85	344

Table 2. The Precision, Recall, and *F*-measure metrics for the *CSOM-CV*, *CSOM*, and *CV* models

Image in	CSOM-CV model			CSOM model			CV model		
	P(%)	R(%)	F-meas.(%)	P(%)	R(%)	F-meas.(%)	P(%)	R(%)	F-meas.(%)
Fig. 2 row 1	99.7	99.8	99.8	93.5	94.7	94	97	88.3	92.5
Fig. 2 row 2	99.8	99.9	99.8	94.7	97.5	96.1	94.2	87.2	90.5
Fig. 2 row 3	48.2	93.9	63.7	16.4	72.6	26.8	12.7	96.4	22.5
Fig. 2 row 4	56.8	97.4	71.1	48.7	96.4	64.7	12.2	100	21.7
Fig. 3 row 1	100	94.3	97.1	99.6	92.1	95.7	92.8	82.9	87.6
Fig. 3 row 2	63.5	89.5	74.3	39.2	95.4	55.6	73	60	65.9
Fig. 3 row 3	95.7	99.8	97.7	46.5	100	63.5	94.9	61.4	74.6
Fig. 3 row 4	91.8	90.3	91	81.9	95.1	88	94.3	88.3	91.2

and *F*-measure metrics (where the "positive" pixels are the foreground pixels) to evaluate quantitatively the segmentation results of all the models, confiming the effectiveness of the *CSOM-CV* model when compared to the *CSOM* and *CV* models (apart from the image in the fourth row of Fig. 3, on which *CSOM-CV* and *CV* produced similar results), thus compensating the usually larger total (training+testing) time required by the *CSOM-CV* model to segment the same image, when compared to the other models. Concluding, the experimental results demonstrate the ability of the *CSOM-CV* model to combine the positive aspects of both the models from which it derives (e.g., the ability of the *CSOM* model to represent the intensity distributions of the foreground/background due to the labeled pixels, and the energy-functional formulation of the *CV* model).

6   Conclusions

In this paper we have proposed a novel *SOM*-based *ACM* model, the Concurrent Self Organizing Map-based Chan-Vese (*CSOM-CV*) model, which relies mainly on a set of prototypes coming from two trained *SOM*s to guide the evolution of the active contour. The *CSOM-CV* model is a supervised and global region-based *ACM*. It has been demonstrated to be efficient and robust to two different kinds of noise. As compared to the *CV* model, our proposed solution consists instead in modeling globally in a supervised way the intensity distributions of the foreground/background (relying on a few supervised pixels) without using parametric models, but relying on a set of prototypes resulting from the training of a *CSOM*. So, the main reasons for which, as shown experimentally in Section 5, the proposed model affects positively the *CV* model in terms of speed-up in the testing phase and robustness to noise are that - differently from the proposed model - the *CV* model refers to Gaussian intensity distributions of the foreground/background, and does not include supervised examples. Moreover, as compared to *CSOM* and in general to *SOM*-like models used in image segmentation, our solution consists in modeling the active contour using a variational level set method and relying at the same time on a few prototypes coming from the learned *CSOM*. In this way, the *CSOM-CV* model is able to produce a final segmentation result characterized by a smooth contour while most *SOM*-like models usually produce segmentations characterized by disconnected boundaries. Of course, in order to be used in practice, the *CSOM-CV* model requires the availability of labeled pixels; an extension to the unsupervised case, as well as the effect of the initialization of the *SOM* neurons, are currently under study.

References

1. Neagoe, V.-E., Ropot, A.-D.: Concurrent self-organizing maps for pattern classification. In: Proc. of the 1st IEEE Int. Conf. on Cognitive Informatics, pp. 304–312 (2002)
2. Venkatesh, Y.V., Kumar Raja, S., Ramya, N.: A novel SOM-based approach for active contour modeling. In: Proc. of the Conf. on Intelligent Sensors, Sensor Networks and Information Processing, pp. 229–234 (2004)
3. Chan, T.F., Vese, L.A.: Active contours without edges. IEEE Trans. on Image Processing 10(2), 266–277 (2001)
4. Abdelsamea, M.M., Tsaftaris, S.A.: Active contour model driven by globally signed region pressure force. In: Proc. of the 18th Int. Conf. on Digital Signal Processing, pp. 1–6 (2013)
5. Middleton, I., Damper, R.I.: Segmentation of magnetic resonance images using a combination of neural networks and active contour models. Medical Enginering & Physics 26(1), 71–86 (2004)
6. Kohonen, T.: Essentials of the self-organizing map. Neural Networks 37, 52–65 (2013)
7. Chen, S., Radke, R.J.: Level set segmentation with both shape and intensity priors. In: Proc. of the 12th IEEE Int. Conf. on Computer Vision, pp. 763–770 (2009)
8. Zhang, K., Song, H., Zhang, L.: Active contours driven by local image fitting energy. Pattern Recognition 43(4), 1199–1206 (2010)

Five-Dimensional Sentiment Analysis of Corpora, Documents and Words

Timo Honkela[1,2], Jaakko Korhonen[3], Krista Lagus[2,4], and Esa Saarinen[3]

[1] University of Helsinki, Department of Modern Languages, Helsinki
[2] Aalto University School of Science
Department of Information and Computer Science, Espoo
[3] Aalto University School of Science
Department of Industrial Engineering and Management, Espoo
[4] National Consumer Research Centre, Helsinki
Finland
first.second@aalto.fi

Abstract. Sentiment analysis has become a widely used approach to assess the emotional content of written documents such as customer feedback. In positive psychology research, the typical one-dimensional analysis framework has been extended to include five dimensions. This five-dimensional model, PERMA, enables a fine-grained analysis of written texts. We propose an approach in which this model, statistical analysis and the self-organizing map are used. We analyze corpora from various genres. A hybrid methodology that uses the self-organizing maps algorithm and human judgment is suggested for expanding the PERMA lexicon. This vocabulary expansion can be useful for English but it is potentially even more crucial in the case of other languages for which the lexicon is not readily available. The challenges and solutions related to the text mining of texts written in a morphologically complex language such as Finnish are also considered.

Keywords: Text mining, natural language processing, self-organizing map, independent component analysis, positive psychology, education, life-philosophical lecturing.

1 Introduction

Computer-based quantitative methods are becoming more and more popular in the study of complex phenomena in social sciences and humanities. This trend has been strengthened by the fact that many modern analysis methods and tools enable non-reductionistic approaches. Quantitative methods may be useful in qualitative analysis if thousands or even larger number of variables are dealt with simultaneously. This idea is reflected in the representation and analysis of texts as large matrices or tensors. Moreover, computational methods enable modeling and simulation that takes into account the systems nature of real world phenomena [3]. Related research areas include systems intelligence [4] and complexity science [1].

T. Villmann et al. (eds.), *Advances in Self-Organizing Maps and Learning Vector Quantization,* Advances in Intelligent Systems and Computing 295,
DOI: 10.1007/978-3-319-07695-9_20, © Springer International Publishing Switzerland 2014

1.1   Background Motivation

In this paper, we take first steps in approaching one highly complex phenomenon related to psychology, education and philosophy, namely what kind of changes begin to happen in the students that have attended a course built on life-philosophical lecturing. By life-philosophical lecturing we refer to a particular kind of oral pedagogical practice that uses the lecture situation for the benefit of providing for the listeners an enhanced possibility of life-philosophical reflection [17,18]. The dominant lecturing practices seek to function as a channel for predetermined knowledge, theories or learning. Then the goal is to make the listeners to adopt the insights, scholarship or philosophy of the lecturer. In contrast, in life-philosophical lecturing "the paramount aim is to facilitate, stimulate and vitalize the participants own life-philosophical thinking in the first-person – his or her use of the reflective mind" [17]. Life-philosophical lecturing is a form of positive philosophical practice and seeks key inspiration from the breakthroughs of the positive psychology movement [22,21]. Our aim is to be eventually able to measure from texts written by students the changes such a lecturing practice stimulates in them. This article describes our first experiments on the matter.

1.2   Sentiment Analysis and the PERMA Model

Sentiment analysis of written documents aims to determine the overall polarity of each document of the attitude of the author(s) regarding some topic. Sentiment analysis has become commonplace and it is widely applied, e.g., in business intelligence and in analyzing social media contents [25,15,14,5]. The sentiment of a document is typically calculated as a synthesis of the sentiments of the words and phrases in the document. A straightforward approach is to manually associate a positive or negative value for those words that indicate sentiment. Turney automated this process by calculating the sentiment of a given phrase by comparing its similarity to a positive reference word ("excellent") with its similarity to a negative reference word ("poor") [25].

A typical approach in sentiment analysis is to estimate the polarity of the documents. This one-dimensional measure can be replaced by analyzing multiple factors simultaneously. A straightforward extension is to measure both valence (positive vs. negative) and arousal (activation vs. deactivation). A more refined category system of emotions could include level of interest, enjoyment, surprise, contempt, anger, fear, distress and shame.

In the context of positive psychology research, Seligman has developed the PERMA model that addresses different aspects of wellbeing [21]. The PERMA model includes five components related to subjective well-being: Positive emotion (P), Engagement (E), Relationships (R), Meaning (M) and Achievement (A) [21]. Researchers have gathered a PERMA lexicon that is a collection of words that are associated with each of the components in a positive or negative manner [19].

We propose a way to apply the PERMA model to the analysis of document collections. Furthermore, we suggest a way for complementing the PERMA vocabulary that can be useful especially for other languages than English. The PERMA analysis of texts can be considered at three main levels:

1. PERMA profiling of document collections. This can provide an overall understanding of the nature of different corpora. We analyze the five-dimensional profile of corpora in six different genres.
2. PERMA profiling of individual documents. The second level of analysis is seen to be useful for the lecturer who is provided tools for familiarizing himself with certain aspects of hundreds of long essays written by the students. A related idea has been presented in the context of MOOCs (massive online open courses) [8] for mining student contributions.
3. Comparison of PERMA and non-PERMA words. This analysis can be conducted, for example, in order to find new PERMA word candidates. In this paper, we use the self-organizing map [11] for this purpose.

One way to look at sentiment analysis is to ask first, which are the sentiments that need to be detected, and second, which features in the text reflect said sentiments. While PERMA model provides a theory-driven proposal for a set of such sentiments, as well as seed lists of features, challenges remain. For example, many texts might not have many PERMA features at all. Moreover, translating the PERMA vocabulary to another language leads to additional challenges, since each language might have quite different ways to express for example positivity, and literal translation of words may not be a sufficient method for capturing these.

1.3   Why Unsupervised Methodology

When using learning methods, information regarding properties of interest (features) or decisions of interest (e.g. class labels) need to be provided to the learning system. Feature selection is generally considered a weak form of importing supervision to a learning system, whereas applying labeled data would constitute a strong form of supervision. One could approach this as a classification problem, by providing a number of manually classified samples to the system.

Instead, in this case we apply a theory-driven perspective for selecting the features, namely the PERMA vocabulary, and then apply an unsupervised learning method, namely the self-organizing map for exploring the outcome. By providing prior knowledge in the feature selection stage the researcher is able to give the learning system information regarding the properties of interest, without having to determine exactly what the outcome should be regarding any specific case, such as a document or a collection. The fact that PERMA vocabulary has been collected already by researchers allows the ready use of unsupervised clustering and visualization methods for any new corpora as well. This suits well in a text mining scenario, where the interest is in finding new, surprising phenomena in the direction of interest of the researcher.

1.4   Preprocessing Morphologically Highly Complex Languages

From a morphological point of view, English is a rather simple language. This means that methods that are based on lists of keywords (e.g. [24]) can rely on the idea that there are only a small number of different surface forms of the same basic form or stem. On the other hand, many other languages have more complex morphology. For instance, in Finnish every noun has about 2,000 different inflections and every verb more than 10,000. In addition to this, compounding is very commonplace. Common multi-word phrases in English are often translated as compounds in Finnish. The outcome of the complex morphology is that Finnish has billions of surface word forms which cannot be simply categorized and listed. Fig. 1 shows the seven (out of 99) most common forms of 'merkitys' (meaning) in our essay corpus. In order to deal with the problem of varying word forms, we have relied on the methodology originally developed by Koskenniemi [12]. We used an open-source implementation of the model [13] to process our corpus. The Omorfi tool transformed each inflected word into its basic form. Due to ambiguities and differences in subtle meanings, the process does not preserve all information when language borders are crossed but details cannot be dealt with here.

Word	Translation	Freq.
merkitystä	meaning (as a partial object)	142
merkitys	meaning	101
merkityksen	of the meaning	65
merkityksellistä	of the meaningful	41
merkityksiä	meanings (as a partial object)	36
merkityksellisyyden	of the meaningfulness	34
merkityksellisiä	meaningful (plural, as a partial object)	32
...	...	..

Fig. 1. Examples of different forms for the word 'merkitys' (meaning) in Finnish with the frequency count in our essay corpus

2   PERMA Profiles of Different Genres

One can compute a PERMA profile for a document by counting the frequencies of the PERMA words in each component. Our hypothesis was that the PERMA profiles would be different for text corpora that represent different genres. We chose the following kinds of corpora: news feeds from Reuters and Finnish news agency STT, Wikipedia articles on topics that start with the letter A, everyday conversations collected at UC Santa Barbara [2], proceedings of European parliament from 1996 [10], English translation of the fairy tales by Grimm brothers, corporate e-mails messages sent in Enron. The Enron corpus was divided into three parts to check whether the inter-corpus variation is smaller than intra-corpus variation. The result of the PERMA profile analysis is shown in Fig. 2. The results indicate that the PERMA vocabulary is able to identify differences

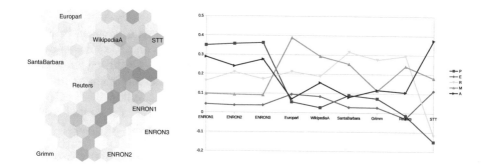

Fig. 2. On the right, relative PERMA profiles for different corpora. Enron e-mails show high values on Positivity and Achievement but low on Meaning, whereas Europarl and Wikipedia are low on Positivity but high on Meaning. On the left, the same PERMA profile information is used to project the corpora on a SOM.

among the document collections in a meaningful and informative way. Firstly, the profiles are markedly different for the various corpora. Secondly, the results clearly make sense.

For instance, the news corpora are markedly negative in their content. On the other hand, the Reuters news corpus also scores high on Relationships whereas the Finnish STT scores very high on Achievement. The latter seems to be due to the large proportion of sports news within the STT corpus.

The European parliament corpus obtains high scores on the Meaning dimension and low on Achievement, which may raise a question regarding whether the parliament is concerned enough about achieving any concrete goals.

In a striking contrast are the Enron discussions, which show low Meaning but high Achievement. The emphasis on achievement of concrete goals can be considered natural in a competitive corporate context. However, one is left to wonder whether the low proportion of meaningfulness might have been indicative of upcoming problems, but this is left here as a question for future exploration.

The PERMA analysis over language borders requires further attention. It is probable that phenomena like linguistic polysemy and cultural contextuality influence the results in such as way that fine-tuning of the methodology is necessary. For instance, the four most common positive PERMA words in Finnish in the STT news articles were "voittaa" (to win), "voitto" (victory), "edustaa" (to represent), and "onnistua" (to succeed). This example reminds that the nature of the corpora needs to be carefully concerned.

3   Self-Organizing Map of Sentiment Words

In the second case study, we explored the possibility of extending the PERMA vocabulary. This goal was motivated further by the observation that translating the vocabulary to another language necessarily introduces errors due to ambiguity, and is likely to result in an incomplete feature set for that particular language. Thus, research on automatic or semi-automatic means of complementing the feature set is important.

The SOM-based process that we suggest is outlined as a diagram in Fig. 3. The sentiment word list by Hu and Bing [7] was used as an external vocabulary. In the experiment, we used the WikipediaA corpus for calculating the context statistics. We formed word-word context matrices so that each element indicates how many times a sentiment word has appeared in the WikipediaA corpus in the vicinity of a context word. The context words were chosen to first exclude the 100 most common words and then to include the next 2000 words in the order of frequency.

The context window was chosen to be seven words to each direction. This can be characterized as an intermediate choice. Very short context windows emphasize syntactic aspects of the words and document-word matrices work relatively best when the documents are different enough from each other. The resulting matrix was analyzed using the self-organizing map algorithm, presented next. The Self-Organizing Map (SOM) has been used to create word clusters automatically from statistical features obtained from corpora [16,6]. The SOM algorithm produces a topological ordering by mapping the input space to an array of nodes. Each node of a SOM consists of a prototype vector m_i of the same dimension as the input vectors x_i. The nodes are typically organized in the form of a lattice. In an organized map, each input is associated with a prototype in a specific location. The basic idea is that if two input are similar they tend to be close to each other on the map. The SOM is rather similar to clustering algorithms but does not produce explicit clusters. It rather creates a diagram that aims to "mirror" the high-dimensional data (usually) in two dimensions as faithfully as possible.

In this case, as the data consists of statistical information on the contextual use of words, the end result is map where similar words are close to each other. In the result, several cases may be discussed qualitatively. The positive achievement words (marked by "A+") have been divided into two clusters in the upper and lower left side of the map. Some nearby words such as "progressive" and "renowned" could clearly be considered candidates for extending the A+ lexicon.

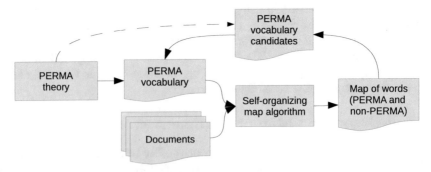

Fig. 3. The process of using the SOM in extending the coverage of a theory-based vocabulary

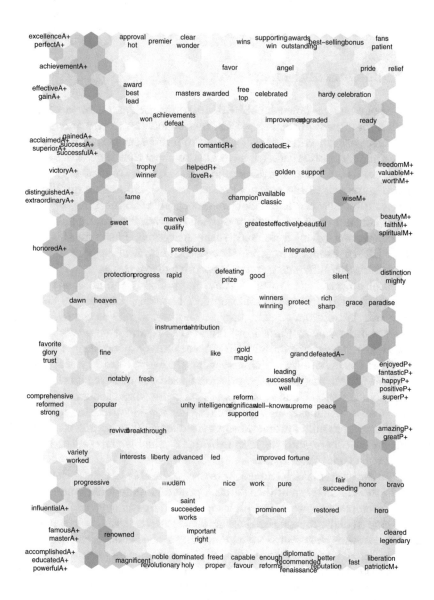

Fig. 4. A map of sentiment words based on context statistics obtained from the WikipediaA corpus. The words that belong to the PERMA lexicon are marked with a label that indicates the category (see Sec. 2 for an explanation).

The number of Relationship (R) and Engagement (E) words in the Hu and Bing word list [7] appeared to be low. The Meaning (M) words form a clear single cluster with the exception of the word "patriotic". These kinds of findings can potentially be used to re-evaluate the PERMA lexicon.

4 Conclusions and Discussion

We have applied a five-dimensional PERMA framework and associated vocabulary on performing sentiment analysis on text collections in two different ways.

In the first case, several text collections from different genres were analyzed and their differences observed. We were able to show that the PERMA profiles of the corpora fit with the intuitions related to the types of genres. In addition, the PERMA analysis of the corpora seemed to raise interesting questions such as did Enron fail because it did not concern itself with meaning, or does EU parliament concern itself relatively too much on overall meaning and too little on the achievement of concrete goals to be successful. Based on this it seems that the PERMA framework is promising on the level of corpora and able to highlight interesting differences in the respective discourses.

Our initial objective was to understand more closely the processes that the students go through during life-philosophical lecturing. Due to the challenges related to translating from one language to another, as well as the high number of different word forms in Finnish we found that it would be advantageous to attempt to complement the initial PERMA vocabulary by additional words. We then explored the possibility of doing so using the Word Category Map methodology, where lexical relations of words were used for ordering both PERMA vocabulary and a set of additional words on a two-dimensional display. The ordering is able to identify new candidates for consideration to be added as PERMA features. A fully automatic supervised learning approach could also be used but, on the other hand, the map provides a valuable view on the relational structure of the conceptual space.

Once the feature set is rich enough we expect to be able to extend the PERMA analysis to the full analysis of the PERMA profiles of individual student essays. Due to the complex nature of the philosophical and psychological contents and cognitive and social processes, reductionistic research methods are not easily applicable. It seems, however, that text mining and visualization methods *can* support traditional qualitative analysis [9]. Development of such tools is useful since manual analysis of student essays is usually limited to rather small numbers of cases. In our case, a qualitative analysis of 304 essays would be a considerable human effort but still manageable. The text mining approach, on the other hand, scales up to thousands and even millions of documents.

Interesting quantitative results can be gained when large corpora are available, collected, e.g., from social media with additional profile information [20]. A lexicon-based approach can be expanded to include sentence-level analysis to take into account context effects and to improve the precision of the analysis [23]. In this paper, we consider some of the problems and solutions related to crossing language borders. One direction is to study more carefully the philosophical and practical aspects related to multilingual and multicultural studies in this area.

Our longer term goal is develop methodology for the text mining and quantitative analysis of texts in the framework of positive psychology. Substantial developments in this field have taken place recently (cf. [20]), partly based on earlier developments (cf., e.g., [24]). Our intention is to experiment with different statistical machine learning and neural-network methods and to facilitate approaches for analyzing corpora written in other languages than English.

Acknowledgments. The authors are highly grateful to Professor Martin Seligman, the Director of the Positive Psychology Center at University of Pennsylvania with his research team for making the PERMA lexicon available. T.H. acknowledges support for the latest stages of the work from the EU Commission through its European Regional Development Fund, and the program "Leverage from the EU 2007-2013".

References

1. Castellani, B., Hafferty, F.W.: Sociology and Complexity Science: A New Field of Inquiry. Springer (2009)
2. Du Bois, J.W.: Santa Barbara Corpus of Spoken American English. University of California, Santa Barbara Center for the Study of Discourse (2000)
3. Goldspink, C.: Methodological implications of complex systems approaches to sociality: Simulation as a foundation for knowledge. Journal of Artificial Societies and Social Simulation 5(1), 1–19 (2002)
4. Hämäläinen, R.P., Saarinen, E.: Systems intelligence – the way forward? a note on Ackoff's "why few organizations adopt systems thinking". Systems Research and Behavioral Science 5(6), 821–825 (2008)
5. Hansen, L.K., Arvidsson, A., Nielsen, F.Å., Colleoni, E., Etter, M.: Good friends, bad news - affect and virality in twitter. In: The 2011 International Workshop on Social Computing, Network, and Services (SocialComNet 2011), pp. 34–43 (2011)
6. Honkela, T., Pulkki, V., Kohonen, T.: Contextual relations of words in Grimm tales, analyzed by self-organizing map. In: Fogelman-Soulié, F., Gallinari, P. (eds.) Proc. of ICANN 1995, vol. II, pp. 3–7. EC2, Nanterre (1995)
7. Hu, M., Liu, B.: Mining and summarizing customer reviews. In: Proceedings of the Tenth ACM SIGKDD International Conference on Knowledge Discovery and Data Mining, pp. 168–177. ACM (2004)
8. Hyman, P.: In the year of disruptive education. Communications of the ACM 55(12), 20–22 (2012)
9. Janasik, N., Honkela, T., Bruun, H.: Text mining in qualitative research application of an unsupervised learning method. Organizational Research Methods 12(3), 436–460 (2009)
10. Koehn, P.: Europarl: A parallel corpus for statistical machine translation. In: MT Summit, vol. 5 (2005)
11. Kohonen, T.: Self-Organizing maps. Springer, Heidelberg (2001)
12. Koskenniemi, K.: A general computational model for word-form recognition and production. In: Proceedings of the 10th International Conference on Computational Linguistics, pp. 178–181. Association for Computational Linguistics (1984)

13. Lindén, K., Silfverberg, M., Pirinen, T.: HFST tools for morphology – an efficient open-source package for construction of morphological analyzers. In: Mahlow, C., Piotrowski, M. (eds.) SFCM 2009. CCIS, vol. 41, pp. 28–47. Springer, Heidelberg (2009)
14. Pak, A., Paroubek, P.: Twitter as a corpus for sentiment analysis and opinion mining. In: Proceedings of LREC 2010. ELRA, Valletta (2010)
15. Pang, B., Lee, L.: Opinion mining and sentiment analysis. Foundations and Trends in Information Retrieval 2(1-2), 1–135 (2008)
16. Ritter, H., Kohonen, T.: Self-organizing semantic maps. Biological Cybernetics 61(4), 241–254 (1989)
17. Saarinen, E.: Life-philosophical lecturing as a systems-intelligent technology of the self. In: The XXIII World Congress of Philosophy, Athens, Greece (2013)
18. Saarinen, E., Lehti, T.: Inducing mindfulness through life-philosophical lecturing. Wiley (to appear, 2014)
19. Schwartz, H.A., Eichstaedt, J.C., Kern, M.L., Dziurzynski, L., Lucas, R.E., Agrawal, M., Park, G.J., Lakshmikanth, S.K., Jha, S., Seligman, M.E.P., Ungar, L.H.: Characterizing geographic variation in well-being using tweets. In: Proceedings of the Seventh International AAAI Conference on Weblogs and Social Media, ICWSM (2013)
20. Schwartz, H.A., Eichstaedt, J.C., Kern, M.L., Dziurzynski, L., Ramones, S.M., Agrawal, M., Shah, A., Kosinski, M., Stillwell, D., Seligman, M.E.: Personality, gender, and age in the language of social media: The open-vocabulary approach. PloS One 8(9), e73791 (2013)
21. Seligman, M.E.: Flourish: A visionary new understanding of happiness and well-being. Free Press, New York (2011)
22. Seligman, M.E., Csikszentmihalyi, M.: Positive psychology: An introduction. American Psychologist, 5–14 (2000)
23. Socher, R., Perelygin, A., Wu, J., Chuang, J., Manning, C.D., Ng, A.Y., Potts, C.: Recursive deep models for semantic compositionality over a sentiment treebank. In: Proceedings of the 2013 Conference on Empirical Methods in Natural Language Processing, pp. 1631–1642. Association for Computational Linguistics, Stroudsburg (2013)
24. Tausczik, Y.R., Pennebaker, J.W.: The psychological meaning of words: LIWC and computerized text analysis methods. Journal of Language and Social Psychology 29(1), 24–54 (2010)
25. Turney, P.D.: Thumbs up or thumbs down? semantic orientation applied to unsupervised classification of reviews. In: Proceedings of the 40th Annual Meeting on Association for Computational Linguistics, pp. 417–424. Association for Computational Linguistics, Stroudsburg (2002)

SOMbrero: An **R** Package for Numeric and Non-numeric Self-Organizing Maps

Julien Boelaert[1], Laura Bendhaiba[1],
Madalina Olteanu[1], and Nathalie Villa-Vialaneix[1,2,3]

[1] SAMM, Université Paris 1 Panthéon-Sorbonne,
90 rue de Tolbiac, 75634 Paris cedex 13, France
[2] UPVD, 52 avenue Paul Alduy, 66860 Perpignan cedex 9, France
[3] INRA, Unité MIAT, BP 52627 31326 Castanet Tolosan cedex, France
{julien.boelaert,laurabendhaiba}@gmail.com,
madalina.olteanu@univ-paris1.fr, nathalie.villa@toulouse.inra.fr

Abstract. This paper presents **SOMbrero**, a new R package for self-organizing maps. Along with the standard SOM algorithm for numeric data, it implements self-organizing maps for contingency tables ("Korresp") and for dissimilarity data ("relational SOM"), all relying on stochastic (i.e., on-line) training. It offers many graphical outputs and diagnostic tools, and comes with a user-friendly web graphical interface, based on the **shiny** R package.

Keywords: Self-Organizing Maps, R, Dissimilarity, Korresp.

1 Introduction

Self-Organizing Maps (SOM), introduced by Teuvo Kohonen [1], are a popular clustering and visualization algorithm. While originally intended for data consisting exclusively of numeric vectors, this prototype-based learning algorithm has been extended to handle other types of data.

One of the oldest attempts to generalize the SOM algorithm to non numeric data is the so-called "Korresp" algorithm (and related methods [2]), that extends standard correspondence analysis: this approach can be used to cluster rows and columns of a contingency table (i.e., values of two categorical variables that are jointly observed) on a topological map. More recently, several extensions of the SOM algorithm to non numeric data have been proposed. The median principle has been used to handle data described by dissimilarity matrices: [3] uses it by replacing the standard computation of prototypes by an approximation in the original data set. Since this approach is very restrictive, it has been improved in [4,5,6] for data described by a kernel, in [7,8] for data described by a dissimilarity matrix and in [9,8] for data described by several dissimilarity matrices or several kernels. In these works, the prototypes are no longer numeric vectors of the input space as in classical SOM, but convex combinations of the observations in a Hilbert or a pseudo-Euclidean vector space. The dissimilarity versions are called "relational SOM" and the kernel versions "kernel SOM". It should be noted

T. Villmann et al. (eds.), *Advances in Self-Organizing Maps and Learning
Vector Quantization*, Advances in Intelligent Systems and Computing 295,
DOI: 10.1007/978-3-319-07695-9_21, © Springer International Publishing Switzerland 2014

that relational SOM is a generalization of kernel SOM when the dissimilarity that describes the data is not Euclidean [8].

Several implementations of the SOM algorithm exist in different mathematical/statistical softwares, most on them usable only for numeric data. The SOM Toolbox[1] is a Matlab library implementing many variants of SOM for numeric data, with graphical outputs, user interfaces and implementations of other clustering algorithms. It is partially based on the original implementation SOM_PAK [10]. SAS Entreprise Miner (current version 12.3) features an implementation of SOM for numeric data. Also, the SAS programs by Patrick Letremy[2] implement standard SOM and several extensions, including SOM for contingency tables as described in [2].

R is one of the most popular statistical software environments, and several R packages implement variants of the SOM algorithm:

- **class** (current version 7.3-9, last updated in August 2013) offers a crude implementation of the SOM algorithm for numeric data with batch training;
- **som** [11] (current version 0.3-5, last updated in April 2010) implements a two-step batch algorithm for numeric data, with basic plotting of the resulting map;
- **popsom** [12] (current version 2.3, last updated in October 2013) is built on the **som** package, with additional diagnostic tools and visualizations;
- **kohonen** [13] (current version 2.0.14, last updated in December 2013) implements the standard SOM for numeric data as well as "super-organized maps", in which the observed variables can be separated into distinct "layers", and two versions of supervised SOM, X-Y fused SOM and Bi-Directional Kohonen maps, in which class information is available for all observations in addition to numeric coordinates. The training is done stochastically and several plot options are available;
- **yasomi** [14] (current version 0.3, last updated in March 2011) implements batch algorithms for standard SOM, relational SOM and kernel SOM (for data consisting of pairwise evaluations of a positive semi-definite kernel function). It features data driven construction methods for the maps, and several plotting options.

In the present article, we describe a new R package for SOM: **SOMbrero**. It implements stochastic versions of the SOM algorithms for numeric data, dissimilarity data, and for data described by contingency tables. To our knowledge, **SOMbrero** is also the R package that proposes the largest number of diagnostic tools (graphics, super-classes and quality measures) designed to help the user understand the outputs of the algorithm. The package uses the S3 object-oriented standard. Its current version (version 0.4-1, last updated in November 2013) is available on R-Forge at http://sombrero.r-forge.r-project.org/. It runs

---

[1] Current version 2.1, last updated in December 2012, available at
   http://research.ics.aalto.fi/software/somtoolbox/

[2] Current version 9.1.3, last updated in December 2005, no longer maintained, available at http://samos.univ-paris1.fr/Programmes-bases-sur-l-algorithme

on R 3.0 or higher, depends on packages including **wordcloud**, **scatterplot3d**, **igraph** and **e1071**, and comes with a **shiny** web graphical user interface (version 0.1) that can be tested at `http://shiny.nathalievilla.org/sombrero/` or directly loading the package into R and running the command:

```
sombreroGUI()
```

The rest of this article is organized as follows : Section 2 presents the main features of **SOMbrero**. Section 3 describes the web graphical user interface and Section 4 gives short examples of applications to various types of data.

2   Main Features of SOMbrero

This section describes the main features of **SOMbrero**: the kind of data the package is able to handle, the available plots and the other diagnostics tools.

Three Types of Self-Organizing Maps: The **SOMbrero** package currently supports three types of stochastic training algorithms, each designed for a specific type of data : "standard" SOM for numeric data, "Korresp" for contingency tables, and "relational SOM" for dissimilarity data.

Numeric SOM. When the data are numeric vectors, **SOMbrero** uses the standard stochastic SOM algorithm [1], which iterates over two steps. In the *affectation step*, a single observation is randomly drawn and affected to the one prototype it is closest to (in terms of Euclidean distance, or some other chosen distance, in the input space). The *representation step* then updates all prototypes, moving the closest unit and its neighbors towards the drawn observation. The algorithm converges empirically towards a minimum of the extended within-class variance. In the particular case of a finite data set, [15] showed that the SOM algorithm trained with a fixed neighborhood radius is equivalent to a gradient-descent minimizing a cost function equal to the extended within-class variance. The stochastic version of the algorithm is preferred over the batch version as it generally provides a better organization, whatever the initialization, at a comparable computational cost [16].

Korresp. When the data consist of a contingency table for two categorical variables, classical correspondence analysis performs a weighted principal components analysis, using the χ^2 distance simultaneously on the row profiles and on the column profiles. The same principle is used in the Korresp algorithm [2], which extends SOM to contingency tables.

Relational SOM. SOMbrero also implements a stochastic version of relational SOM, described in [8], which is an extension of SOM to dissimilarity data. This is useful when the data are not naturally described by a fixed set of numerical attributes (e.g. categorical variables or relations between objects), but when a measure of resemblance between observations (i.e., a similarity or a dissimilarity) can nevertheless be constructed. In this approach, prototypes are expressed as a symbolic convex combination of the observations that is justified by a pseudo-Euclidean framework [7].

Graphical Outputs: The main advantage of SOM over other clustering algorithms is that it combines clustering with a nonlinear projection since the clusters are organized on a grid, while preserving the topology of the original data. **SOMbrero** offers a wide range of plots aimed at giving a comprehensive overview of the resulting clusters.

All available plots are obtained using a single function `plot.somRes` (or `plot.somSC` if the visualization must be combined with a super-clustering as described below) and just two arguments handle the type of the output plot in a handy way: `what` and `type`. Argument `what` must be one of: `obs` for plotting a graphic based on the original observations, `prototypes` for plotting a graphic related to the prototypes, `energy` for plotting the evolution of the extended within-class variance during training and `add` for combining the clustering with one or several additional variables. Argument `type` sets which type of graphic is to be plotted (colors, pie charts, bars, distances...). Table 1 summarizes all available graphics for the three implemented SOM algorithms.

Table 1. Summary of plots available in **SOMbrero** 0.4-1

type	numeric			korresp			relational		
	obs	proto	add	obs	proto	add	obs	proto	add
color	x	x	x		x				x
3d		x			x				
lines	x	x	x		x			x	x
barplot	x	x	x		x			x	x
radar	x	x	x		x			x	x
boxplot	x		x						x
poly.dist		x			x			x	
umatrix		x			x			x	
smooth.dist		x			x			x	
mds		x			x			x	
grid.dist		x			x			x	
hitmap	x			x			x		
names	x		x	x			x		x
words			x						x
pie			x						x
graph			x						x

Plot types fall into two main categories: they either display the distances between prototypes (only when `what="prototypes"`), or the actual values, for each cluster, of the prototypes, observations or additional variables (for `what="prototypes"`, `what="obs"` or `what="add"`, respectively).

The plots that display the values (values of the prototypes, or average values of the observations or additional variables) are listed below:

– `color` shows the value of a single variable using a gradient of colors;
– `3d` is similar to `color` but shows the values as a three-dimensional surface plot;

- `lines` shows the values of all variables together, with lines;
- `barplot` is similar to `lines` but uses vertical bars and is limited to 5 variables at most;
- `radar` is similar to `lines` but uses radar plots;
- `boxplot` is similar to `lines` but uses boxplots and is limited to 5 variables at most. This option is used when the user wants to visualize the distribution of the variables and not only their means.

The plots that display distances between prototypes are listed below:

- `poly.dist` represents the distances between neighboring prototypes with polygons plotted for each cell of the grid [17]. The smaller the distance between a polygon's vertex and a cell border, the closer the pair of prototypes. The polygons are filled with colors indicating the number of observations in each cell;
- `umatrix` is the well known "u-matrix" [18] that plots the grid and fills the cells with colors according to the mean distance between a prototype and the neighboring prototypes;
- `smooth.dist` depicts the average distance between a prototype and its neighbors using smooth color changes;
- `mds` plots a two-dimensional Multi Dimensional Scaling projection of the prototypes;
- `grid.dist` plots all two-way grid distances (computed on the grid) against the corresponding prototype distances (computed in the input space).

Other Plots Are also Available: `type="hitmap"` displays the distribution of the observations on the map with rectangles; each cluster is represented by a rectangle with an area proportional to the number of observations it contains, [19]. Using `type="names"` or `type="words"` (the latter only available for additional variables) displays a grid of word clouds, either of observation names or of words related to the observations of each cluster. Finally, `type="pie"` or `type="graph"` can be used to display pie charts (for an additional categorical variable) or graphs (with nodes corresponding to the observations that have been clustered on the map).

Diagnostic Tools: **SOMbrero** offers additional diagnostic tools: super-clustering (the result of which can be plotted on most graphical outputs) and quality measures. The SOM algorithm often results in a large number of classes, which is not very handy for interpretation. Therefore, a common practice is to run an ascending hierarchical clustering algorithm on the prototypes of the trained map. **SOMbrero** implements this in function `superClass`. A dendrogram and a scree-plot can be drawn using function `plot.somSC`, which can guide the user's choice.

Furthermore, apart from the energy plots, **SOMbrero** offers two measures of quality for a trained SOM, through function `quality` [20]:

- the topographic error, which is the average frequency (over all observations) with which the prototype that comes second closest to an observation is not

in the direct neighborhood (on the grid) of the winner prototype. It is a real number between 0 and 1, a value close to 0 indicating good quality.
- the quantization error, which is the distance (in the input space) of observations to their assigned prototype, averaged over all observations. It is a positive real number, with a value close to 0 when projection quality is good.

3   Graphical User Interface

SOMbrero comes with a user-friendly graphical interface, which makes most of its options available in a few clicks, without resorting to the command line. The interface is programmed using the R package **shiny** [21]. It can be tested using a simple web browser (some of the features may not work with Internet Explorer or Chrome; Firefox must be preferred), and can be accessed on-line at `http://shiny.nathalievilla.org/sombrero` or in R using the command

```
sombreroGUI()
```

It is shown in Figure 1.

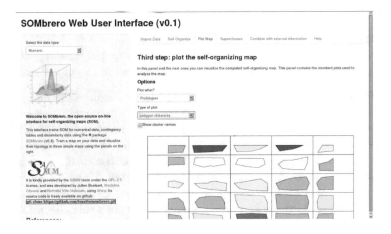

Fig. 1. Screenshot of the **SOMbrero** web user interface

The interface consists of seven panels: the left hand side panel allows the user to choose the type of SOM and gives general information and references. An "Import Data" panel is used to import a data file in csv or text format, and to set formatting options for the importation. If the data are properly imported, a preview table is shown in this panel. The "Self-Organize" panel is used to select the SOM options and train the algorithm. The "Plot Map" panel provides the different graphical outputs implemented in **SOMbrero**. The "Superclasses" panel is used to compute and to display super-classes. The "Combine with external information" panel can be used to import additional data and to display them on the map and thus to combine the results of SOM with external information. Finally, the "Help" panel contains indications about how to use the interface.

4   Examples

SOMbrero also provides five vignettes (documentation files accessible from within the package) that detail the use of the package for different types of data. Three datasets, provided with **SOMbrero**, are used to illustrate the numeric case, the Korresp case and the relational case. The present section provides a few examples of the analyses that can be performed using **SOMbrero**; we refer the reader to the package's vignettes for comprehensive illustrations.

Numeric SOM: the `iris` data set. The numeric SOM case is illustrated with the Fisher's famous *iris* dataset [22]. The following two command lines are used to train the SOM and assess the quality of the map[3]:

```
iris.som <- trainSOM(iris[,1:4])
quality(iris.som)
# $topographic          $quantization
# [1]  0.06             [1]  0.1933871
```

The results of the algorithm can be combined with the categorical variable `Species`, plotted as pie-charts in Figure 2. This graphic shows a good organization and separation of the different species on the map.

```
plot(iris.som, what= "add", variable= iris$Species)
```

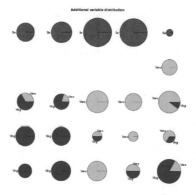

Fig. 2. Pie charts of `Species` for a SOM trained on the four numeric variables of the *iris* dataset

Korresp: French Presidential Election Data. The second example illustrates the Korresp algorithm using the **presidentielles2002** data set[4]. The data consist of a contingency table containing the number of votes for each region

---

[3] The values of these quality measures may vary between maps because of the stochastic nature of SOM training procedure.

[4] Source: "Ministère de l'Intérieur, France", http://www.interieur.gouv.fr/
content/download/1789/18734/file/Presidentielle-2002-departements.zip

(département, in rows) and each candidate (columns) in the first round of the 2002 French presidential elections.

The names plot shows the codes of the départements and the names of the candidates in their assigned cells, as in Figure 3. Departments with similar vote results are plotted close together, along with the candidates most representative of their voting profiles:

```
presi.som <- trainSOM(presidentielles2002,
                      type= "korresp")
plot(presi.som, what= "obs", type= "names")
```

The top left side of the map corresponds to the extreme right candidates, and corresponding "départements", that attracted much attention during these elections. More information about the interpretation of these results is given in the package's corresponding vignette.

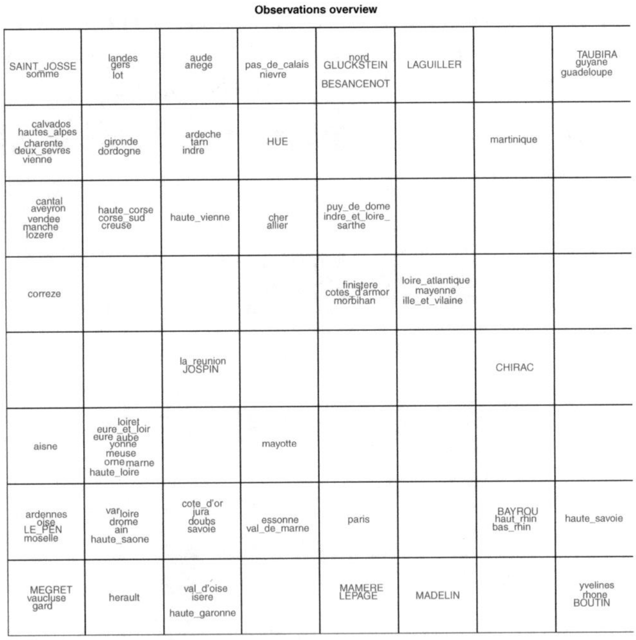

Observations overview

SAINT_JOSSE somme	landes gers lot	aude ariege	pas_de_calais nievre	nord GLUCKSTEIN BESANCENOT	LAGUILLER		TAUBIRA guyane guadeloupe
calvados hautes_alpes charente deux_sevres vienne	gironde dordogne	ardeche tarn indre	HUE			martinique	
cantal aveyron vendee manche lozere	haute_corse corse_sud creuse	haute_vienne	cher allier	puy_de_dome indre_et_loire sarthe			
correze				finistere cotes_d'armor morbihan	loire_atlantique mayenne ille_et_vilaine		
		la_reunion JOSPIN				CHIRAC	
aisne	loiret eure_et_loir eure aube yonne meuse orne marne haute_loire		mayotte				
ardennes oise LE_PEN moselle	var loire drome ain haute_saone	cote_d'or jura doubs savoie	essonne val_de_marne	paris		BAYROU haut_rhin bas_rhin	haute_savoie
MEGRET vaucluse gard	herault	val_d'oise isere haute_garonne		MAMERE LEPAGE	MADELIN		yvelines rhone BOUTIN

Fig. 3. Names plot for a Korresp analysis of French presidential elections

Relational SOM: Data Set "Les Misérables". Data set lesmis[5], is the co-appearance network of the characters in Victor Hugo's novel "Les Misérables". A dissimilarity matrix has been derived from this graph (using the shortest path lengths), which can be used as input for relational SOM.

Figure 4 shows a projection of the original graph of characters onto the SOM grid: each node in this figure represents a cluster and has an area proportional

---

[5] Source: http://people.sc.fsu.edu/~jburkardt/datasets/sgb/jean.dat

to the number of characters classified inside. The edge widths are proportional to the number of connections between the characters of the two clusters. Colors highlight superclasses computed by hierarchical clustering on the SOM proto-types. The SOM clustering, the super classes and the graph of Figure 4 are obtained with just three command lines:

```
lesmis.som  <-trainSOM(dissim.lesmis,type="relational")
lesmis.SC <- superClass(lesmis.som, k=6)
plot(lesmis.SC, what="add", type="graph", var=lesmis)
```

Fig. 4. Projection of the graph of characters' links on a relational SOM grid

5   Conclusion

The **SOMbrero** R package implements on-line algorithms of SOM for three types of data (numeric vectors, contingency tables and dissimilarity data), and provides multiple diagnostic tools, graphical outputs, as well as a handy graphical user interface. Further versions intend to include support for other types of data (multiple categorical data, multiple dissimilarity SOM), and to add training options, such as different grid topologies and weighting of the observations.

References

1. Kohonen, T.: Self-Organizing Maps, 3rd edn., vol. 30. Springer, Heidelberg (2001)
2. Cottrell, M., Letremy, P., Roy, E.: Analyzing a contingency table with Kohonen maps: a factorial correspondence analysis. In: Mira, J., Cabestany, J., Prieto, A.G. (eds.) IWANN 1993. LNCS, vol. 686, pp. 305–311. Springer, Heidelberg (1993)
3. Kohohen, T., Somervuo, P.: Self-organizing maps of symbol strings. Neurocomput-ing 21, 19–30 (1998)
4. Mac Donald, D., Fyfe, C.: The kernel self organising map. In: Proceedings of 4th International Conference on Knowledge-Based Intelligence Engineering Systems and Applied Technologies, pp. 317–320 (2000)
5. Andras, P.: Kernel-Kohonen networks. International Journal of Neural Systems 12, 117–135 (2002)

6. Villa, N., Rossi, F.: A comparison between dissimilarity SOM and kernel SOM for clustering the vertices of a graph. In: 6th International Workshop on Self-Organizing Maps (WSOM), Bielefield, Germany, Neuroinformatics Group, Bielefield University (2007)
7. Hammer, B., Hasenfuss, A.: Topographic mapping of large dissimilarity data sets no access. Neural Computation 22(9), 2229–2284 (2010)
8. Olteanu, M., Villa-Vialaneix, N.: On-line relational and multiple relational som. Neurocomputing (forthcoming, 2014)
9. Olteanu, M., Villa-Vialaneix, N., Cierco-Ayrolles, C.: Multiple kernel self-organizing maps. In: Verleysen, M. (ed.) XXIst European Symposium on Artificial Neural Networks, Computational Intelligence and Machine Learning (ESANN), Bruges, Belgium, pp. 83–88. d-side publications (2013)
10. Kohonen, T., Hynninen, J., Kangas, J., Laaksonen, J.: Som_pak: The self-organizing map program package. Technical Report A31, Helsinki University of Technology, Laboratory of Computer and Information Science (1996)
11. Yan, J.: som: Self-Organizing Map. R package version 0.3-5 (2010)
12. Hamel, L., Ott, B., Breard, G.: popsom: Self-Organizing Maps With Population Based Convergence Criterion. R package version 2.3 (2013)
13. Wehrens, R., Buydens, L.: Self- and super-organising maps in r: the kohonen package. J. Stat. Softw. 21(5) (2007)
14. Rossi, F.: yasomi: Yet Another Self Organising Map Implementation. R package version 0.3/r39 (2012)
15. Ritter, H., Martinetz, T., Shulten, K.: Neural computation and Self-Organizing Maps, an Introduction. Addison-Wesley (1992)
16. Fort, J., Letremy, P., Cottrell, M.: Advantages and drawbacks of the batch kohonen algorithm. In: Verleysen, M. (ed.) Proceedings of 10th European Symposium on Artificial Neural Networks (ESANN 2002), Bruges, Belgium, pp. 223–230 (2002)
17. Cottrell, M., de Bodt, E.: A Kohonen map representations to avoid misleading interpretations. In: Verleysen, M. (ed.) Proceedings of ESANN 1996, D Facto, Bruxelles, pp. 103–110 (1996)
18. Ultsch, A., Siemon, H.: Kohonen's self organizing feature maps for exploratory data analysis. In: Proceedings of International Neural Network Conference, INNC 1990 (1990)
19. Vesanto, J.: Data Exploration Process Based on the Self–Organizing Map. PhD thesis, Helsinki University of Technology, Espoo (Finland), Acta Polytechnica Scandinavica, Mathematics and Computing Series No.115 (2002)
20. Polzlbauer, G.: Survey and comparison of quality measures for self-organizing maps. In: Paralic, J., Polzlbauer, G., Rauber, A. (eds.) Proceedings of the Fifth Workshop on Data Analysis (WDA 2004), Sliezsky dom, Vysoke Tatry, Slovakia, pp. 67–82. Elfa Academic Press (2004)
21. RStudio, Inc.: shiny: Web Application Framework for R. R package version 0.6.0 (2013)
22. Becker, R., Chambers, J., Wilks, A.: The New S Language. Wadsworth & Brooks/Cole (1988)

K-Nearest Neighbor Nonnegative Matrix Factorization for Learning a Mixture of Local SOM Models

David Nova, Pablo A. Estévez, and Pablo Huijse

Department of Electrical Engineering and Advanced Mining Technology Center,
University of Chile, Casilla 412-3, Santiago, Chile
dnovai@ug.uchile.cl, pestevez@ing.uchile.cl, pablo.huijse@gmail.com

Abstract. In this work we present a modified Nonnegative Matrix Factorization (NMF) method for learning a mixture of local SOM models. The proposed method approximates a data point with a linear combination of its k-nearest neighbor prototypes. This allows obtaining a low quantization error and at the same time keeping the interpretability of the prototypes. The results of the new method are compared with those obtained using non-negative least squares, NMF and SOM, using four benchmark data sets. Two metrics are used to assess the performance of the different approaches. The proposed k-nn NMF method obtained the lowest relative local quantization error, while keeping a global quantization error similar to the best alternative methods.

Keywords: Self Organizing Map, Nonnegative Matrix Factorization, k-Nearest Neighbors, Linear Mixture, Relative Quantization Error.

1 Introduction

The Self-Organizing Map (SOM) is a popular unsupervised neural network method widely applied to data exploration and clustering [7,9]. Let $X \in R^{m \times n}$ be a data matrix, where m is the dimension and n the number of samples; and let $W \in R^{m \times r}$ be a dictionary matrix, where r represents the number of prototypes and each column vector is defined as $\mathbf{w}_i = [w_1, ..., w_m]_i^T$. The best matching unit (BMU) for a given input $\mathbf{x}$ is defined as follows:

$$i^* = \arg \min_{i=1,...,r} = \|\mathbf{x} - \mathbf{w}_i\|^2. \tag{1}$$

The SOM performs a vector quantization, representing the input data by a finite set of prototypes (models). In addition, the prototypes are associated with the nodes of an output grid (usually 2D), with the aim of preserving the topographic relationships of the data. SOM gives visualization and topological interpretability to the underlaying structure of the data [9]. The quantization error (QE) of an input vector is the Euclidean norm of the difference of the input vector and the BMU, $\|\mathbf{x} - \mathbf{w}_{i*}\|$. As the sum of the squared QEs is directly

T. Villmann et al. (eds.), *Advances in Self-Organizing Maps and Learning Vector Quantization*, Advances in Intelligent Systems and Computing 295,
DOI: 10.1007/978-3-319-07695-9_22, © Springer International Publishing Switzerland 2014

minimized by the k-means algorithms, but not by SOM, the latter algorithm usually gets higher QE than k-means. Some exceptions to this rule have been studied by Kohonen [10]. SOM may obtain a lower QE than k-means only if the number of training samples per model is small and if the effective dimensionality of the data vectors is sufficiently high [10].

In [8], Kohonen extended the use of SOM, by approximating the input data more accurately using an optimized linear mixture of prototypes, instead of using a single BMU. Kohonen's approach employs a mixture of any subsets of models (prototypes), i.e., it does not consider the topological structure of the data. As a consequence the topological interpretability of the fitting is lost. Kohonen's linear mixtures of SOM models is useful for certain application such as text analysis, where the task is to find out whether a text comes from different sources. We are interested here in local approximations of data points using their k-nearest neighbor prototypes. A possible application is modeling and predicting complex time series. In [17,18] the SOM is trained to minimize the prediction error and local models are used to improve its performance.

In this work, we introduce a method to represent data samples by local linear combinations of SOM prototypes based on Nonnegative Matrix Factorization (NMF). NMF has been successfully applied to clustering and dimensionality reduction tasks [19]. This technique finds a lower rank approximation of a nonnegative matrix X, such as

$$X \approx WH^T, \tag{2}$$

where W is defined as the dictionary matrix and H is the coefficient matrix. The dictionary and coefficient matrices are element-wise nonnegative. The nonnegative constraint forces the decomposition to be purely additive, part-based and sparse, achieving a compact representation of the data. NMF is related to clustering techniques such as K-means [3] for nonnegative data, and to dimensionality reduction methods such as Principal Component Analysis (PCA) [21]. As we know, PCA finds global and dense decompositions that might be hard to interpret. NMF enhances interpretability due to the nonnegativity constraint, which is crucial in some real-world applications such as: image pixel values, chemical compound concentration, web patterns, text analysis, just to name a few.

A related work is the Sparse Coding Neural Gas (SCNG) [11,12], where overcomplete data representations are learned. In a first step, the original neural gas algorithm is combined with Oja's rule [16], in order to represent each training sample by a single basis vector. Then this algorithm is generalized, by using Orthogonal Matching Pursuit to represent each training sample by a linear combination of basis elements. The SCNG algorithm was applied to natural images, obtaining band-pass like basis elements localized in space and orientation.

In this work, we introduce a method which represents the data point by employing a subset of prototypes of the dictionary matrix, which are the k-nearest neighbor prototypes in the data space. This approach keeps the quality of SOM of preserving the structure of the data, and gives intuitive descriptions of the patterns by using nonnegative constraints. These properties allow us to interpret the resulting representation of the data points.

The remainder of this paper is organized as follows: In section 2, a background on NMF and the linear mixture of SOM models is presented. In section 3, the proposed method is introduced. In section 4, the results using four benchmark data sets are shown. Finally, in section 5 the conclusions are drawn.

2   Nonnegative Matrix Descomposition

In this section a brief background on nonnegative matrix decompositions is provided.

2.1   Linear Mixture of Models

In [8] Teuvo Kohonen introduced a method to describe an input pattern as a mixture of SOM models (prototypes). This method uses the prototypes determined by SOM as a fixed dictionary matrix $W \in R^{m \times r}$ where r is the number of prototypes and m is the dimension of the input pattern. A matrix of coefficients $H \in R^{n \times r}$, where n is the number of samples, is adjusted in order to find linear mixtures of models that reduce the fitting error (3). Both W and H are nonnegative matrices. In practice, a separate optimization procedure is performed for each data sample. The objective function to minimize is:

$$E = \sum_{\mathbf{x}} \|\mathbf{x} - W\mathbf{h}\|_2^2, \tag{3}$$

where $\mathbf{x} \in R^m$ is a column vector corresponding to a given data sample, $\mathbf{h} = [h_1, h_2, ..., h_r]^T$ is a column vector of coefficients to be adjusted. A least square optimization is done for each data point which corresponds to a quadratic optimization problem based on the *Kuhn-Tucker theorem* [3]. This may be solved by using the Matlab[1] function *lsqnonneg*.

2.2   Nonnegative Matrix Factorization

Nonnegative Matrix Factorization [19] provides a low rank approximation of a nonnegative matrix of data $X \in R^{m \times n}$, where each row represents a feature and each column represents an observation or sample. Let r be an integer such that $r < \min\{m, n\}$. This method finds two nonnegative matrices: $W \in R^{m \times r}$ which corresponds to a dictionary matrix and $H \in R^{n \times r}$ which is the coefficient matrix. The NMF approach solves the following optimization problem:

$$\min_{W,H} f_r(W, H) \equiv \frac{1}{2} \|X - WH^T\|_F^2, \tag{4}$$

subject to $W, H \geq 0$, i.e., W and H are element-wise nonnegative, $\|\cdot\|_F$ denotes Frobenius norm, and the index r in f_r stands for the number of dictionary elements used for obtaining a compact representation of the data matrix X.

---

[1] Available on: http://www.mathworks.com/

2.3   Sparse Nonnegative Matrix Factorization

In order to achieve sparsity in the solution of eq. (4) , the L_1-norm regularization [4,6] in the H coefficient matrix can be used:

$$\min_{W,H} \frac{1}{2} \left[\left\| X - WH^T \right\|_F^2 + \beta \sum_{j=1}^{n} \left\| H(:,j) \right\|_1^2 \right], \tag{5}$$

where β is the parameter control of the trade-off between the accuracy of the approximation and the sparseness of H. Note that for $\beta = 0$ the formulation is equal to eq. (4), while a large value of β implies high sparseness.

3   K-Nearest Neighbor Nonnegative Matrix Factorization

Here we propose a method called k-nearest neighbor Non-negative Matrix Factorization (k-nn NMF), where the k-nearest neighbor prototypes of each sample are used to learn a mixture of local models. The set of prototypes is determined previously by using conventional SOM. These prototypes are kept fixed and used as the dictionary matrix W in the training procedure. The objective function is as follows:

$$E = \left\| X - WS^T \right\|_F^2, \tag{6}$$

where $X \in R^{m \times n}$ is the data matrix, $W \in R^{m \times r}$ is the fixed dictionary matrix (or matrix of SOM prototypes), $S = H \otimes \Gamma$, where $\otimes$ stands for the point-wise multiplication; and $H \in R^{n \times r}$ is the coefficient matrix that weights the prototypes in the linear combination. Γ is a mask matrix having $1's$ only for the k-nearest neighbor prototypes of each sample. Usually, an Alternate Nonnegative Least Square [5] is used to solve the optimization problem in NMF to adjust both W and H. But, in our case a multiplicative update rule was used [13], in order to adjust the matrix H by solving:

$$\min_{H \geq 0} \left\| X - WS^T \right\|_F^2,$$

while the dictionary matrix W and the mask matrix Γ are fixed. The update rule for $S = H \otimes \Gamma$ is obtained as follows:

$$S \leftarrow S - \eta_H \nabla f_H(W, S),$$
$$= S - \frac{S}{\nabla_+} \cdot (\nabla_+ - \nabla_-)$$
$$= +S \cdot \frac{\nabla_-}{\nabla_+}$$
$$= S \otimes \left(W^T X \right) \oslash \left(W^T W S \right), \tag{7}$$

where $\nabla_+ = W^T W S$, $\nabla_- = W^T X$ and $\eta_H = \frac{S}{\nabla_+}$. The notation $\oslash$ stands for the point-wise division. In our algorithm, a threshold of the error variance was used as a stop criterion.

The algorithm is summarized as follows:

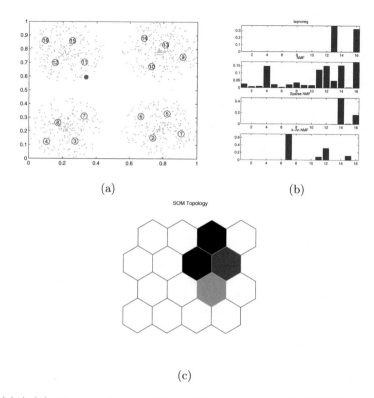

(a)   (b)

SOM Topology

(c)

Fig. 1. (a) A didactic example of the Simple Cluster data set with 16 SOM prototypes, the selected data point is represented by a red dot at the top left cluster. (b) The coefficient column vector **h** obtained for each method : lsqnonneg, NMF, Sparse NMF and k-nn NMF, from top to bottom. (c) SOM topology (magnitudes of coefficients are shown in gray scale) of the k-nearest neighbor used for reconstructing the selected data point, the darker the node the higher its relevance in the linear combination. The colored cells in the SOM map represent the k-nearest prototypes in the data space.

1. Execute the SOM algorithm to obtain a dictionary matrix W (set of proto-types).
2. Initialize randomly the elements of the nonnegative coefficient matrix H in the range $[0, 1]$.
3. Compute the mask matrix Γ, by finding the k-nearest neighbor prototypes for each data point **x**.
4. Solve $\min_{H \geq 0} \left\| X - W S^T \right\|_F^2$
   (a) Update the matrix S according to eq. (6),
   (b) $t \leftarrow t + 1$.
   (c) If $t > t_{max}$ or the stop criterion is reached go to step 5, otherwise go back to step 4.
5. End

In Fig. 1 a didactic example is shown where the different methods are compared in a data set having four clusters. In Fig. 1(a) a red dot in the top left cluster represents a selected data point. Fig. 1(b) shows the coefficient vectors obtained to represent the data point using four different methods (lsqnonneg, NMF, Sparse NMF and k-nn NMF). In Fig. 1(c) a SOM map is shown where the colored cells represent the k-nearest prototypes to a selected point. The k-nn NMF gives a local approximation of the data point using the k-nearest neighbor prototypes to represent it. With $k = 4$ the nearest neighbor prototypes are 7, 11, 12 and 15. Note that although the lsqnonneg and sparse NMF obtained sparse representations, they included prototypes that are far away in the map, and without using the closest prototype (BMU).

4   Results

In order to compare the different methods previously mentioned (SOM, lsqnonneg, NMF, Sparse NMF and k-nn NMF), four benchmark clustering data sets were used. The first data set is Simple Cluster which consists of four two-dimensional Gaussian distributions with 1000 samples. The next three data sets are taken from [20]: Engy Time which consists of 2 two-dimensional overlapped Gaussian distributions with 4096 samples; Two Diamonds which consists of 2 two-dimensional diamond shape distributions with 800 samples; and L-sun which consists of three distributions with different variances and inter cluster distances with 400 samples.

For comparison purposes two metrics are used. The first one is associated with the quantization error which allows measuring the quality of the L-2 norm reconstruction. The metric uses the following expressions:

$$SC_{within} = \sum_{x} \left\| \mathbf{x} - W\mathbf{h}_{c(\mathbf{x})} \right\|^2, \tag{8}$$

and

$$SC_{total} = \sum_{x} \left\| \mathbf{x} - \bar{\mathbf{x}} \right\|^2, \tag{9}$$

where W is the dictionary matrix, $\mathbf{h}_{c(\mathbf{x})}$ is the coefficient column vector belonging to the coefficient matrix H which better approximates the data point $\mathbf{x}$, and $\bar{\mathbf{x}}$ is the mean of the data points. Thus, the relative quantization error [1] (RQE) is defined as

$$RQE = \frac{SC_{within}}{SC_{total}}. \tag{10}$$

In addition, another metric is defined in order to measure the quantization error when using the k-nearest neighbor prototypes to approximate every data point x. This allows measuring the quality of the local reconstruction. This metric is defined as follows:

$$SC_{loc} = \sum_{\mathbf{x}} \frac{1}{|\mathcal{V}(c(\mathbf{x}))|} \left\| \mathbf{x} - \sum_{k \in \mathcal{V}(c(\mathbf{x}))} \mathbf{w}_k \mathbf{h}_{c(\mathbf{x})} \right\|^2, \tag{11}$$

Table 1. Mean (Standard Deviation) of the Relative Quantization Error and Relative Local Quantization Error using 5 runs for: Simple Cluster, Engy Time, Two Diamonds and Lsun data sets, with $k = 4$ nearest neighbor prototypes.

Data Set	Metric	SOM	lsqnonneg	NMF	sNMF	k-nn NMF
Simple Cluster	RQE	0.0076	**4.4871e-4**	**4.4872e-4**	0.0024	9.3523e-4
		(5.6304e-6)	**(4.5444e-5)**	**(4.5439e-5)**	(2.1923e-5)	(1.3659e-4)
	RQE_{loc}	-	0.0252	0.0532	0.0240	**2.3381e-4**
		-	(9.7696e-4)	(7.181e-4)	(3.2107e-4)	**(3.4148e-5)**
Engy Time	RQE	0.0092	**5.8137e-4**	**5.8137e-4**	0.0059	8.7729e-4
		(3.0939e-7)	**(8.3724e-6)**	**(8.3724e-6)**	(2.1399e-5)	(2.4427e-5)
	RQE_{loc}	-	0.1216	0.1127	0.0971	**2.1932e-4**
		-	(0.0035)	(3.5757e-4)	(7.0013e-4)	**(6.1067e-6)**
Two Diamonds	RQE	0.0085	**1.8693e-4**	**1.8693e-4**	0.0036	5.2390e-4
		(4.0711e-5)	**(1.0951e-5)**	**(1.0951e-5)**	(2.3328e-4)	(4.5007e-5)
	RQE_{loc}	-	0.1096	0.0758	0.0561	**1.3098e-4**
		-	(0.0032)	(0.0011)	(0.0051)	**(1.1252e-5)**
Lsun	RQE	0.0065	**3.8397e-4**	**3.8397e-4**	0.0026	5.4354e-4
		(3.7278e-5)	**(3.5907e-5)**	**(3.5908e-5)**	(2.4131e-5)	(2.4881e-5)
	RQE_{loc}	-	0.321	0.394	0.0324	**1.3588e-4**
		-	(0.0037)	(0.0011)	(0.0043)	**(6.2202e-6)**

where $\mathbf{w}_k$ is the k-th nearest neighbor prototype of the data point x, and $\mathcal{V}$ is the set of the k-nearest neighbor prototypes of the data point $\mathbf{x}$. The relative local quantization error for reconstruction is defined as follows:

$$RQE_{loc} = \frac{SC_{loc}}{SC_{total}}. \tag{12}$$

A low RQE_{loc} value indicates that the reconstruction prioritizes the use of neighboring prototypes. This allows incorporating the topological knowledge obtained with the SOM procedure used to get W. The following methods are compared: SOM, lsqnonneg, NMF, sparse NMF (sNMF), k-nn NMF. The number of prototypes is set to $N = 16$, a maximum number of epochs $t_{max} = 1000$ and a sparsity parameter $\beta = 0.001$ were used in all experiments.

Table 1 shows the RQE and RQE_{loc} values (mean and standard deviation) obtained with the four data sets for $k = 4$ nearest neighbor prototypes. In all cases the original SOM has the highest RQE since this method uses only the closest prototype to represent a data point. It has no meaning to compute RQE_{loc} for SOM because $k = 1$ always. The RQE value obtained by SOM can be used as a reference value against which to compare all other figures. It can be seen that all the different methods of mixture models are able to reduce the RQE with respect to SOM. This means that linear mixture models are by far more accurate than single prototype methods.

In all cases both lsqnonneg and NMF obtained the lowest RQE, followed closely by our proposed method k-nn NMF. A multi-comparison statistical test with a Bonferroni correction was performed for $\alpha = 0.005$. This test reveals that

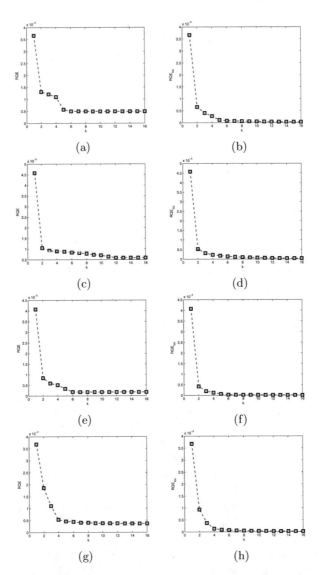

Fig. 2. Relative Quantization Error RQE (a, c, e, g), and Relative Local Quantization Error RQE_{loc} (b, d, f, h) using k-nn NMF as a function of k. Each row from top to bottom represents a different data set: Simple Cluster, Engy Time, Two Diamonds and Lsun.

for all four experiments both lsqnonneg and NMF mean RQE values are not statistically different. In the case of sNMF a trade-off exists between accuracy and sparseness, which yields a relatively high RQE with respect to lsqnonneg and NMF. On the other hand, k-nn NMF obtained slightly higher RQE values than those obtained by lsqnonneg and NMF, and the mean difference is statistically significant. This result is expected because lsqnonneg and NMF use all prototypes in the linear combination, but k-nn NMF restricts it to only the k-nn prototypes. In Table 1, the results for $k = 4$ are shown, but it is easy to check that when k grows larger the RQE value of k-nn NMF approaches the RQE value of NMF.

As expected the lowest RQE_{loc} values were obtained by k-nn NMF in all four experiments, and these values are significantly different than those of all other methods according to the multi-comparison statistical test with a 95% degree of confidence. By design k-nn NMF optimizes the linear combination of the k-nn prototypes to approximate data, while all the other methods are global.

In Fig. 2, the first column of subfigures show the RQE obtained by k-nn NMF as a function of k for the four data sets. These subfigures illustrate how the global fitting error is reduced when k increases. It can be seen that as k grows, the k-nn NMF RQE values converge to those obtained by NMF (Table 1). In Figure 2, the second column of subfigures show the RQE_{loc} values obtained by k-nn NMF as a function of k. These subfigures show the local fitting error as a function of k.

5 Conclusions

In all the tests knn-NMF demonstrated a good performance in terms of using the k-nearest neighbor prototypes for reconstructing the data, while achieving on-par performance to lsqnonneg and NMF in terms of the global reconstruction of the data. The k-nn NMF method allows a better topological interpretability because only the nearest neighbors provide information for reconstructing the data. This is in line with an ideal SOM map representation. The proposed method could be easily combined with prototypes obtained by Neural Gas [14] or Growing Neural Gas [2]. In addition, this method could be extended to supervised learning, where we can select the appropriate elements of the dictionary matrix (e.g. Learning Vector Quantization prototypes [15]) to represent particular data points, e.g., data points near the border between classes. Also, one might not always want to get the lowest reconstruction error of the data, as this can lead to an over-fitting of the model, losing the ability to generalize and failing to classify new points. The proposed method may help to understand better the representation of the data using the knowledge of its topological structure.

Acknowledgments. This research was supported by CONICYT-Chile under grant Fondecyt 1140816.

References

1. Côme, E., Cottrell, M., Gaubert, P.: Professional trajectories of workers using disconnected self-organizing maps. In: Estevez, P.A., Principe, J.C., Zegers, P. (eds.) Advances in Self-Organizing Maps. AISC, vol. 198, pp. 295–304. Springer, Heidelberg (2013)
2. Fritzke, B., et al.: A growing neural gas network learns topologies. Advances in Neural Information Processing Systems 7, 625–632 (1995)
3. Hartigan, J.A., Wong, M.A.: Algorithm as 136: A k-means clustering algorithm. Journal of the Royal Statistical Society. Series C 28(1), 100–108 (1979)
4. Kim, H., Park, H.: Sparse non-negative matrix factorizations via alternating non-negativity-constrained least squares for microarray data analysis. Bioinformatics 23(12), 1495–1502 (2007)
5. Kim, H., Park, H.: Nonnegative matrix factorization based on alternating nonnegativity constrained least squares and active set method. SIAM Journal on Matrix Analysis and Applications 30(2), 713–730 (2008)
6. Kim, J., Park, H.: Sparse nonnegative matrix factorization for clustering. Technical Report Technical report GT-CSE-08-01, Georgia Institute of Technology (2008)
7. Kohonen, T.: Self-organizing maps. Springer-Verlag New York, Inc., Secaucus (1997)
8. Kohonen, T.: Description of input patterns by linear mixtures of som models. In: Proceedings of WSOM, Bielefeld, Germany (2007)
9. Kohonen, T.: Essentials of the self-organizing map. Neural Networks 37, 52–63 (2013)
10. Kohonen, T., Nieminen, I.T., Honkela, T.: On the quantization error in SOM vs. VQ: A critical and systematic study. In: Príncipe, J.C., Miikkulainen, R. (eds.) WSOM 2009. LNCS, vol. 5629, pp. 133–144. Springer, Heidelberg (2009)
11. Labusch, K., Barth, E., Martinetz, T.: Sparse coding neural gas for the separation of noisy overcomplete sources. In: Kůrková, V., Neruda, R., Koutník, J. (eds.) ICANN 2008, Part I. LNCS, vol. 5163, pp. 788–797. Springer, Heidelberg (2008)
12. Labusch, K., Barth, E., Martinetz, T.: Sparse coding neural gas: Learning of overcomplete data representations. Neurocomputing 72(7), 1547–1555 (2009)
13. Lin, C.J.: On the convergence of multiplicative update algorithms for nonnegative matrix factorization. IEEE Trans. on Neural Networks 18(6), 1589–1596 (2007)
14. Martinetz, T.M., Berkovich, S.G., Schulten, K.J.: Neural-gas' network for vector quantization and its application to time-series prediction. IEEE Trans. on Neural Networks 4(4), 558–569 (1993)
15. Nova, D., Estévez, P.A.: A review of learning vector quantization classifiers. Neural Computing and Applications (2014), doi:10.1007/s00521-013-1535-3
16. Oja, E.: Simplified neuron model as a principal component analyzer. Journal of Mathematical Biology 15(3), 267–273 (1982)
17. Principe, J.C., Wang, L.: Non-linear time series modeling with self-organization feature maps. In: Proceedings of the 1995 IEEE Workshop on Neural Networks for Signal Processing V, pp. 11–20. IEEE (1995)
18. Principe, J.C., Wang, L., Motter, M.A.: Local dynamic modeling with self-organizing maps and applications to nonlinear system identification and control. Proceedings of the IEEE 86(11), 2240–2258 (1998)
19. Seung, D., Lee, L.: Algorithms for non-negative matrix factorization. Advances in Neural Information Processing Systems 13, 556–562 (2001)
20. Ultsch, A.: Clustering with som: U*c. In: Proceedings of WSOM, pp. 75–82 (2005)
21. Wold, S., Esbensen, K., Geladi, P.: Principal component analysis. Chemometrics and Intelligent Laboratory Systems 2(1), 37–52 (1987)

Comparison of Spectrum Cluster Analysis with PCA and Spherical SOM and Related Issues Not Amenable to PCA

Masaaki Ohkita[1], Heizo Tokutaka[1],
Kazuhiro Yoshihara[2], and Matashige Oyabu[3]

[1] SOM Japan Inc.,
[2] Omicron Nano Technology Inc.
[3] Kanazawa Institute of Technology, Japan
mohkita111@yahoo.co.jp,
tokuhema@hal.ne.jp

Abstract. A chemical spectral data set was analyzed with the PCA method and the SSOM method. The results are in agreement albeit that the PCA method is only valid when the number of spectral dimensionalities is small. This is not the case with the SSOM method. In the present paper, the data of the AES depth profile, where Sn was plated on Cu, and which has a high dimensionality, is also analyzed. These results show the excellence of the SSOM method.

Keywords: PCA, Spherical SOM, cluster analysis.

1   Introduction

Multi-dimension data can be visualized in a low (usaually two) dimensional space with an array of techniques such as Principal Component Analysis (PCA), Self-Organizing Map (SOM), Multidimensional scaling (MDS), etc. In the present paper, we use PCA and SOM to analyze and compare the characteristics of TOF-SIMS spectrum data and high-dimensional Auger Electron Spectroscopy (AES) data. The Time-of-Flight Secondary Ion Mass Spectrometry (TOF-SIMS) is a surface-sensitive analytical method that relies on a pulsed beam of a primary ion (such as Ar^+, Cs^+, $O2^+$, $C60^+$, or microfocused Ga^+, In^+, Aun^+, Binx) with several kiloelectronvolts of energy to eject and ionize material from the uppermost layers of the sample. The actual desorption or sputtering of the material from the surface is the result of collision cascades or correlated atomic motions in the solid, initiated by the primary ion imprinting on the sample surface. A small fraction of the sputtered material is ionized during the emission process. The resulting atomic and molecular secondary ions, which are characteristic of the surface chemistry, are accelerated into a mass spectrometer, where they become mass analyzed by measuring their time-of-flight from the sample surface to the detector [1]. When organic matter is analyzed by TOF-SIMS, many molecular species will be detected. In that case, it is both important and

T. Villmann et al. (eds.), *Advances in Self-Organizing Maps and Learning Vector Quantization*, Advances in Intelligent Systems and Computing 295,
DOI: 10.1007/978-3-319-07695-9_23, © Springer International Publishing Switzerland 2014

practical to classify the latter smoothly and quickly. To this end, a multivari-able evaluation method is used to analyze the different molecular species. In the present paper, in support of a multivariable evaluation, the Principal Component Analysis (PCA) and the Self-Organizing Maps (SOM) [2–6] are evaluated and clusters identified. The data of the PET film, which was obtained by TOF-SIMS [1, 2], is visualized with both techniques and their clustering results compared. A PET film consists of polyethylene terephthalate a thin, stiff, thermally stabilized polyester material with many different applications. A clustering method that also visualizes the multidimensional data set has been proposed [3–7]. There, the phase distance of the labeled data was computed on the spherical surface and a dendrogram created by calculating the distance between the labels. Then, based on this dendrogram, the members of the same cluster were represented on the spherical surface. Also position relations of the clusters can be visualized by the U-matrix, the spherical surface with the different Glyph values and by coloring the clusters. Besides, we can emphasize that the dendrogram can be constructed by using the distance among the labels on the distorted spherical surface via the group average method. At first this method is applied to the TOF-SIMS spectrum analysis for comparison with PCA. Next this method is applied to the AES depth profile data where Sn was plated on Cu. This time, the divisions of the spectrum are quite large leading to a 151 dimensional data set. Therefore, with the PCA method it is impossible to visualize the data only by using, e.g., the 1st to the 3rd principal components.

2 TOF-SIMS Spectrum Analysis with PCA and Spherical SOM Method

2.1 TOF-SIMS Spectrum Analysis with PCA

There are 26 molecular species identified. Several typical spectra are shown in Fig. 1

When some factors can be interpreted, a principal component analysis (PCA) does not treat them independently, i.e., one by one, but considers them as a whole. When there are p variables, then maximally p principal components can be taken. The PCA is a technique to represent the data in a new coordinate system (z) of which the axes are mutually uncorrelated. When considering less than p axes, a subspace is obtained. The number of axes is chosen so as to minimize the loss of information when representing the original data (x) in that subspace. For example, as shown in Fig. 2, when two dimensional data are given, a one dimensional straight line can be drawn so that the amount of information loss is as small as possible, and the original two dimensional data can be denoted by their values on the straight line. Thus, two dimensional data can be taken as one dimensional data. In order to have the number of dimensions as small as possible, a variance-covariance matrix is made based on the dispersion of data. And the axis of the corresponding eigenvector is made into the principal components (the 1st and 2nd ...) in order with a large eigenvalue of the matrix. Then, the axes can be selected to minimize the loss of the amount of information.

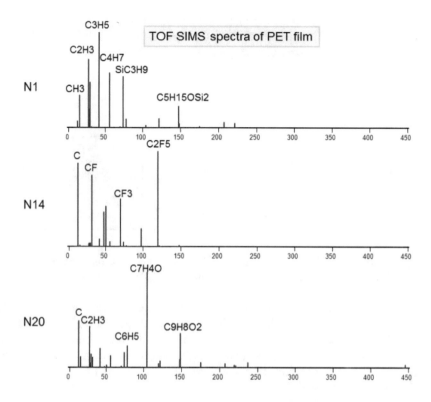

Fig. 1. The original data of TOF-SIMS

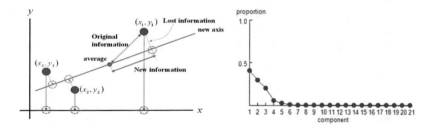

Fig. 2. Left panel: logic behind the principal component ; right panel: eigenvalues for our case, ranked according to their magnitude

2.2   Result of TOF-SIMS Spectrum Analysis with PCA

The value on the axis of the 1st principal component is calculated from the following expression 0.1910[C] 0.1302[CH3] 0.2489[C2H3] 0.1581[Si] 0.1792[C2H5] 0.2633[CF] with [ ] denoting the intensity of each variable of each sample. The

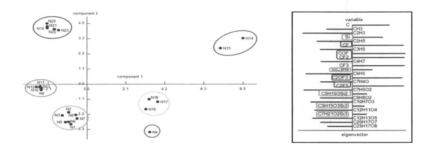

Fig. 3. The result by the 1st, and the 2nd principal component analyses. The N*i the left panel represent data names(thus, not chemical component names).

coordinates of the 1st principal component is shown on the horizontal line in the figure. It was found that the 1st principal component gives a high score for the spectrum in that the molecules containing Si and CF were observed. The 2nd principal component gives the spectrum which contains Si a high score and, conversely, demotes the spectrum that contains CF. The values of the 1st and 2nd principal components are calculated based on the previous expression for every sample, and the result displayed on two dimensions in Fig. 3. It is clear that 26 samples can be classified into six clusters. However, judging from the graph that shows the magnitude of the eigenvalues, three principal components suffice. Hence, three principal components are displayed in Fig.4. When considering the 3rd principal component, N16, N17 and N18 are separated into two groups. Then, it can be found that the spectra are classified into seven groups. Incidentally, the 3rd principal component gives a high score for the spectra that contains Si and a reduced score for the spectra that contains C-H -. The number of axes in PCA is determined by compromising between the number of components and the incurred loss of information. In this experiment, the first 3 components had quite large proportions compared to those of the other components. Therefore, 3 axes were selected for display.

2.3   Result of TOF-SIMS Spectrum Analysis with Self-Organizing Map (SOM)

The originally 26 components of Fig.1 are represented by line spectra. We utilize the spherical SOM for spectrum analyses because of its improved learning accuracy and clustering performance. We trained the SSOM with the tool described in [2–4] considering the following conditions: The number of learning nodes is

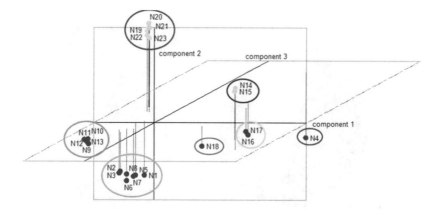

Fig. 4. Result when considering also the 3rd principal component

642 in SSOM. The number of learning iterations is 500 times the number of input data. The neighborhood function is a Gaussian of which the radius is reduced linearly during learning. The normalized spectra and the original data are shown in Fig.5. Analyzing the normalized data of Fig.5 with the spherical SOM method leads to the results shown in Figs.6 and 7. The color-display for each group is shown below.

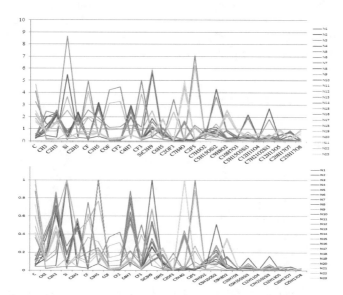

Fig. 5. Top: original 23 data of Fig 1 displayed by broken line graph. The vertical scale is intensity. Bottom: The same data but normalized.

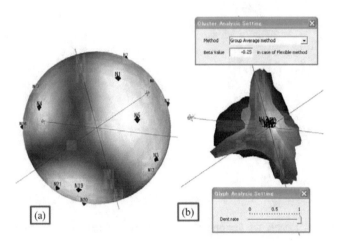

Fig. 6. (a) The learning result of the data in Fig.5 (bottom) (b) When a spherical surface was distorted at Glyph value 1 and by the group average-method using the distance of the label coordinates, the clustering result is shown in Fig.7

It can be seen that N1, N2, N3, N5, N6, N7 and N8 belong to the same group for the spherical SOM (upper left). Next, it is found that N9, N10, N11, N12, and N13 belong to the same group as well. N4 is isolated from the spherical SOM as shown in the upper right. It can be seen that N14 and N15 belong to the same group in the lower right. And so on. The spectra of G11, G12, G21, and G22 only roughly correspond to each other.

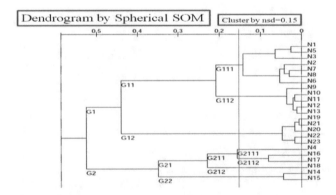

Fig. 7. The clustering outcome is selected at a distance (nsd=0.15) for which the same number of clusters is obtained as with PCA; nsd means the non-similar distance

The spectra of four groups from G11 to G22 in Fig.7 were compared in Fig. 8 The precision for the matching of the grouping by the spectra is as follows: G21¿G11¿G22=G12.

2.4   Comparison of PCA and Spherical SOM Results

The result of PCA is shown in Fig.4 When not considering the 3rd principal component, N16, N17 and N18 are not separated. For the SOM, grouping means that a dendrogram is obtained without being concerned with the issue how to select the number of principal components. When considering the dendrogram of Fig.7, for cluster distance 0.15, the clustering results agrees with that of PCA. SOM gathers resembling data and can bundle them as the clusters. Then, the clusters can be carried out by considering a difference among the groups. The spherical SOM software which was used here emphasizes distance among the

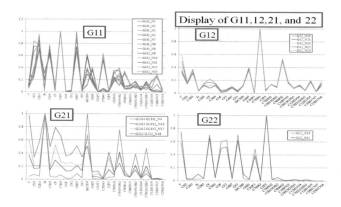

Fig. 8. Comparison among the spectra of groups G11, G12, G21 and G22 in Fig.7

clusters and displays it as shown in Fig.6. The separation of each cluster can be clearly seen. The user judges the distance among the clusters. Thus, a final grouping can be completed.

2.5   Summary of Subsection 2

- The principal component analysis (PCA) and Self-Organizing Maps (SOM) are valid tools for clustering the spectra of TOF-SIMS.
- As for the clustering of the spectra, almost identical results were obtained with PCA and SOM.
- PCA differs in the information that is obtained for different numbers of principal components considered but there is not such the difference with the SOM. This becomes even more of an issue when the number of principal components becomes equal to four or more.
- Seven groups which were obtained in PCA of Fig.4 agreed very well with the cluster labels with nsd=0.15 of the dendrogram of Fig.7 which was obtained in SSOM.
- The spherical SOM is excellent for displaying results without being influenced by the purpose of the analysis. It can easily classify a spectrum.

3   Clustering of Other Spectral Data with Spherical SOM

3.1   Data Preparation and Analysis Procedure

The spectra [6] in 410-440 eV of the energy range with 151 divisions were used for
the analysis. Each spectrum was normalized at the line with the maximum and
the minimum in the above energy range as shown in Fig.9. Then, the analyses
are detailed.

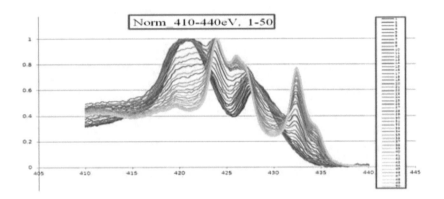

Fig. 9. Normalized spectra of 410 - 440 eV of energy ranges numbered from 1-50

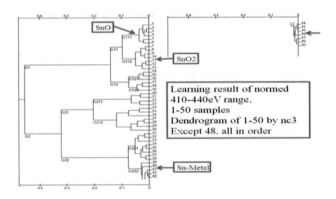

Fig. 10. The dendrogram is arranged sequentially in 1-50 spectra only, except for the
48th spectrum (right panel, position marked by an arrow). The positions of SnO, SnO2,
and Sn-Metal as the standard material are shown by an arrow.

There is only here a portion for which the order is garbled. After SOM learning 1-50 spectra, the positions of SnO, SnO2 and Sn-Metal are located on the
spherical surface. SnO is in the G111 group. SnO2 agrees with the 11th spectrum.

Sn-Metal agrees with the 41st spectrum. The normalized spectra of 1-50 and a standard one of SnO, SnO2 and Sn-Metal are the spherical surface and further described.

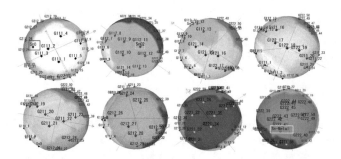

Fig. 11. From the upper left to the lower right, the each colored group from G111 to G222 is sequetially arranged. SnO belongs to G111 and SnO2 is on the sample number 11. Also, Sn Metal belongs to G222 in the lower right and is shown for sample 41.

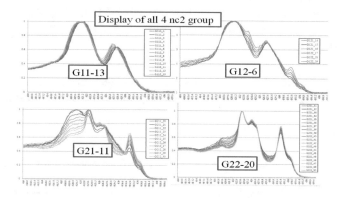

Fig. 12. The results of four groups from G11 to G22 in Fig.10 are shown

The rectangular figures are shown on the right of each group in Fig.12 according to the number of spectra that comprise each group. The distance order is G21¿G11¿G12=G22 according to the dendrogram of Fig.10. G22 seems that the distance from G11separates considerably by the observation. The spectrum number is as large as 20. Then, the distance per spectrum would become roughly small.

3.2   Summery of Subsection 3

- The energy range was chosen as 410-440 eV. Given the other energy ranges, the spectrum was the smoothest. The spectra were normalized to have 0-1. Using the spectra of 1-50, the sequentially arranged dendrogram, however, could have been constructed. They are separated into eight groups from G111 to G222. The spectra in each group resembled quite well.

- The standard spectrum of SnO is in G111 group. And that of SnO2 was in G112 group and agreed with sample 11.
- The standard spectrum of Sn-Metal is in G222 group and agreed with sample number 41.
- This data could not be analyzed with PCA in reasonable way since some of the dominant terms would disappear.

4   Conclusion

TOF-SIMS spectra data have been clustered with PCA and SOM. When the 1st to 3rd principal components were used in the PCA, the result agreed with that of the SOM. The cluster analysis disagreed for more than 3 principal components. But the SOM method can be applied to arbitrary number of dimensions. We have shown this for the case of 151 dimensions. As shown in Fig. 12, the spectral data was classified into four groups in an excellent way. The results were also compared with the group average-method of Multi-variate analysis. As a result, the SSOM method showed far better clustering results for the TOf-SIMS data. The SSOM method also agrees with the result of PCA. Finally, for the AES case, the SSOM method also yielded a nice clustering result as in Fig.12 also with four groups.

Acknowledgement. The authors are very obliged to Prof. M. Van Hulle of KU Leuven, Belgium, for kindly reading and correcting our manuscript.

References

1. Benninghoven, A., Rudenauer, F.G., Werner, H.W.: Secondary Ion Mass Spectrometry, p. 671. John Wiley (1987)
2. Yoshihara, K., Tokutaka, H.: The analysis of TOF-SIMS spectra by PCA and the spherical SOM methods. In: Proceeding of the 41st Meeting of Surface Analysis Society of Japan (SASJ), Nagoya, Japan, June 16 (2013) (in Japanese)
3. Tokutaka, H., Fujimura, K., Ohkita, M.: Cluster Analysis using Spherical SOM. Journal of Biomedical Fuzzy Systems Association 8(1), 29–39 (2006) (in Japanese)
4. Tokutaka, H., Fujimura, K., Ohkita, M.: Cluster Analysis using Spherical SOM. In: WSOM 2007, Bielefeld, Germany, September 3-6 (2007)
5. Tokutaka, H., Ohkita, M., Hai, Y., Fujimura, K., Oyabu, M.: Classification using topologically preserving Spherical Self-Organizing Maps. In: Laaksonen, J., Honkela, T. (eds.) WSOM 2011. LNCS, vol. 6731, pp. 308–317. Springer, Heidelberg (2011)
6. Ohkita, M., Tokutaka, H., Fujimura, K., Gonda, E.: Self-Organizing Maps and the tool. Springer Japan Inc. (2008)
7. http://www.somj.com
8. Tokutaka, H.: The classification by the spherical SOM method for the AES depth profile data where the sample was plated Sn on Cu and the results were compared with identifying by the standard-sample. In: Proceeding of the 41st Meeting of Surface Analysis Society of Japan (SASJ), Nagoya, Japan, June 16 (2013)

Exploiting the Structures of the U-Matrix

Jörn Lötsch[1,2] and Alfred Ultsch[3]

[1] Institute of Clinical Pharmacology, Goethe - University, Theodor-Stern-Kai 7,
D-60590 Frankfurt am Main, Germany
[2] Fraunhofer Institute of Molecular Biology and Applied Ecology - Project Group
Translational Medicine and Pharmacology (IME-TMP), Theodor-Stern-Kai 7,
D-60590 Frankfurt am Main, Germany
[3] DataBionics Research Group, University of Marburg, Hans-Meerwein-Strae,
D-35032 Marburg, Germany

Abstract. The U-matrix has become a standard visualization of self-organizing feature maps (SOM). Here we present the abstract U-matrix, which formalizes the structures on a U-matrix such that distance calculations between best-matching units w.r.t. the height structures of a U-matrix are precisely defined (U-cell distance). This enables the assessment of the topological correctness of the SOM and the implementation of clustering algorithms that take the structures seen on the U-matrix into account. A weighted Delaunay graph of the U-cell distances allows the calculation of a dendrogram corresponding to the structures of the U-matrix. The method is shown to detect and visualize meaningful cluster structures on difficult artificial and real-life data.

1 Introduction

Self-organizing feature maps (SOM) [2] are often visualized by using the U-matrix [3]. A trained SOM represents a topology pre-serving mapping of n high-dimensional data points $x_i \in \mathbb{R}^D$ onto a two dimensional grid of neurons. A neuron n and the neurons in its Moore neighborhood $N(n)$ on the output grid of the SOM represent points in the data space. The sum of distances between n and the neurons in $N(n)$ in the high-dimensional space is shown on a U-matrix as a height value (U-height) at neuron n. Large U-heights mean that there is a large gap in the data space. Low U-heights mean that the points in $\{n \cup N(n)\}$ are close to each other within the data space. On a 3D-display of U-matrix valleys, ridges and basins can be seen (Figure: 1). If the best matching units (BMUs) of data points are located in a valley surrounded by large walls (water-basin), then these data points are within a distance-induced cluster structure in the data space. Water-sheds, respectively water-basins, on a U-matrix allow for emergence in SOM-based algorithms [3]. Emergent algorithms have the property that novel, formerly unseen structures on a macroscopic level (e.g., valley ridges, clusters) become visible on top of the only locally defined U-heights. The described usage of a SOM and its U-matrix can be used to visualize the distance structures in the high dimensional data space. If clustering of the data space is sought, additional

T. Villmann et al. (eds.), *Advances in Self-Organizing Maps and Learning
Vector Quantization,* Advances in Intelligent Systems and Computing 295,
DOI: 10.1007/978-3-319-07695-9_24, © Springer International Publishing Switzerland 2014

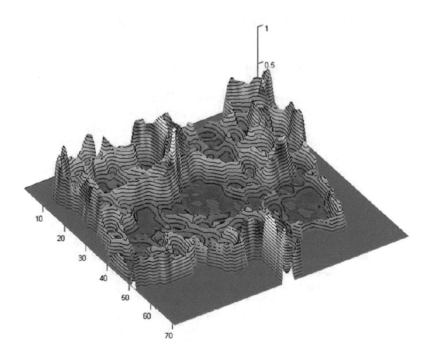

Fig. 1. U-matrix of the pain data described below [1]

clustering methods need to be applied such as a second SOM layer [4] , Fuzzy clustering [5] or spectral clustering [6]. An alternative to these is the usage of visual observation to identify coherent valleys on the U-matrix, i.e. clusters in the data. In this work, we present the abstract U-matrix (AU-matrix), which formalizes the structures on a U-matrix such that distance calculations between BMUs become meaningful. This enables assessing the topological correctness of the SOM and the implementation of clustering algorithms that take structures on the U-matrix into account.

2   Definitions

We assume that n high-dimensional data points $x \in \mathbb{R}^D$ are projected (topology preserving) onto a two-dimensional grid of neurons trough a sufficiently trained SOM. The output grid of neurons (units) is embedded in $O \subset \mathbb{R}^2$ (output space). The images (projections) of the points are the corresponding best-matching units (BMU). We assume that the size of the grid is large enough to map sufficiently distinct points of the data space to distinct BMU coordinates on the grid. For this type of SOM, called emergent SOM (ESOM), the size of the output grid is such that the Voronoi cells of a Voronoi tessellation [11] of the BMUs are sufficiently large. If a Voronoi-cell V_i has a cell V_j as neighbor, then there is an edge in the corresponding Delaunay graph D [7]. Let b_i and b_j be BMUs of data points x_i and x_j, and b_i and b_j are connected by an edge in D. Define a U-cell as follows: a

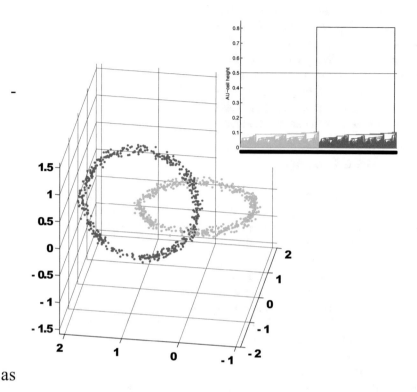

as

Fig. 2. The chain link data set and a dendrogram induced by the AU-cell distances, from which two classes,colored as either red or blue, clearly emerge (top right)

U-cell has a floor shaped by the border lines of the Voronoi cell of the BMU. On each borderline there is a vertical plane. If the borderline is between b_i and b_j, the height of the U-cell on this borderline (AU-height) is the distance $d(x_i, x_j) > 0$ of the data points in the data space. The abstract U-matrix, (AU-matrix) is then the set of all U-cells on a SOM grid. This gives a geometric structure on top of the output space O which is analog to the U-matrix. Height values on the AU-matrix are displayed on top of the Voronoi cell lines and have a clear meaning: the distance in data space of the corresponding BMUs. The usual U-matrix can be regarded as a quantized visualization of the AU-matrix (see below). A U-matrix corresponds to its AU-matrix, if for all pairs of BMUs having an edge in the Delaunay graph the sums of the U-heights on suitable paths between the pairs of BMU correlate to the AU-heights. Let AUH denote the Delaunay graph D induced by the BMUs and weighted by the AU-cell distances. If the edges with weights above a threshold $\min(AUH) < t < \max(AUH)$ are removed from AUH the graph may be separated into different connected components. Using all possible values of t results in an ordered set of critical threshold $t_0, t_1, \ldots, t_c$ such that for $t_i < t < t_{i+1}$ the clustering is the same. Following the method proposed by Carlsson et al. [8], a dendrogram can be constructed that shows the threshold along with the number and the sizes of the resulting clusters. Using the dendrogram the data can be clustered by providing either the number of clusters or the threshold for the maximal AU-cell distance. This is called AU-cell clustering. The adjustment of a suitable threshold respectively the number of clusters is the same problem as in hierarchical clustering. A political map of a U-matrix, resp. AU-matrix, is a top view of the AU-matrix where Voronoi cells of BMU b_i and b_j have the same color if the corresponding data points of b_i and b_j are assigned to the same cluster.

3   A First Example

Data sets from the Fundamental Clustering Problems Dataset (FCPS,) are used to demonstrate the application of the AU matrix. For the three-dimensional Chainlink data set of 1000 data points (Figure 2) an ESOM of grid size 80×50 was trained using the Databionic ESOM software [9].

A top view of the U-matrix using physical-map analogy for color-coding of the distances separates the two classes visually by a ridge between two valleys (Figure 3). An overlay of the U-matrix with a top view of the AU-matrix, for each BMU its Voronoi cell can be seen (Figure 3). The ridge on the U-matrix coincides with the Voronoi-cells borders having large AU-heights. The dendrogram for AU-cell clustering clearly indicates a definition of two classes. These two clusters are the two separate rings in the data. On the Chainlink data set, AU-cell clustering provides complete accuracy (100%). On other data sets from FCPS, the AU matrix method outperforms common cluster algorithms such as k-means and Ward clustering by obtaining always the correct cluster membership of a data point (100% accuracy), whereas the classical methods often provides lower accuracies with more difficult data, up to occasional complete failure (Table 1).

Table 1. Comparative of performance (accuracy [%] of data point assignment to the correct cluster) the AU based and other (Ward, k-means) clustering methods for identifying the cluster structure of data sets with different degrees of difficulty selected form the Fundamental Clustering Problems Dataset (FCPS http://www.uni-marburg.de/fb12/datenbionik/data)

Data set	Main problem	Accuracy [%] of cluster member-ship assignment		
		AU clustering	Ward	k-means
Hepta	Easy	100%	100 %	100 %
Lsun	Standard	100 %	50 %	50 %
Tetra	Small inter distances	100 %	90 %	100 %
Chainlink	Linearly not separable	100 %	50 %	50 %
Atom	Variance differences	100 %	50 %	50 %
Target	Outlier	100 %	25 %	25 %
Golf ball	Equidistant points	100 %	50 %	0 %

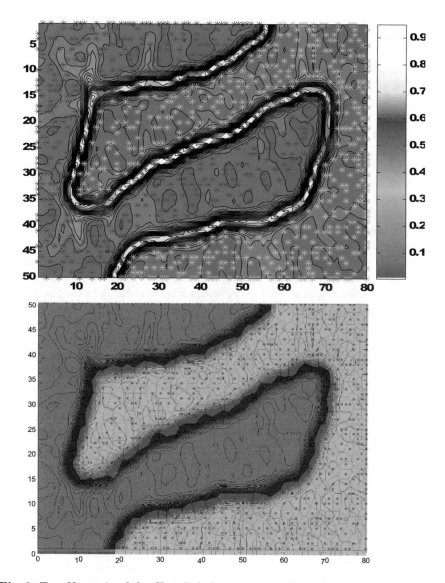

Fig. 3. Top: U-matrix of the Chainlink data, where a visual separation emerges from the ridge in the physical map. Bottom: Political map superimposed on the U- matrix. The two classes are colored in either red or green and their clear visual separation follows the edges of Voronoi cells.

4   Application of the Abstract U-Matrix to Real-Life Data

Pain and its genetic background is a complex problem in biology. Pain is a trait defined as an unpleasant sensory and emotional experience associated with actual or potential tissue damage, or described in terms of such damages. Its sensory, affective, motor, vegetative and emotional components [10] are associated with a complex pathophysiology [11] reflected in the large network of molecular nociceptive pathways [12]. A genetic basis of pain and analgesia has been well established. Today, more than 410 genes have been recognized to contribute to the individual sensitivity of pain [13]. For example, red-haired women displayed greater pain relief following administration of a kappa-opioid receptor specific analgesic (pentazocine) than women without this phenotype [14]. Another example is the hereditary insensitivity to pain due to a loss-of-function genetic mutation, which was found in a single family whose members work as fakirs using the absence of pain professionally [15]. Such mutations are today a valuable source of targets of new analgesic drugs. However, the utility of genetic markers to predict pain sensitivity in the average population and to guide personalized analgesic therapy has remained modest [16] due to the complexity in both, the phenotype and the genotype of pain [17]. Initial approaches using clustering of patients with similar sensitivities to particular pain stimuli have so far not provided reproducible predictions of pain phenotypes and associations of underlying pain-relevant genotypes. These approaches used mainly k-means clustering [18]. However, as shown above, k-means clustering may provide poor cluster associations depending on the distribution of the data.

A data set [1] was obtained following administration of defined pain stimuli to 214 (105 men) healthy volunteers (approval of the Ethics Committee of the Medical Faculty of the Goethe University and informed written consent from each participant obtained). The pain phenotype was assessed by means of measuring pain thresholds to four different pain stimuli (heat, cold, blunt pressure, electricity $0 - 20mA$). After appropriate preprocessing (for details, see [1]) the data were projected onto a ESOM of $50 \times 82 = 4200$ neurons and an U-Matrix was generated (Figure 4). The dendrogram of the AU-cell distances suggested eight clusters in the pain data resulting in a political map that can be overlaid on the U-matrix (Figure 5). The cluster identification using the U-matrix provided a suitable basis for the desired genotype-phenotype association. That is, on the basis of a combined genotype, consisting of 10 variants in four genes (plus gender), subjects with a high pain sensitivity phenotype were predicted with an accuracy of 78% [1]. For comparison, among single genetic markers and gender, only the latter provided a prediction better than guessing. Similarly, for a pain phenotype called *stoics with a selective high sensitivity to heat*, a genetic association with a genotype composed of seven variants in three genes provided a mean cross-validated classification accuracy of $88 \pm 12\%$. These examples clearly demonstrate the utility of AU-cell clustering for real-life pain phenotype genotype associations, suggesting that this method indeed may be essential to advance personalized therapy approaches.

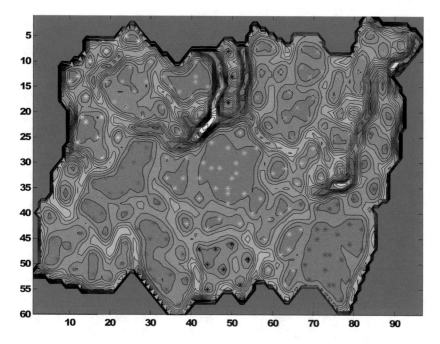

Fig. 4. U-matrix of the Pain data set

5 Discussion

In this contribution we shed insights onto what can be seen on a U-matrix and how this can be used to identify structures and or clusters in high dimensional data. The AU-matrix can be seen as theoretical model to explain a given U-matrix. It can be used for the assessment of the topological correctness of the underlying SOM and the implementation of clustering algorithms which take the structures seen on the U-matrix into account. A dendrogram as known from hierarchical clustering algorithms which closely correspond to the structures seen on a U-matrix can be constructed. The political map of a U-matrix is a very flexible tool to visualize the result of possible clusterings. It allows to easily identify outliers and critical distance structures where the membership of data points to the same or different clusters is debatable. Sometimes cluster structures on high dimensional data are not defined by distance structures (alone) [19]. Local densities of the data space must be taken into account. DBSCAN is an example of a distance and density based clustering algorithm [20]. The CONNvis approach recently proposed [21] integrates density information into a Delaunay graph on the high dimensional data points. In our approaches density information is regarded separately using the P- and/or U* matrix methods [22]. A corresponding technique for AU-matrices is subject to further research. So far, the here presented method provides accurate clustering in model data sets and seems to provide the necessary clustering of real-life data sets, with promising results to provide the

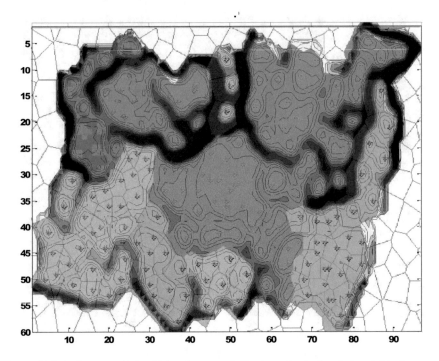

Fig. 5. Political map of the Pain data set following clustering into eight classes of subjects with similar pain sensitivity patterns and overlaid onto the U-matrix (Figure 4)

necessary methods to identify, for example, sub-populations for individualized treatments and drug discovery and development.

References

1. Lötsch, J., Ultsch, A.: A machine-learned knowledge discovery method for associating complex phenotypes with complex genotypes. Application to pain. Journal of Biomedical Informatics 46, 921–928 (2013)
2. Kohonen, T.: Self-organized formation of topologically correct feature maps. Biol. Cybernet. 43, 59–69 (1982)
3. Ultsch, A.: Emergence in Self-Organizing Feature Maps. In: International Workshop on Self-Organizing Maps (WSOM 2007). Neuroinformatics Group (2007)
4. Vesanto, J., Alhoniemi, E.: Clustering of the self-organizing map. IEEE Transactions on Neural Networks / A Publication of the IEEE Neural Networks Council 11, 586–600 (2000)
5. Sarlin, P., Eklund, T.: Fuzzy Clustering of the Self-Organizing Map: Some Applications on Financial Time Series. In: Laaksonen, J., Honkela, T. (eds.) WSOM 2011. LNCS, vol. 6731, pp. 40–50. Springer, Heidelberg (2011)
6. Taşdemir, K.: Spectral Clustering as an Automated SOM Segmentation Tool. In: Laaksonen, J., Honkela, T. (eds.) WSOM 2011. LNCS, vol. 6731, pp. 71–78. Springer, Heidelberg (2011)

7. Delaunay, B.: Sur la sphère vide. Izvestia Akademii Nauk SSSR 7, 793–800 (1934)
8. Carlsson, G., Mémoli, F.: Characterization, Stability and Convergence of Hierarchical Clustering Methods. J. Mach. Learn. Res. 11, 1425–1470 (2010)
9. Ultsch, A., Moerchen, F.: ESOM-Maps: tools for clustering, visualization, and classification with Emergent SOM (2005)
10. Tracey, I., Mantyh, P.W.: The cerebral signature for pain perception and its modulation. Neuron 55, 377–391 (2007)
11. Cross, S.A.: Pathophysiology of pain. Mayo Clin. Proc. 69, 375–383 (1994)
12. Julius, D., Basbaum, A.I.: Molecular mechanisms of nociception. Nature 413, 203–210 (2001)
13. Lötsch, J., Doehring, A., Mogil, J.S., Arndt, T., Geisslinger, G., Ultsch, A.: Functional genomics of pain in analgesic drug development and therapy. Pharmacology & Therapeutics 139, 60–70 (2013)
14. Mogil, J.S., Wilson, S.G., Chesler, E.J., Rankin, A.L., Nemmani, K.V., Lariviere, W.R., Groce, M.K., Wallace, M.R., Kaplan, L., Staud, R., Ness, T.J., Glover, T.L., Stankova, M., Mayorov, A., Hruby, V.J., Grisel, J.E., Fillingim, R.B.: The melanocortin-1 receptor gene mediates female-specific mechanisms of analgesia in mice and humans. Proc. Natl. Acad. Sci. U S A 100, 4867–4872 (2003)
15. Cox, J.J., Reimann, F., Nicholas, A.K., Thornton, G., Roberts, E., Springell, K., Karbani, G., Jafri, H., Mannan, J., Raashid, Y., Al-Gazali, L., Hamamy, H., Valente, E.M., Gorman, S., Williams, R., McHale, D.P., Wood, J.N., Gribble, F.M., Woods, C.G.: An SCN9A channelopathy causes congenital inability to experience pain. Nature 444, 894–898 (2006)
16. Mogil, J.S.: Are we getting anywhere in human pain genetics? Pain 146, 231–232 (2009)
17. Lötsch, J., Flühr, K., Neddermayer, T., Doehring, A., Geisslinger, G.: The consequence of concomitantly present functional genetic variants for the identification of functional genotype-phenotype associations in pain. Clin. Pharmacol. Ther. 85, 25–30 (2009)
18. Baron, R., Binder, A., Wasner, G.: Neuropathic pain: diagnosis, pathophysiological mechanisms, and treatment. Lancet Neurol. 9, 807–819 (2010)
19. Ultsch, A., Moutarde, F.: U*F Clustering: a new performant Cluster-mining method based on segmentation of Self-Organizing Maps. In: International Workshop on Self-Organizing Maps, WSOM 2005 (2005)
20. Ester, M., Kriegel, H.-P., Sander, S., Xu, X.: A density-based algorithm for discovering clusters in large spatial databases with noise, pp. 226–231 (1996)
21. Tasdemir, K., Merényi, E.: SOM-based topology visualization for interactive analysis of high-dimensional large datasets. University of Bielefeld, Germany (2012)
22. Ultsch, A.: The U-Matrix as Visualization for Projections of high-dimensional data. In: Proc. 11th IFCS Biennial Conference (2003)

Partial Mutual Information for Classification of Gene Expression Data by Learning Vector Quantization

Mandy Lange, David Nebel*, and Thomas Villmann

University of Applied Sciences Mittweida - Dept. of Mathematics
Mittweida, Germany

Abstract. Gene expression data analysis is frequently performed using correlation measures whereas unsupervised and supervised vector quantization methods are usually designed for Euclidean distances. In this paper we summarize recent approaches to apply correlation measures to those vector quantization algorithms for analysis of microarray gene expression data. Additionally, we consider k-th order partial correlations as a natural extension if pseudo-correlations should be avoided. Further, we draw the focus to mutual information as powerful alternatives to correlation measures. Related to this we provide the concept of k-th order partial mutual information as counterpart to partial correlations. We apply these methods to an exemplary but real classification problem in gene expression analysis for detection of diabetic patients.

1 Introduction

Gene expression analysis based on microarray data is a challenging task in biology and medicine. It is one way of molecular biology and medicine to understand and to investigate biological processes, diseases and evolutions. During the last years a rapidly increasing amount of data is available although a careful data analysis is still difficult due to the huge number of genes, which are parallely considered in a single experiment. This number frequently is in the order of thousands whereas the number of experiments is quit low compared to this, usually only a few hundred or less. Thus, microarrays deliver high-dimensional data to be analyzed with only a few data samples for model generation available. This problem is also known as 'curse of dimensionality'.

Prototype based methods for clustering and classification have been established as powerful methods in high-dimensional data analysis as an alternative to classical multi-variate statistics. Robust machine learning variants were inspired by neural computing. Self-organizing maps (SOMs, [17]) and neural gas (NG, [24]) for unsupervised vector quantization and clustering as well as learning vector quantizers (LVQ, [18,27,34]) for supervised training (classification learning) belong to those algorithms. For example, these vector quantization approaches were successfully applied for hyper-spectral data analysis in remote sensing [25,46], non-invasive biochemical analysis of food [15], mass-spectrometry [35,36,47] and fMRI-analysis [22] all to be known as high-dimensional problems. One key observation in this context is that prototype based vector quantizer reduce the risk to get affected by the curse of dimensionality. In fact,

---

* Supported by the European Social Fund (ESF), Saxony.

T. Villmann et al. (eds.), *Advances in Self-Organizing Maps and Learning Vector Quantization*, Advances in Intelligent Systems and Computing 295,
DOI: 10.1007/978-3-319-07695-9_25, © Springer International Publishing Switzerland 2014

prototypes are low-noise representations of data [1,13]. Therefore, they are particularly suitable also for analysis of gene expression data.

Because of the huge data dimension of the raw gene expression data, usually, the data are preprocessed in advance using dimensionality reduction techniques like principal component analysis (PCA), cluster analysis and other [4,8,7,12]. Among them, correlation analysis is a standard method frequently applied with subsequent selection schemes based on correlation ranks [33]. Yet , frequently the remaining number of genes to be considered is still high.

Most of the vector quantization algorithms have in common that the Euclidean distance is used for dissimilarity evaluation of the data and prototypes. For analysis of gene expression data, this can lead to moderate problems: the Euclidean distance is sensitive to normalization like centralization and variance normalization, which may cause difficulties when merging several data sets from different investigations [6,3]. Therefore, correlation measures are preferred for gene expression analysis [42,39,41,40,16]. However, correlation measures can be affected by pseudo-correlations. To reduce these influences, partial correlations are appropriate. While these quantities are well-known, their application in LVQ-approaches is not considered so far. In this paper we give a respective framework.

A more general alternative to correlation measures are information theoretic quantities like divergences and closely related mutual information [44], which take into account also higher-order correlations. We present in this paper several definitions of mutual information according to the underlying divergence and entropy types. Further, we provide the theoretical framework for partial mutual information of k-th order as counterpart to the k-th order partial correlation to avoid pseudo-dependencies. We proof the the suitability for those quantities for an exemplary gene expression data set to distinguish diabetic profiles from profile of healthy volunteers.

The paper is structured as follows: First we briefly review generalized learning vector quantization for classification. One particular aspect here is how to process if the dissimilarity measure between data is not differentiable. Second we revisit k-th order partial correlations and derive an analog quantity for mutual information. Thereafter we present the exemplary application in gene expression analysis.

2   Learning Vector Quantization Based on Cost Functions

Classification by learning vector quantization (LVQ) is the supervised counterpart of neural vector quantization by self-organizing maps (SOMs) and are heuristically motivated by KOHONEN [17]. A cost function based version, keeping the basic ideas as well as the Hebbian enhancement learning paradigm from original LVQ, is known as generalized LVQ (GLVQ) [34].

For classification learning each training data vector $\mathbf{v} \in V \subset \mathbb{R}^n$ is equipped with a class label $x_{\mathbf{v}} \in \mathcal{C} = \{1, 2, 3, ..., C\}$. Now, the task is to distribute the set $W = \{\mathbf{w}_k\}_{k \in A} \subset \mathbb{R}^n$ of prototypes such that the classification accuracy is maximized. For this purpose each prototype is also equipped with a class label y_k such that $\mathcal{C}$ is covered by all y_k. After LVQ training a data point is assigned to the class y_s of that prototype $\mathbf{w}_s \in W$ which has minimum distance, i.e.

$$s\left(\mathbf{v}\right) = argmin_k d(\mathbf{v}, \mathbf{w}_k).$$ (1)

A gradient based GLVQ scheme proposed by SATO AND YAMADA uses the following energy function:

$$E(W) = \frac{1}{2} \sum_{\mathbf{v} \in V} f(\mu_W(\mathbf{v}))$$ (2)

where the classifier function

$$\mu_W(\mathbf{v}, \kappa) = \frac{d^+(\mathbf{v}) - d^-(\mathbf{v})}{d^+(\mathbf{v}) + d^-(\mathbf{v})} + \kappa$$ (3)

approximates the non-differentiable classification error depending on W and the constant κ frequently set to zero. The function $f \colon \mathbb{R} \to \mathbb{R}$ is monotonically increasing, usually chosen as sigmoid. Further, $d^+(\mathbf{v}) = d(\mathbf{v}, \mathbf{w}^+)$ denotes the distance between the data point $\mathbf{v}$ and the nearest prototype $\mathbf{w}^+$, which has the same label like $x_{\mathbf{v}} = y_{\mathbf{w}^+}$. In the following we abbreviate $d^+(\mathbf{v})$ simply by d^+. Analogously d^- is defined as the distance to the nearest prototype of all other classes.

In case of a differentiable distance measure $d(\mathbf{v}, \mathbf{w})$, or more general dissimilarity measure, stochastic gradient descent learning can be applied. The respective stochastic gradients of $E(W)$ are

$$\frac{\partial_s E}{\partial \mathbf{w}^+} = \frac{\partial_s E}{\partial d^+(\mathbf{v})} \cdot \frac{\partial d^+(\mathbf{v})}{\partial \mathbf{w}^+}$$

and

$$\frac{\partial_s E}{\partial \mathbf{w}^-} = \frac{\partial_s E}{\partial d^-(\mathbf{v})} \cdot \frac{\partial d^-(\mathbf{v})}{\partial \mathbf{w}^-}$$

where $\frac{\partial_s}{\partial}$ denotes the stochastic gradient with

$$\frac{\partial_s E}{\partial d^+(\mathbf{v})} = \left(2 \cdot g^-(\mathbf{v}, W) - \kappa\right) \cdot f'(\mu_W(\mathbf{v}, \kappa))$$

and

$$, \frac{\partial_s E}{\partial d^-(\mathbf{v})} = \left(2 \cdot g^+(\mathbf{v}, W) - \kappa\right) \cdot f'(\mu_W(\mathbf{v}, \kappa))$$

Further, we introduce the *positive* quantities

$$g^+(\mathbf{v}, W) = \frac{\kappa}{2} - \frac{d^+(\mathbf{v})}{d^+(\mathbf{v}) + d^-(\mathbf{v})}$$ (4)

$$g^-(\mathbf{v}, W) = \frac{\kappa}{2} + \frac{d^-(\mathbf{v})}{d^+(\mathbf{v}) + d^-(\mathbf{v})}.$$ (5)

Obviously, in case of the (squared) Euclidean distance we have to calculate $\frac{\partial d^\pm(\mathbf{v})}{\partial \mathbf{w}^\pm} = -2(\mathbf{v} - \mathbf{w})$, which still refers to Hebbian-like learning. If more general dissimilarity measures $d(\mathbf{v}, \mathbf{w})$ are in use, the respective gradients have to be applied [11,14,45].

Algorithm 1. gEM-algorithm of M-GLVQ

---

1. Initialize W^{old}
2. **E-Step:** set $\gamma(W|\mathbf{v}) \leftarrow p(W^{\mathrm{old}}|\mathbf{v})$
3. **M-Step:** for fixed $\gamma(W|\mathbf{v})$ determine $W^{\mathrm{new}} = \arg\max_W \mathcal{L}$, which improves $\mathcal{L}$
4. If $W^{\mathrm{new}} = W^{\mathrm{old}}$ then STOP, else set: $W^{\mathrm{new}} \leftarrow W^{\mathrm{old}}$ and go to step 2.

If the dissimilarity measure $d(\mathbf{v}, \mathbf{w})$ is not differentiable or too complex for numerically efficient derivative calculations, a recently proposed median variant of GLVQ (M-GLVQ, [26]) can be applied: Introducing the *formal probabilities*

$$p^+(W|\mathbf{v}) = \frac{g^+(\mathbf{v}, W)}{g^+(\mathbf{v}, W) + g^-(\mathbf{v}, W)} \tag{6}$$

$$p^-(W|\mathbf{v}) = \frac{g^-(\mathbf{v}, W)}{g^+(\mathbf{v}, W) + g^-(\mathbf{v}, W)} \tag{7}$$

and the functions $\gamma^+(W|\mathbf{v}) \geq 0$ and $\gamma^-(W|\mathbf{v}) \geq 0$, which play the role of generating models for the prototypes for correct and incorrect classification of a given data point $\mathbf{v}$ with the additional constraint $\gamma^+(W|\mathbf{v}) + \gamma^-(W|\mathbf{v}) = 1$. For this purpose, we rewrite the cost function (2) as

$$K(V, W) = \sum_{\mathbf{v} \in V} \log \left(g^+(\mathbf{v}, W) + g^-(\mathbf{v}, W) \right) \tag{8}$$

specifying the transfer function f to be the logarithm. This is done to meet the requirements for a *generalized* EM-learning (gEM). In particular, we can decompose (8) into

$$K(V, W) = \sum_{\mathbf{v} \in V} \left[\mathcal{L}_{\mathbf{v}}(\gamma||g) - \mathcal{K}_{\mathbf{v}}(\gamma||p) \right]. \tag{9}$$

with the Kullback-Leibler-divergence (KLD)

$$\mathcal{K}_i(\gamma||p) = \gamma^+(\mathbb{W}|\mathbf{v}) \cdot \log\left(\frac{p^+(W|\mathbf{v})}{\gamma^+(W|\mathbf{v})} \right) + \gamma^-(W|\mathbf{v}) \cdot \log\left(\frac{p^-(W|\mathbf{v})}{\gamma^-(W|\mathbf{v})} \right) \tag{10}$$

and $\mathcal{L}_{\mathbf{v}}(\gamma||g)$ is a *generalized* KLD (see [5,44])

$$\mathcal{L}_i(\gamma||g) = \gamma^+(W|\mathbf{v}) \cdot \log\left(\frac{g^+(\mathbf{v}, W)}{\gamma^+(W|\mathbf{v})} \right) + \gamma^-(W|\mathbf{v}) \cdot \log\left(\frac{g^-(\mathbf{v}, W)}{\gamma^-(W|\mathbf{v})} \right) \tag{11}$$

with $g = \{g^+(\mathbf{v}, W), g^-(\mathbf{v}, W)\}$, $p = \{p^+(\mathbf{v}, W), p^-(\mathbf{v}, W)\}$ and both, $\gamma^+(W|\mathbf{v})$ and $\gamma^-(W|\mathbf{v})$, form together the *formal probability density function* $\gamma(W|\mathbf{v})$. In fact, (9) has the structure of a maximum likelihood problem with the first term $\mathcal{L} = \sum_{\mathbf{v} \in V} \mathcal{L}_{\mathbf{v}}(\gamma||g)$ being a lower bound.

At this point we emphasize that for this variant of the gEM-approach we do not search for a set W^* in the M-step, which would maximize the cost function $K(V, W)$. We only assume that the cost function is not decreasing for W^{new}.

Obviously, the algorithm only requires the distances $d(\mathbf{v}, \mathbf{w}_k)$ between data and prototypes. If we restrict the prototypes to be data points, which corresponds exactly to the *median* principle, only the dissimilarities between the data are needed.

3   Appropriate Dissimilarity Measures for Gene Analysis in Microarrays

As mentioned above, GLVQ and SOMs were introduced originally using the Euclidean distance to calculate the dissimilarities between data and prototype vectors. Yet, gene expression analysis is frequently done applying correlations in microarrays [37]. Higher order correlations are taken into account if entropy based methods are applied in more sophisticated schemes like divergences and related mutual information measures [50]. In the following we review correlation measures and consider mutual information.

3.1   Correlations

Following the approach in [39], the *linear* Pearson correlation can be applied in gradient based vector quantization. Pearson correlation implicitly undertakes the data a centralization and, therefore, is well suited for analysis of gene expression analysis [42,43], where individually calibrated biomedical measuring devices are in common [31,42]. The Pearson correlation between a data vector $\mathbf{v} \in \mathbb{R}^n$ and a prototype $\mathbf{w} \in \mathbb{R}^n$ is defined as

$$\varrho_P(\mathbf{v}, \mathbf{w}) = \frac{\sum_{k=1}^{n} (v_k - \mu_\mathbf{v}) \cdot (w_k - \mu_\mathbf{w})}{\sqrt{\sum_{k=1}^{n} (v_k - \mu_\mathbf{v})^2 \cdot \sum_{k=1}^{n} (w_k - \mu_\mathbf{w})^2}} \tag{12}$$

with $\mu_\mathbf{v}$ and $\mu_\mathbf{w}$ are the means of $\mathbf{v}$ and $\mathbf{w}$, respectively.

Spearman's rank correlation ϱ_S is a *non-linear* correlation measure. However, due to its rank based computation scheme, it is not differentiable at hand. In a first step we calculate the ranks in terms of sums of Heaviside functions

$$H(x) = \begin{cases} 0 \ if \ x \leq 0 \\ 1 \ else \end{cases} \tag{13}$$

and express in this way the $\varrho_S(\mathbf{v}, \mathbf{w})$ between vectors $\mathbf{v}$ and $\mathbf{w}$ by the Pearson correlation $\varrho_P(\mathbf{v}, \mathbf{w})$. For that purpose we define an indicator matrix $\mathbf{R}(\mathbf{x})$ of a vector $\mathbf{x}$ as

$$\mathbf{R}(\mathbf{x}) = \begin{pmatrix} H(x_1 - x_1) & \cdots & H(x_1 - x_n) \\ \vdots & & \vdots \\ H(x_n - x_1) & \cdots & H(x_n - x_n) \end{pmatrix} \tag{14}$$

with row vectors $\mathbf{R}_i(\mathbf{x})$, which determine the rank function

$$rnk(\mathbf{x}) = \sum_{i=1}^{n} \mathbf{R}_i(\mathbf{x}). \tag{15}$$

Using this indicator matrix, the Spearman rank correlation between a data vector $\mathbf{v}$ and a prototype vector $\mathbf{w}$ can be expressed in terms of the Pearson correlation (12) by

$$\varrho_S(\mathbf{v}, \mathbf{w}) = \varrho_P(rnk(\mathbf{v}), rnk(\mathbf{w})) \tag{16}$$

using the rank vectors (15).

It turns out that soft variants of Spearman rank correlation as proposed in [2,38] can be related to other variants of soft and fuzzy rank correlations, the latter ones based on t-norms and t-conorms [2]. Moreover, rank-based approaches frequently benefit from the robustness of this paradigm to achieve high performance.

3.2 Mutual Information

Correlation do not consider statistical independence. Therefore, we propose to consider the mutual information in gene expression analysis based on microarray data to keep higher order correlations in dissimilarity determination. For this reason we suppose positive gene expression vectors $\mathbf{v}$ with expression levels $v_i \geq 0$ and normalization $\sum_{i=1}^n v_i = 1$. Then the mutual information is defined as

$$I_S(\mathbf{v}, \mathbf{w}) = H_S(\mathbf{v}) + H_S(\mathbf{w}) - H_S(\mathbf{v}, \mathbf{w}) \qquad (17)$$

with

$$H_S(\mathbf{x}) = -\sum_{i=1}^n x_i \ln(x_i) \qquad (18)$$

being the Shannon-entropy and $H_S(\mathbf{v}, \mathbf{w})$ is the joint Shannon-entropy [23]. $I_S(\mathbf{v}, \mathbf{w})$ is a symmetric quantity, which is closely related to the KLD [20]. Yet, there exist several divergence types [5]. Their use in vector quantization is extensively investigated in [44]. These divergences are based on different entropy definitions. A robust alternative to the Shannon-entropy is the *Rényi-entropy*

$$H_R^\alpha(\mathbf{v}) = \frac{1}{1-\alpha} \log\left(\sum_{i=1}^n v_i^\alpha\right) \qquad (19)$$

depending on the parameter α [32]. The respective mutual information writes as

$$I_R^\alpha(\mathbf{v}, \mathbf{w}, \alpha) = H_R^\alpha(\mathbf{v}) + H_R^\alpha(\mathbf{w}) - H_R^\alpha(\mathbf{v}, \mathbf{w}) .$$

Easy computation of $H_R^\alpha(\mathbf{v})$ is achieved for $\alpha = 2$, which is well studied for respective Rényi-divergences in information theoretic learning (ITL) by J. PRINCIPE [30].

The estimation of the joint entropies $H_S(\mathbf{v}, \mathbf{w})$ and $H_R^\alpha(\mathbf{v}, \mathbf{w})$ contributing to the mutual information is generally difficult. Successful estimators for mutual information were proposed for both types by KRASKOV ET AL. and PÁL ET AL. in [19] [28].

3.3 Partial Correlation and Partial Mutual Information

Partial Correlations were developed to eliminate the influence of pseudo-correlations in correlation analysis. For a given correlation measure ϱ, the partial correlation of first order is defined by the quotient

$$\varrho(\mathbf{v}, \mathbf{w}|\mathbf{z}) = \frac{\varrho(\mathbf{v}, \mathbf{w}) - \varrho(\mathbf{v}, \mathbf{z}) \cdot \varrho(\mathbf{w}, \mathbf{z})}{\sqrt{1 - \varrho^2(\mathbf{v}, \mathbf{z})} \cdot \sqrt{1 - \varrho^2(\mathbf{w}, \mathbf{z})}} \qquad (20)$$

whereas the partial correlation of second order

$$\varrho\left(\mathbf{v},\mathbf{w}|\mathbf{z},\mathbf{x}\right)=\frac{\varrho\left(\mathbf{v},\mathbf{w}|\mathbf{z}\right)-\varrho\left(\mathbf{v},\mathbf{x}|\mathbf{z}\right)\cdot\varrho\left(\mathbf{w},\mathbf{x}|\mathbf{z}\right)}{\sqrt{1-\varrho^2\left(\mathbf{v},\mathbf{x}|\mathbf{z}\right)}\cdot\sqrt{1-\varrho^2\left(\mathbf{w},\mathbf{x}|\mathbf{z}\right)}}$$

is recursively defined taking into account the partial correlations of first order. Partial correlations higher order are defined correspondingly.

The counterpart for mutual information is partial mutual information (PMI). Given a mutual information I, the PMI can be written in terms of the underlying entropy H as

$$I\left(\mathbf{v},\mathbf{w}|\mathbf{z}\right)=H\left(\mathbf{v},\mathbf{z}\right)+H\left(\mathbf{w},\mathbf{z}\right)-H\left(\mathbf{z}\right)-H\left(\mathbf{v},\mathbf{w},\mathbf{z}\right)$$

taking into account all mutual information between $\mathbf{v}$ and $\mathbf{w}$, which is not contained in $\mathbf{z}$ [10]. This concept can also be extended to higher order PMIs [21,49]: Let S_k be a set of k vectors $\mathbf{v}_k$. Then the k-th order PMI is given by

$$I\left(\mathbf{v},\mathbf{w}|S_k\right)=H\left(\mathbf{v},S_k\right)+H\left(\mathbf{w},S_k\right)-H\left(S_k\right)-H\left(\mathbf{v},\mathbf{w},S_k\right)$$

being also a recursive definition.

The numerical estimation for the k-th order PMI is by means not trivial, because the calculation of the higher order joint entropies are required. As it is explained in [21], for the Shannon-entropy based PMI the estimator provided by KRASKOV ET AL. in [19] can be extended to the general k-th order case. A kernel-based estimator was proposed in [49]. Unfortunately, an easy transfer of the methods for the usual joint entropies to higher orders in case of Rényi-entropies is not known so far. The methods applied in [28,29] seem to be not easily extendible to higher orders.

4 Experiments

For a numerical experiment we selected a data set studying the insulin effects on gene expression in skeletal muscle and muscle biopsies. The data were obtained from 20 insulin sensitive individuals before and after euglycemic hyperinsulinemic clamps as described in [48]. Additionally, data records of 15 patients suffering from type-2-diabetes were considered as offered in [9]. This result in 6 classes with overall 110 data samples. The data were generated by an Affymetrix Human Genome U95A Array. The microarray delivered 12626 genes to be considered.

We trained a M-GLVQ with one prototype per class. As similarities we used the Pearson-correlation and the Shannon-entropy based mutual information as well as partial Pearson correlation and partial mutual information of second order. The conditional variables were $S_2=\{\mathbf{m}_b,\mathbf{m}_a\}$ being the mean vectors $\mathbf{m}_b$ and $\mathbf{m}_a$ of the gene expression vectors for all individuals and patients *before* and *after* euglycemic hyperinsulinemic clamps, respectively. For comparison we also performed M-GLVQ using the Euclidean distance d_E. The result depicted in Tab.1 are obtained by 5-fold cross-validation.

We observe that, taking into account pseudo-correlations, significantly improvements in classification accuracy are achieved compared to application of only standard correlation or mutual information. Further, we detect a slight worsening in case of partial

Table 1. Classification rates obtained for the several dissimilarity measures in use for M-GLVQ

	$d_E\,(\mathbf{v},\mathbf{w})$	$\varrho_P\,(\mathbf{v},\mathbf{w})$	$\varrho_p\,(\mathbf{v},\mathbf{w}\vert S_2)$	$I_S\,(\mathbf{v},\mathbf{w})$	$I_S\,(\mathbf{v},\mathbf{w}\vert S_2)$
train	0.91	0.91	0.99	0.89	0.91
test	0.78	0.75	0.97	0.75	0.83

mutual information in comparison to partial correlation. This may be dedicated to the crucial estimation procedures for the joint entropies needed for calculation of the partial mutual information.

5   Conclusion

In this contribution we provided the theoretical framework for application of partial correlations and partial mutual information in learning vector quantization. These techniques are of particular interest in gene-expression analysis, where usually correlations play the role of the dissimilarity measure between data instead of the Euclidean distance. The utilization of these quantities eliminate the influence of pseudo-correlations and generally should lead to better performances. As an example we investigated a real world application studying the insulin effects on gene expression in skeletal muscle and muscle biopsies in type-2-diabetes as well as healthy but sensitive individuals. The proposed methods achieve a considerably improvement of the classification rate. From machine learning point of view, the application of the recently developed M-GLVQ was essential, because derivatives of PMI and partial correlations are numerically not available.

References

[1] Biehl, M., Hammer, B., Villmann, T.: Distance measures for prototype based classification. In: Petkov, N. (ed.) Proceedings of the International Workshop on Brain-Inspired Computing 2013, Cetraro, Italy. Springer (2014)

[2] Bodenhofer, U., Klawonn, F.: Robust rank correlation coeffcients on the basis of fuzzy orderings: Initial steps. Mathware & Soft Computing 15, 5–20 (2008)

[3] Chelloug, S., Meshoul, S., Batouche, M.: Clustering microarray data within amorphous computing paradigm and growing neural gas algorithm. In: Ali, M., Dapoigny, R. (eds.) IEA/AIE 2006. LNCS (LNAI), vol. 4031, pp. 809–818. Springer, Heidelberg (2006)

[4] Chiaromonte, F., Martinelli, J.: Dimension reduction strategies for analyzing global gene expression data with a response. Mathematical Biosciences 176, 123–144 (2002)

[5] Cichocki, A., Zdunek, R., Phan, A., Amari, S.-I.: Nonnegative Matrix and Tensor Factorizations. Wiley, Chichester (2009)

[6] Covell, D., Wallqvist, A., Rabow, A., Thanki, N.: Molecular classification of cancer: unsupervised self-organizing map analysis of gene expression microarray data. Molecular Cancer Therapeutics 2(36), 317–332 (2003)

[7] da Costa, J.F.P., Alonso, H., Roque, L.: A weighted principal component analysis and its application to gene expression data. IEEE/ACM Transactions on Computational Biology and Bioinformatics 8(1), 246–252 (2011)

[8] Dai, J., Lieu, L.: Dimension reduction for classification with gene expression microarray data. Statistical Applications in Genetics and Molecular Biology 5(1), 1–19 (2006)
[9] Frederiksen, C., Højlund, K., Hansen, L., Oakeley, E., Hemmings, B., Abdallah, B., Brusgaard, K., Beck-Nielsen, H., Gaster, M.: Transcriptional profiling of myotubes from patients with type 2 diabetes: no evidence for a primary defect in oxidative phosphorylation genes. Diabetologia 51, 2068–2077 (2008)
[10] Frenzel, S., Pompe, B.: Partial mutual information for coupling analysis of multivariate time series. Physical Review Letters 99, 204101-1–204101-4 (2007)
[11] Hammer, B., Villmann, T.: Generalized relevance learning vector quantization. Neural Networks 15(8-9), 1059–1068 (2002)
[12] Han, X.: Nonnegative principal component analysis for cancer molecular pattern discovery. IEEE/ACM Transactions on Computational Biology and Bioinformatics 7(3), 537–549 (2010)
[13] Kaden, M., Lange, M., Nebel, D., Riedel, M., Geweniger, T., Villmann, T.: Aspects in classification learning - Review of recent developments in Learning Vector Quantization. In: Foundations of Computing and Decision Sciences (accepted, 2014)
[14] Kästner, M., Hammer, B., Biehl, M., Villmann, T.: Functional relevance learning in generalized learning vector quantization. Neurocomputing 90(9), 85–95 (2012)
[15] Kästner, M., Nebel, D., Riedel, M., Biehl, M., Villmann, T.: Differentiable kernels in generalized matrix learning vector quantization. In: Proc. of the Internacional Conference of Machine Learning Applications (ICMLA 2012), pp. 1–6. IEEE Computer Society Press (2012)
[16] Kästner, M., Strickert, M., Labudde, D., Lange, M., Haase, S., Villmann, T.: Utilization of correlation measures in vector quantization for analysis of gene expression data - a review of recent developments. Machine Learning Reports, 6 (MLR-04-2012), 5–22 (2012), http://www.techfak.uni-bielefeld.de/~fschleif/mlr/mlr_04_2012.pdf, ISSN:1865-3960
[17] Kohonen, T.: Self-Organizing Maps, Springer Series in Information Sciences, vol. 30. Springer, Heidelberg (1995) (2nd extended edn., 1997)
[18] Kohonen, T., Kangas, J., Laaksonen, J., Torkkola, K.: LVQ_PAK: A program package for the correct application of Learning Vector Quantization algorithms. In: Proc. IJCNN 1992, International Joint Conference on Neural Networks, vol. I, pp. 725–730. IEEE Service Center, Piscataway (1992)
[19] Kraskov, A., Stogbauer, H., Grassberger, P.: Estimating mutual information. Physical Review E 69(6), 66–138 (2004)
[20] Kullback, S., Leibler, R.: On information and sufficiency. Annals of Mathematical Statistics 22, 79–86 (1951)
[21] Lange, M.: Partielle Korrelationen und Partial Mutual Information zur Analyse von fMRT-Zeitreihen. Master's thesis, University of Applied Sciences Mittweida, Mittweida, Saxony, Germany (2012)
[22] Lange, M., Kästner, M., Villmann, T.: About analysis and robust classification of searchlight fMRI-data using machine learning classifiers. In: Proceedings of International Joint Conference on Neural Networks, Dallas, Texas, USA, pp. 2026–2033. IEEE Press (2013)
[23] Mackay, D.: Information Theory, Inference and Learning Algorithms. Cambridge University Press (2003)
[24] Martinetz, T.M., Berkovich, S.G., Schulten, K.J.: 'Neural-gas' network for vector quantization and its application to time-series prediction. IEEE Trans. on Neural Networks 4(4), 558–569 (1993)

[25] Merényi, E., Villmann, T.: Self-organizing neural network approaches for hyperspectral images. In: Tolba, M., Salem, A. (eds.) Intelligent Computing and Information Systems, Ain Shams University Cairo, Fac. of Computer and Information Science, pp. 33–42 (2002) ISBN 977-237-172-3

[26] Nebel, D., Hammer, B., Villmann, T.: A median variant of generalized learning vector quantization. In: Lee, M., Hirose, A., Hou, Z.-G., Kil, R.M. (eds.) ICONIP 2013, Part II. LNCS, vol. 8227, pp. 19–26. Springer, Heidelberg (2013)

[27] Nova, D., Estévez, P.: A review of learning vector quantization classifiers. In: Neural Computation and Applications (2013)

[28] Pál, D., Póczos, B., Szepesvári, C.: Estimation of Rényi entropy and mutual information based on generalized nearest-neighbor graphs. In: Proc. of the Workshop on Neural Information Processing Systems, NIPS (2010)

[29] Póczos, B., Kirshner, S., Szepesvári, C.: REGO: Rank based estimation of Rényi information using Euclidean graph optimization. In: Proc. of the 13th International Conference on Artificial Intelligence and Statistics (AISTATS). Journal of Machine Learning Research (JMLR), vol. 9 (2010)

[30] Principe, J.: Information Theoretic Learning. Springer, Heidelberg (2010)

[31] Raghava, G., Han, J.H.: Correlation and prediction of gene expression level from amino acid and dipeptide composition of its protein. BMC Bioinformatics 6, 59 (2005)

[32] Rényi, A.: On measures of entropy and information. In: Proceedings of the Fourth Berkeley Symposium on Mathematical Statistics and Probability, University of California Press (1961)

[33] Saeys, Y., Inza, I., Larra-Naga, P.: A review of feature selection techniques in bioinformatics. Bioinformatics 23(19), 2507–2517 (2007)

[34] Sato, A., Yamada, K.: Generalized learning vector quantization. In: Touretzky, D.S., Mozer, M.C., Hasselmo, M.E. (eds.) Advances in Neural Information Processing Systems 8. Proceedings of the 1995 Conference, pp. 423–429. MIT Press, Cambridge (1996)

[35] Schleif, F.-M., Villmann, T., Hammer, B.: Prototype based fuzzy classification in clinical proteomics. International Journal of Approximate Reasoning 47(1), 4–16 (2008)

[36] Schleif, F.-M., Villmann, T., Kostrzewa, M., Hammer, B., Gammerman, A.: Cancer informatics by prototype networks in mass spectrometry. Artificial Intelligence in Medicine 45(2-3), 215–228 (2009)

[37] Sharma, A., Paliwal, K.: Cancer classification by gradient LDA technique using microarray gene expression data. Data & Knowledge Enginneering 66, 338–347 (2008)

[38] Strickert, M.: Enhancing M|G|RLVQ by quasi step discriminatory functions using 2nd order training. Machine Learning Reports, 5(MLR-06-2011), 5–15 (2011), http://www.techfak.uni-bielefeld.de/~fschleif/mlr/mlr_06_2011.pdf, ISSN:1865-3960

[39] Strickert, M., Schleif, F.-M., Seiffert, U., Villmann, T.: Derivatives of Pearson correlation for gradient-based analysis of biomedical data. Inteligencia Artificial, Revista Iberoamericana de Inteligencia Artificial (37), 37–44 (2008)

[40] Strickert, M., Schleif, F.-M., Villmann, T., Seiffert, U.: Unleashing pearson correlation for faithful analysis of biomedical data. In: Biehl, M., Hammer, B., Verleysen, M., Villmann, T. (eds.) Similarity-Based Clustering. LNCS, vol. 5400, pp. 70–91. Springer, Heidelberg (2009)

[41] Strickert, M., Seiffert, U., Sreenivasulu, N., Weschke, W., Villmann, T., Hammer, B.: Generalized relevance LVQ (GRLVQ) with correlation measures for gene expression analysis. Neurocomputing 69(6-7), 651–659 (2006) ISSN: 0925-2312.

[42] Strickert, M., Sreenivasulu, N., Usadel, B., Seiffert, U.: Correlation-maximizing surrogate gene space for visual mining of gene expression patterns in developing barley endosperm tissue. BMC 8, 165 (2007)

[43] Strickert, M., Sreenivasulu, N., Villmann, T., Hammer, B.: Robust centroid-based clustering using derivatives of Pearson correlation. In: Encarnação, P., Veloso, A. (eds.) Proceedings of the First International Conference on Biomedical Electronics and Devices, BIOSIGNALS 2008, Funchal, Madeira, Portugal, vol. 2, pp. 197–203. INSTICC - Institute for Systems and Technologies of Information, Control and Communication (2008)

[44] Villmann, T., Haase, S.: Divergence based vector quantization. Neural Computation 23(5), 1343–1392 (2011)

[45] Villmann, T., Haase, S., Kaden, M.: Kernelized vector quantization in gradient-descent learning. In: Neurocomputing (in press, 2014)

[46] Villmann, T., Merényi, E., Hammer, B.: Neural maps in remote sensing image analysis. Neural Networks 16(3-4), 389–403 (2003)

[47] Villmann, T., Schleif, F.-M., Kostrzewa, M., Walch, A., Hammer, B.: Classification of mass-spectrometric data in clinical proteomics using learning vector quantization methods. Briefings in Bioinformatics 9(2), 129–143 (2008)

[48] Wu, X., Wang, J., Cui, X., Maianu, L., Rhees, B., Rosinski, J., So, W., Willi, S., Osier, M., Hill, H., Page, G., Allison, D., Martin, M., Garvey, W.: The effect of insulin on expression of genes and biochemical pathways in human skeletal muscle. Endocrine 31, 5–17 (2007)

[49] Yuan, C., Zhang, X., Xu, S.: Partial mutual information for input selection of time series prediction. In: Proceedings of the 2011 Chinese Control and Decision Conference, CCDC, Mianyang, pp. 2010–2014. IEEE Press (2011)

[50] Zhu, S., Wang, D., Yu, K., Li, T., Gong, Y.: Feature selection for gene expression using model-based entropy. IEEE/ACM Transactions on Computational Biology and Bioinformatics 7(1), 25–36 (2010)

Composition of Learning Patterns Using Spherical Self-Organizing Maps in Image Analysis with Subspace Classifier

Nobuo Matsuda[1], Fumiaki Tajima[2], and Hedeaki Sato[3]

[1] Dept. Electr. and Mech. Eng., Oshima National College of Maritime Technology
1091-1, Komatsu, Suo-oshima-cho, Oshima-gun, 742-2193 Yamaguchi-ken, Japan
[2] Education and Human Science, Yokohama National University
[3] Federation of National Public Service Personnel Mutual Aid Association
matsuda@oshima-k.ac.jp,
tajima@ynu.ac.jp

Abstract. This paper describes a composition of learning patterns based on the visualization of a Spherical SOM for improving the performance of image analysis using the Subspace classifier. We have applied the Subspace classifier to image analysis because it has fewer parameters and higher performance. Then we have experienced that the selection of features and learning patterns influence greatly the classification performance through examinations. The Spherical SOM has no border in the array of nodes and eliminates the *Border effect* problem. Comparing the performance of the image analysis, we show that visualization of the Spherical SOM allows the composition of learning patterns to improve more performance and degree of its reliability than those without the composition.

Keywords: Subspace Classifier, Spherical Self-Organizing Map, Learning Pattern, Fundus Image, Visualization.

1 Introduction

The Self-Organizing Map (SOM) by Kohonen [1] is a kind of neural network algorithm that projects high dimensional data onto a low dimensional space. Several Spherical SOMs based on a geodesic dome [2] or a Toroidal SOM have been proposed as a remedy against the *Border effect* problem in the traditional SOM algorithm. To show its potential effectiveness, Tokutaka *et al.*[3] have proposed a highly accurate cluster analysis using the Spherical SOM.

We have already reported the cluster analysis using the Spherical SOM for real medical fundus data [4]. It was very difficult for input data to determine the class which should be classified in performance evaluation of the real medical data. In such a case, there was a possibility to make a subjective interpretation for the dendrogram.

As a popular classification method, whereas, the Support vector machine (SVM) [5] is often adapted in recent research papers because of its high performance. For example, a glaucoma diagnosis using a data mining technique has

T. Villmann et al. (eds.), *Advances in Self-Organizing Maps and Learning Vector Quantization*, Advances in Intelligent Systems and Computing 295,
DOI: 10.1007/978-3-319-07695-9_26, © Springer International Publishing Switzerland 2014

been proposed by Nishiyama *et al.*[6]. In the real medical data with considerably overlapped distributions and noise, even if the high recognition technique such as the SVM was employed, the application to optimal parameters and its performance assessment would be difficult since its performance generally depends on the choice of parameters, such as Kernel functions and optimal margin, and data distributions.

Hence we aimed at the fact that the Subspace classifier [7] has the simplicity of parameter selection as well as high classification performance. We have proposed an image analysis using the Subspace classifier [8]. We have applied the Subspace classifier to the fundus image analysis and then have experienced that the selection of feature and learning patterns influence greatly the performance through examinations. Therefore on the composing learning patterns or training datasets, we focused on the potential effectiveness with the accurate visualization of Spherical SOM. We propose a composition for learning pattern effectively using the Spherical SOM, and show the effectiveness of the proposed composition through experiments of three cases of composition.

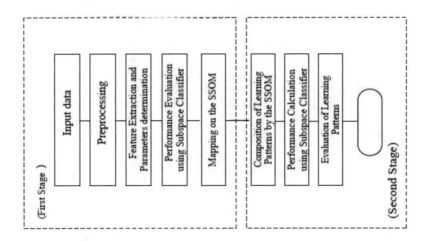

Fig. 1. Overview of proposed method using the Subspace classifier and composition with visualization by the Spherical SOM

2   Composition of Learning Patterns by Spherical SOM

2.1   Spherical SOM

The arrangement of nodes in the SOM is originally free, several arrangements have been proposed. Ritter [9] suggested the use of tessellated platonic polyhedra as the underlying lattices and also pointed out the advantage of Spherical

SOM. In order for grid units to be uniformly distributed on the surface of sphere, every grid unit needs the same number of immediate neighbors and the distances between a unit and its immediate neighbors are the same. However, such uniformity cannot be achieved on the sphere except for the five platonic polyhedra; tetrahedron, cube, octahedron, icosahedron and dodecahedron. These polyhedra can be tessellated into different frequencies: the geodesic domes.

2.2   Subspace Classifier

Let N be the number of pattern space dimension which is the number of pattern vector elements. Let $\varphi = (\varphi_1, \varphi_2, \ldots, \varphi_i, \ldots, \varphi_r)$ be reference vectors for each class which are normal and orthogonal. Let r be the number of reference vectors and let $\boldsymbol{x}$ be an input vector. The similarity S or the squared length of the orthogonal projection of the input vector $\boldsymbol{x}$ on each class subspace is defined as

$$S = \mathbf{x}^T \varphi \mathbf{x} = \sum_{i=1}^{r} \left(\mathbf{x}^T \varphi_{\mathbf{i}} \right)^2 \tag{1}$$

There are several ways to construct the vector φ_i . Simply the vector φ_i for each class can be constructed by using Karhunen-Loéve expansion. The reference vectors are defined for each class or category and the similarities S are also calculated for each category. The category with maximum similarity from their ones should be determined as the answered category for unknown data. Here, note that the number r is the dimension of the spanned space by ϕ_i, while the number N is the total space's dimension.

2.3   Overview of Proposed Method

The proposed method shown in Fig. 1 distinguishes two stages: the first stage of image analysis using the Subspace classifier and mapping by Spherical SOM, and the second stage of composition of learning patterns and its evaluation. In the first stage the experiments on feature extraction are conducted from input datasets and a suitable feature among several features is selected by the highest accuracy. After determining the best feature for classification, the experiments on classification performance are conducted by the Cross-validation, and then in order to obtain the visualization information for composition learning patterns the Spherical SOM applies to the suitable feature datasets.

In the next stage several kinds of learning patterns are reconstructed based on the visualization information by the Spherical SOM and then are evaluated by the Cross-validation. The operation of this stage is repeated until accuracy and/or sensitivity is satisfied. Because the performance of the Subspace classifier is originally high, the number of repetition required here is only several times. The details concerning the composition of learning patterns will be described in Section 4.

3   Experimental Data and Analytical Method

3.1   Input Data

A series of experiments was conducted with fundus images produced by a clinical doctor. The total number of images was 133: 91 normal subjects (labels o1 to o91) and 42 abnormal ones (labels x1 to x42). Colored fundus photographs of 24 bit RGB bitmaps as shown in Figure 1a were acquired with a scanner.

The data used in our experiments was intensity values of these images. The intensity plane of the 2-D image was partitioned into 24 channels as shown in Figure 1b. The mean was computed in each of these 24 channels in the intensity domain. The intensity plane partitioning was uniform along the angular direction (equal step size of 15 degrees) and uniform along the radial direction (equal step size of 10 dots). When the input data were prepared in this way, the minimum dimensionality of input data was 24 and its maximum was 120.

The input data was divided into three groups of datasets with three different dimensionalities. The first group is the *ring* region, and there are 5 rings. Each ring contains 24 channels and is marked in the clockwise direction from inside outward. The second group is called the *zone* region with 48 channels; the zone group is made from two adjacent rings. There are 4 zones, and each zone is also numbered in the direction from inside outward. The last group is called as the *all* region, containing all five rings and thus, 120 channels.

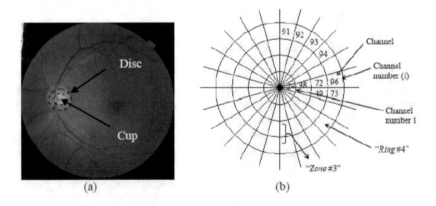

Fig. 2. (a) Fundus image: excavation cup and optic nerve disc, and (b) Channel configuration (the origin of the coordinate is taken as the center of the Cup)

3.2   Feature Extraction and Classification Performance

To prevent the performance degradation and unreliability on the classification, specific essential features required for classification can be extracted from an original dataset with various kinds of information. The required classifier designed using the training data should reduce the average of misclassification rates to a minimum.

Firstly the experiments on feature extraction are conducted from the intensity value data with element numbers of three kinds: 24-D, 48-D and 120-D. The best ones among these features are then selected by the highest accuracy. After determining the feature for the classification, the experiments on classification performance are conducted by Cross-validation. The Cross-validation is performed in order to determine the accuracy for the test datasets, which is a measure of classification performance for unknown data.

The Cross-validation for evaluating classification performance applies to 10-fold Cross-validation. The 10-fold Cross-validation is calculated by the following ways. Firstly all data $\mathbf{X}$ is divided into ten groups $\mathbf{x}_1, \mathbf{x}_2, \ldots, \mathbf{x}_{10}$ randomly. One group $\mathbf{x}_i$ $(i = 1, 2, \ldots, 10)$ is included 10% of all the data and defined as testing set. Another group except $\mathbf{x}_i$ are included 90% of all data and defined as training set. Learning rules are obtained by training, and the performance evaluation for testing set is conducted with the learning rules. This process is repeated 10 times and the evaluated values are averaged.

4   Results and Discussion

4.1   Feature Extraction

Eigenvalues of two classes and the corresponding cumulative proportions for the 48-dimensional input dataset are shown in Figure 3. The eigenvalues of each class reduce rapidly with the increasing of dimensionality r and are very small for $r > 3$. A cumulative proportion reaches a value of 0.99 when the dimensionality r is about 4.

The graph of top left in the Fig. 3 shows the results of classification for learning datasets of zones when spanned space dimensionality r varying from 1 to 48. Table 1 summarizes the results of maximum recognition rate for the training sets when the dimension r of spanned space was varied from 1 to maximum value. The figure and the table show that classification accuracy for the training data is up as the increase of dimensionality of spanned space in any regions. The feature data with 48 channels is suitable for classification from these experiments of feature extraction. From above results, the Cross-validation test was performed in the four zones.

4.2   Cross-Validation

The three graphs from top right to bottom right in Fig. 4 shows the results of the Cross-validation for testing data of four zones. The result of the Cross-validation presents that the variation of accuracy with respect to the dimensionality r is small. Almost all zone data attains the maximum of accuracy when the dimensionality r is a value of 2. Table 2 lists the maximum of accuracy for all zone data.

By contrast the values of the specificity and sensitivity depend on values of dimensionality considerably. It can be seen that a large dimensionality reduces the value of sensitivity remarkably and it also reduces generalization ability. It

Table 1. Accuracies [%] and number of channel in each region

Channels (name)		Region number				
		1	2	3	4	5
24	(*ring*)	87.2	87.2	88.7	87.2	89.5
48	(*zone*)	97.0	97.0	97.0	96.2	-
120	(*all*)			97.0		

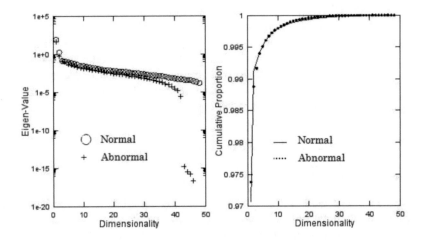

Fig. 3. Eigenvalues and Cumulative proportions

is possible that a lack of training data of abnormal subjects causes the singular matrix, or an over training may take place during learning process. Since a detection of abnormal subjects rather than normal ones is important in real clinical diagnosis, the number of spanned space dimension should be the value of the range of 2 to 6.

Table 2. Accuracies of each zone (%)

Zone	r	Accuracy	Specificity	Sensitivity
1	2	70.7	66.7	72.5
2	2	71.4	69.1	72.5
3	2	74.4	73.8	77.7
4	6	71.4	52.4	80.2

4.3 Comparison with Other Methods

In order to evaluate the classification performance using the Subspace classifier, these results were compared with the Cross-validation results by other methods. The Learning vector quantization method LVQ was consequently selected as a compared one. That is why LVQ has a high computational performance and a

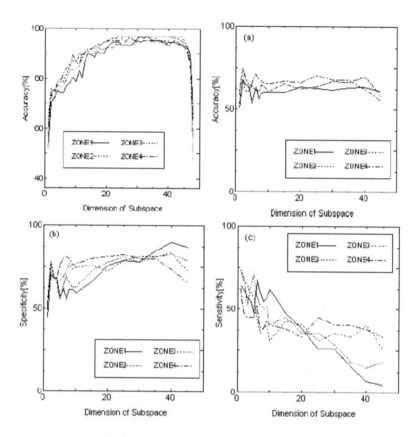

Fig. 4. Accuracies for training data of the zones and the Cross-validation for testing data: (a) accuracy, (b) specificity, and (c) sensitivity

good handling of parameters through our experiments of fundus images. For the classification, we used LVQ_PAK program package Version 3.1 [10], from which we used the program LVQ2. The main parameters used in the experiments were the same parameters in reference [8].

We conducted the classification using the LVQ2 for testing datasets of the third zone under conditions for the number of prototype NOC varying from 2 to 10. Maximum accuracy was 74.1%, and specificity and sensitivity were 88.1% and 43.3%, respectively, at NOC=7. In the Subspace classifier, the best accuracy, specificity and sensitivity were 74.4%, 77.2%, and 73.8%, respectively, at $r = 2$ from Table 2. Comparing the two methods, it can be seen that the performance of Subspace classifier is superior in the accuracy and sensitivity; the value of sensitivity was especially high.

4.4 Effects of Composition for Learning Patterns on Classification

Table 3 lists Cross-validation of each subset of testing patterns for dimensionality $r = 2$ in the third zone. In subsets g4 and g8, the less sensitivity and accuracy can be seen from this table. We found that from the similarity of the Subspace the extremely less sensitivity and accuracy were probably caused by four data: numbers 35 and 36 of normal data, the numbers 40 and 41 of abnormal ones. Therefore, in order to know how these data were arranged on the map, visualization was made using the blossom [11] as a Spherical SOM tool.

Figure 5 shows the numbers 35 of normal data and 41 of abnormal data for Glyph value 1.0 on the spherical map. The symbol * marked on the map is used for emphasis. The Glyph value is used to express the degree of the modification of the nodes of the blossom. From this map, it is shown that a normal data and an abnormal one were placed on the same node and two data are consequently inconsistent with each other.

Table 3. Cross-validation [%] for subsets of testing data in the third zone

Subset	g0	g1	g2	g3	g4	g5	g6	g7	g8	g9	Av.
Accuracy	84.6	84.6	61.5	84.6	53.9	46.2	68.2	84.6	53.9	50.0	74.4
Specificity	88.9	100.0	77.8	88.9	66.7	33.3	77.8	77.8	77.8	40.0	73.8
Sensitivity	75.0	50.0	75.0	25.0	25.0	75.0	50.0	100.0	0.0	66.7	77.2

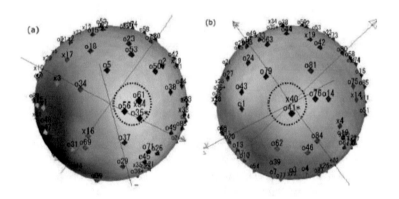

Fig. 5. Two normal data 35 and 41 for Gliph = 1 on the Spherical SOM

Although the image analysis using the Subspace classifier attains very high performance in spite of the small numbers of parameter (only a dimension of subspace), in order to improve further the accuracy of testing data such as data of table 3, the result of Cross-validation was corresponded with the map of Spherical SOM, and several combinations of learning pattern were composed. As a combination for the composition, we prepared the compositions four cases of learning patterns and evaluated their performance; the first composition was exchange of a pair of data between two subsets of testing patterns, this composition was further divided into two cases; exchange of a pair of normal data (case 1), and exchange for abnormal ones (case 2). The second composition, the case

3, was an alteration of class category within the same subset of testing patterns and the third composition was the exclusion of conflicting pattern from testing patterns.

Of the 4 cases shown in Tables 4 and 5, the three cases of the Cross-validation for $r = 2$ to 5. In the case 1, two numbers 35 and 41 of normal data in the subsets g3 and g4 were exchanged with the numbers 9 and 19 of normal ones in the subsets g0 and g1, respectively. Similarly in the case 2, two numbers 34 and 40 of abnormal data in the subsets g8 and g9 were also exchanged with the numbers 24 and 29 of abnormal ones in the subsets g5 and g7, respectively.

For the reliability of accuracy, the evaluation of performance improved by the composition is not the highest accuracy, and the average values of the accuracy for the range of dimensionality $r = 2$ and 5. From tables it was shown that composition of the learning patterns using visualization of Spherical SOM was effective in improving the performance of the analysis by the Subspace classifier; the operation of exchange of the learning data improved the averaged classification accuracy from 2.2 to 2.8% as compared with not performing the operation. As seen from Tables 4 and 5, the operation of composition of the learning pattern prevented the fall of accuracy in a wide range of dimensionality r and provided the high reliability of the performance.

Table 4. Accuracies [%] for the compositions from the case 1 to 3

r	Non composition	Case 1	Case 2	Case 3
2	74.44	74.23	74.23	72.09
3	67.67	68.80	68.70	68.46
4	64.66	69.00	67.45	67.58
5	63.91	66.54	66.54	67.47
Av	67.67	69.64	69.23	68.90

Table 5. Averaged performances and their standard deviations [%] for each case

Case No.	Accuracy Av.	Std. dev.	Specificity Av.	Std.dev.	Sensitivity Av.	Std.dev.
Non composition	67.76	4.80	70.33	5.23	61.91	4.73
1	69.64	3.26	72.53	3.70	61.91	3.37
2	69.23	3.45	71.98	4.06	61.91	3.37
3	68.90	2.17	71.35	2.80	61.91	1.43

5   Conclusion

We have proposed a composition of learning patterns based on the visualization of Spherical SOM for improving classification performance in the image analysis using the Subspace classifier. Without a composition of learning patterns from visual information, the Cross-validation for testing datasets was attained 74.4 % of maximum accuracy, 73.8% of specificity and 77.7% of sensitivity, respectively.

Furthermore we examined the classification performance on the image analysis using the proposed composition. With the proposed composition of learning patterns, the classification accuracy was improved from 2.2 to 2.8%. The potency with the accurate visualization of Spherical SOM allows the composition of the learning patterns to improve more performance and its reliability than those without the composition in the image analysis using the Subspace classifier.

References

1. Kohonen T.: Self-Organizing Maps. Springer Series in Information Sciences, vol. 3.0 (2001)
2. Wu, Y., Takatsuka, M.: Geodesic self-organizing map. In: Proceedings of Conference on Visualization and Data Analysis IS&T/SPIE (2005)
3. Tokutaka, H., Fujimura, K., Ohkita, M.: Cluster Analysis using Spherical SOM. Journal of Biomedical Fuzzy Systems Association 8(1), 29–39 (2006) (in Japanese)
4. Matsuda, N., Tokutaka, H., Laaksonen, J., Tajima, F., Miyatake, N., Sato, H.: Spherical SOM and its Application to Fundus Image Analysis. Journal of Biomedical Soft Computing and Human Sciences 11(1), 29–34 (2009) (in Japanese)
5. Cortes, C., Vapin, V.N.: Support vector networks. Machine Learning 20, 273–295 (1995)
6. Nishiyama, H., Hiraishi, H., Iwase, A., Mizoguch, F.: Design of Glaucoma Diagnosis System by Data Mining. In: 3A1-4 The 20th Annual Conference of the Japanese Society for Artificial Intelligence (2006) (in Japanese)
7. Watanabe, S., Pakvasa, N.: Subspace method of pattern recognition. In: 1st International Joint Conference of Pattern Recognition Proceeding, pp. 25–32 (1973)
8. Matsuda, N., Laaksonen, J., Tajima, F., Miyatake, N., Sato, H.: Fundus Image Analysis using Sub-space Classifier and its Performance. In: Proceedings of the Joint 5th International Conference on Soft Computing and Intelligent Systems and 11th International Symposium on Advanced Intelligent Systems, pp. 146–151 (2010)
9. Ritter, H.: Self-Organizing Maps on non-euclidean Spaces, Kohonen Maps. In: Oja, E., Kaski, S. (eds.), pp. 95–110. Elsevier (1999)
10. Kohonen, T., Kangas, J., Laaksonen, J., Torkkala, K.: LVQ-PAK: The Learning Vector Quantization Program Package. Helsinki Univ. of Tech., Finland (1995)
11. Blossom software tool, http://www.somj.com/

Self-Organizing Map for the Prize-Collecting Traveling Salesman Problem

Jan Faigl[1] and Geoffrey A. Hollinger[2]

[1] Czech Technical University in Prague,
Department of Computer Science and Engineering,
Technická 2, 166 27 Prague, Czech Republic
[2] Oregon State University,
School of Mechanical, Industrial, and Manufacturing Engineering,
204 Rogers Hall, Corvallis, OR 97331

Abstract. In this paper, we propose novel adaptation rules for the self-organizing map to solve the prize-collecting traveling salesman problem (PC-TSP). The goal of the PC-TSP is to find a cost-efficient tour to collect prizes by visiting a subset of a given set of locations. In contrast with the classical traveling salesman problem, where all given locations must be visited, locations in the PC-TSP may be skipped at the cost of some additional penalty. Using the self-organizing map, locations for the final solution may be selected during network adaptation, and locations where visitation would be more expensive than their penalty can be avoided. We have applied the proposed self-organizing map learning procedure to autonomous data collection problems, where the proposed approach provides results competitive with an existing combinatorial solver.

1 Introduction

The self-organizing map (SOM) is a two-layered artificial neural network that can be considered as a non-linear transformation (map) of a high-dimensional input space into a lower dimensional discrete output space. Its main feature is that it preserves topological properties of the input space in the output space. Although SOM was originally proposed as a data visualization technique, it has also been applied to solve NP-hard routing problems. The first such attempt was SOM for the Traveling Salesman Problem (TSP), which was proposed in 1988 by Angéniol and Fort.

The traveling salesman problem can be formulated as follows. Having a set of locations (cities) in a plane and a distance function between them, the TSP stands to find a shortest tour connecting all the given cities, such that each city is visited exactly once and the tour returns to the origin city. This problem arises from many practical applications [2], and the TSP is a well-studied problem in the operational research domain where efficient heuristics have been proposed [12].

SOM for the TSP has the output layer organized in a one-dimensional array of units representing a Peano curve that fills the input space. During unsupervised

T. Villmann et al. (eds.), *Advances in Self-Organizing Maps and Learning
Vector Quantization,* Advances in Intelligent Systems and Computing 295,
DOI: 10.1007/978-3-319-07695-9_27, © Springer International Publishing Switzerland 2014

learning, the cities are presented to the input layer, and the network is adapted by the means of moving neuron weights towards each presented city. Then, a solution is found as a sequence of cities retrieved by traversing the output layer where each winning neuron has an associated city in the input space.

Even though SOM has improved its performance in the TSP over the last decades [14,7,8], combinatorial heuristics still provide better results in classical graph-based instances of the TSP. On the other hand, SOM exhibits interesting results in planning problems where the locations to be visited are not explicitly prescribed [9] or where each location is represented by a set of possible points to be visited (i.e., the traveling salesman problem with neighborhoods (TSPN) [10]).

In this paper, we follow the recent advancements of SOM and propose a new adaptation rule to solve a variant of the TSP called the Prize Collecting Traveling Salesman Problem (PC-TSP) [6]. This problem is an extension of the standard TSP, where each city represents a prize that might be collected and where each prize also has an associated penalty cost if it is not collected. Thus, in the case where the penalty is significantly lower than the travel cost to the city, it is more suitable to avoid visitation of the city by the tour. The problem is to find a cost-efficient tour collecting the most important prizes (high penalty cities), i.e., to find a tour with the minimal total cost that is computed as the sum of the tour length (cost) in addition to the sum of penalties of all cities that are not visited by the tour.

The herein proposed SOM-based approach for the PC-TSP is based on the self-adjusting structure of the neural network proposed in [10] for solving the TSPN that has been extended to select and adapt the network to the most promising cities. To the best of our knowledge, the proposed method is the first application of SOM to the PC-TSP. The approach extends the application domain of SOM to additional optimization problems in which SOM can provide new ways of solving routing problems. The PC-TSP represents a class of problems where it is not sufficient to just alternate cities, but where it is also desirable to learn the underlying problem domain to select the most important cities to visit. Finding a solution to the PC-TSP consists of 1) a selection of the most promising cities and those that should be avoided and 2) finding a tour visiting the selected cities. SOM can be used to simultaneously address both of these problems together.

The paper is organized as follow. The problem motivation and the problem definition are presented in the next section. The proposed method is introduced in Section 3. Results of the proposed approach evaluation and a comparison with other methods are presented in Section 4 together with a discussion of found insights. Section 5 is dedicated to concluding remarks.

2   Problem Statement

The addressed problem is motivated by autonomous data collection, where it is requested to collect data from a number of sampling stations to create a model of a spatial phenomena as quickly as possible. Due to the spatial distribution of the sampling stations, the information retrieved from one station can also be included in measurements provided by other stations; thus, it is not necessary to retrieve data from all stations to acquire the desired model of the phenomena. The problem is formulated as a simplified variant of the PC-TSP [5], where each city has an associated penalty representing an importance of the measurement provided by the station [13].

In robotics, the TSP-based routing problems are alternatively called multi-goal path planning problems because the cities in the TSP represent goals towards which the robot navigates [13]. Therefore, to emphasize the robotic motivation of the studied problem, the term goal (or goal location) is used in the rest of this paper to denote the equivalent of cities in the TSP.

2.1 Problem Definition

Having n possible goal locations $\mathbf{G} = \{g_1, \ldots, g_n\}$, $g_i \in \mathbb{R}^2$, where each goal has associated penalty $\zeta(g_i) \geq 0$, and distance between two goals g_i and g_j is $c(g_i, g_j) \geq 0$, the problem is to find a tour T visiting a subset of the goals $G_T \subseteq \mathbf{G}$ such that the total cost of the tour $\mathcal{C}(T)$ is minimal

$$\mathcal{C}(T) = \sum_{(g_{s_i}, g_{s_{i+1}}) \in T} c(g_{s_i}, g_{s_{i+1}}) + \sum_{g \in \mathbf{G} \setminus G_T} \zeta(g). \qquad (1)$$

The tour T is a sequence of the selected goals $T = (g_{s_1}, \ldots, g_{s_{k-1}}, g_{s_k})$, where $g_{s_j} \in G_T$, $s_j \geq 1$, and $s_j \leq n$. The sequence represents a closed tour over the selected goals $g_{s_1} = g_{s_k}$, and all the selected goals, except the first (and last) visited goal, are included in the tour at most once.

For simplicity, which is also in line with the considered motivational application of autonomous data collection [13], the travel cost between two goals is computed as the Euclidean distance $c(g_i, g_j) = |(g_i, g_j)|$; i.e., the problem is considered in a planar environment without obstacles.

2.2 Performance Metric

The PC-TSP is NP-complete because it includes the TSP for very high penalties. Several approximation algorithms have been proposed for variants of the problem [4]. Although these algorithms provide guaranteed approximation factors relative to optimal, the approximation factors are relatively high (e.g., factor of 2.5 for the approach [6] based on the Christofides' algorithm) or somewhat better for a more complex algorithm [3].

In this paper, we consider a solution quality metric using the ratio of the PC-TSP solution to the related TSP (i.e., the shortest tour visiting all the goals and thus with zero penalty term in (1)). A solution to the TSP is available out-of-the-box using the Concorde solver [1], which provides an optimal solution for TSP problems (up to several hundred goals) within a reasonable time. Moreover, this ratio allows us to aggregate results for different problem instances and consider

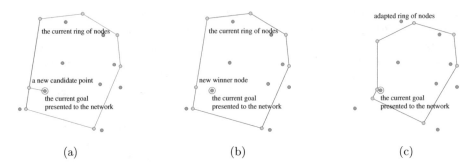

Fig. 1. Visualization of winner selection: (**a**) the closest point of the ring's segment to the presented goal; (**b**) new neuron added to the ring at the position of the closest point; (**c**) the current ring after adaptation of the winner and its neighbouring nodes. The goals are represented by green discs, while the blue discs denote the neuron weights.

statistical evaluation of the algorithmic performance using several hundreds of trials. The ratio is computed as

$$R = \frac{\mathcal{C}(T)}{\mathcal{C}(T_{TSP})},\tag{2}$$

where T_{TSP} is the optimal solution of the underlying TSP.

3   Proposed SOM for the PC-TSP

The proposed adaptation schema for the PC-TSP is based on two relatively straightforward properties that are not included in standard SOM-based approaches for the TSP. The first is a mechanism to adjust the number of neurons according to the currently selected number of goals to be visited. Otherwise, a poor solution will result because a low number of neurons will not provide convergence guarantees, and too many neurons will result in an inefficient adaptation of the winner neighborhood. The second property is the selection of the goals itself.

The first property is addressed by considering the self-adjusting adaptation rules proposed in [10]. The selection of goals is addressed in winner selection, where the neuron is considered to be a winner candidate only if its distance to the goal currently presented to the network is smaller than the goal's penalty. These two ideas are the foundation of the proposed solution.

The learning procedure consists of a two-layered competitive neural network similar to the one used for the TSP. The input layer represents a two-dimensional input vector, and the second layer consists of the output units organized into a uni-dimensional structure, where the neuron weights represent coordinates in a plane. The output units are defined by a sequence of straight line segments (called the ring) in $\mathbb{R}^2$ that represents the tour connecting the selected goals.

The main idea of the self-adjusting neural network is based on winner selection using the shortest distance of the presented goal to the network (i.e., the shortest distance of the goal to the ring). Having the current ring of neuron weights as a sequence of the straight line segments, the winner candidate is found as the closest point of the ring to the presented goal. If such a point corresponds to a neuron already presented in the ring, and the neuron was not a winner in the current learning epoch, the neuron is considered the winner candidate. Otherwise, a new neuron is created, and its weights are set to the position of the point. If such a winner candidate becomes the winner, the winner neuron is added to the ring; otherwise it is deleted whenever a new winner candidate is determined. The winner selection is schematically visualized in Fig. 1.

The winner selection provides a mechanism to create new neurons. The reduction of neurons is performed by the ring regeneration at the end of each learning epoch. At this time, the current winner neurons are preserved, and all other neurons are removed from the network. An additional neuron is inserted between each preserved winner to support spreading neurons during the adaptation. The weights of the neurons are set to the center of the segment connecting the two winners. The learning procedure is summarized in the following eight steps:

1. *Initialization:* Create $2n$ neurons around the first goal for the input set of n goals **G**. Set the neighbouring function variance $\sigma \leftarrow 10$ and the learning rate $\mu \leftarrow 0.99$. Set the current number of learning epoch $i \leftarrow 1$.
2. *Randomizing:* Create a random permutation of goals $\Pi(\mathbf{G}) \leftarrow \text{permute}(\mathbf{G})$.
3. *Winner Selection:* If $i == 1$
   - **Then:** Select the closest point p of the current ring to the presented goal $g \in \Pi(\mathbf{G})$;
   - **Otherwise:** Select the closest point p of the current ring to $g \in \Pi(\mathbf{G})$ such that $|(p, g)| < \zeta(g)$.
4. *Adapt:* If p is selected
   - Determine the appropriate winner neuron $\nu^\star$ (select or create new one).
   - Adapt the winner $\nu^\star$ and its neighbouring nodes ν_j within the distance d (in the number of nodes) using the neighbouring function $f(\sigma, d) = \mu e^{(-d^2/\sigma^2)}$ for $d < 0.2m$ and $f(\sigma, 0) = 0$ otherwise, where m is the current number of neurons, i.e., move ν closer towards g about $|(\nu, g)| f(\sigma, d)$.
   Remove g from the permutation, $\Pi(\mathbf{G}) \leftarrow \Pi(\mathbf{G}) \setminus \{g\}$, and If $|\Pi(\mathbf{G})| > 0$ go to Step 3.
5. *Ring regeneration:* Create a new ring using only the winner nodes of the current learning epoch, and add a new neuron between each two consecutive winner nodes ν_i and ν_j with the weights set to $\nu_i + (\nu_i - \nu_j)/2$.
6. *Update the number of the learning epoch and neighbouring function variance:* $i \leftarrow i + 1$; $\sigma \leftarrow (1 - 0.0005i)\sigma$.
7. *Termination condition:* If the maximal distance of the winner to its goal is less than 10^{-3}, stop the adaptation. Otherwise go to Step 2.
8. *Final tour construction:* Traverse the ring and use the associated goals to the last winners to construct the final goal tour.

Computational complexity of the learning procedure can be derived from the number of comparisons needed to find the best matching neuron for each presented goal to the network, which is the most time-consuming operation. The number of goals is n, and the number of neurons can be bounded by $3n$ at every moment of the adaptation, which gives $3n^2$ operations. The values of μ and $f(\sigma, d)$ are always less than 1, and thus the adaptation rule is stable [15]. Besides, the neighbors of the winner are effectively moved only for a sufficiently high value of $f(\sigma, d)$, and since σ does not depend on n, the network is stabilized in a constant number of learning epochs. Thus, the overall computational complexity can be bounded by $O(n^2)$.

The required memory only depends on the representation of the goals and neurons, which are basically coordinates in the plane accompanied by the penalty and associated goal for the winners. Hence, the space can be bounded by $O(n)$.

4   Results

The proposed SOM for the PC-TSP was first validated in simple use cases to verify feasibility of the proposed principle of goal selection. Then, its performance was evaluated in problem instances proposed in [13]. Results from these two evaluation scenarios are presented in the following sections.

4.1   Simple Use Cases

The initial feasibility test was performed using 8 goals forming two concentric squares with a side length 10 and 8. According to the value of the penalties, the solution of the PC-TSP is a tour connecting all the goals (for penalties 10 and 2) and a single connected square of the outer (penalties 10 and 0) or inner (penalties 0 and 10) goals. The found solutions are visualized in Fig. 2, where the red discs denote goals with higher penalties (more important goals), and goals with zero or small penalties are shown in blue. The solutions of these simple problems correspond to optimal solutions and thus provide validation of the proposed adaptation rules.

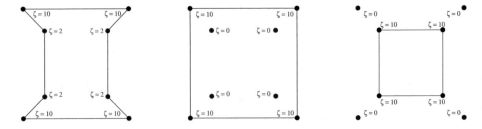

Fig. 2. Example of found solutions for concentric squared goals problems

Table 1. Average ratios R of the solution cost to the optimal solution of the related TSP for 50 problems in the 100 km × 100 km area and 2500 SOM trials

Problem	f	SOM TSP [10]	PC-TSP [11],[1]	SOM PC-TSP Proposed	SOM PC-TSP+TSP Proposed + [1]
area 20×20	0.001	1.03/0.02	1.00/<0.01	1.04/0.02/(−)	**1.00/<0.01/(+)**
area 20×20	0.100	1.03/0.02	1.00/0.01	1.03/0.02/(−)	**1.00/<0.01/(+)**
area 20×20	1.000	1.03/0.02	1.02/0.02	1.02/0.02/(=)	**0.99/0.01/(+)**
area 20×20	2.000	1.03/0.02	1.02/0.03	**1.01/0.02/(+)**	**0.98/0.02/(+)**
area 20×20	5.000	1.03/0.02	1.03/0.05	**0.97/0.05/(+)**	**0.95/0.05/(+)**
area 20×20	7.000	1.03/0.02	1.00/0.04	1.00/0.07/(=)	**0.99/0.07/(=)**
area 20×20	10.000	1.03/0.02	0.84/0.08	**0.79/0.06/(+)**	**0.79/0.06/(+)**

Values in columns are: average / standard deviation / (statistical comparison).
(+) - the algorithm provides statistically better solutions than the PC-TSP [11,1].

4.2 Performance Evaluation

The performance evaluation is based on the solution to data collection problems consisting of 100 randomly placed sensors within a 100 km × 100 km large area, where each goal penalty is randomly drawn from the range 0 to 25. Similarly to [13], the vehicle speed is assumed to be 5 km per hour. Thus, the goals are effectively placed in a 20 × 20 large square, and the cost between goals is directly computed as their Euclidean distance. In addition, more problem instances are created from this setup by dividing the penalty by the value f, which allows us to study the algorithm's performance for different penalties. Notice, that for very high penalties (e.g., $f \leq 0.1$), the problems become close to the TSP.

The proposed SOM approach is compared with the combinatorial deterministic approach considered in [13]. The prior combinatorial approach was based on the heuristic determination of the goals to ignore [11] and a consecutive optimal solution of the TSP for the remaining goals using the Concorde solver [1]. The SOM approach is considered in two variants. First, it is used for simultaneous selection of the goals together with the tour connecting them. The second variant uses the goals determined by the first variant that are connected by the optimal tour found by [1].

For each problem scenario, 50 random instances are created, and the algorithm's performance is measured as the average value of the ratio R of the found solution of the PC-TSP to the optimal solution of the related TSP introduced in (2). Due to stochastic nature of SOM, 50 trials are solved for each problem instance, which gives 2500 solutions to compute the average ratio for a particular problem scenario.

Based on the statistical data, a comparison of the SOM-based algorithms with the reference algorithm was performed. This tests the null hypothesis that the random variables R describing the quality of the provided solutions are from the some distribution. For each scenario and particular factor, the ratio R is considered to be a random variable over the 50 problem instances. For SOM, the average C from all 50 trials is considered to be the solution quality of the scenario. These random variables are considered to be drawn from normal

Table 2. Average ratios R of the solution cost to the optimal solution of the related TSP for 50 problems in the 200 km × 200 km area and 2500 SOM trials

Problem	f	SOM TSP [10]	PC-TSP [11],[1]	SOM PC-TSP Proposed	SOM PC-TSP+TSP Proposed + [1]
area 40×40	0.001	1.03/0.02	1.00/<0.01	1.04/0.02/(−)	1.00/<0.01/(=)
area 40×40	0.100	1.03/0.02	1.00/<0.01	1.04/0.02/(−)	1.00/<0.01/(=)
area 40×40	1.000	1.03/0.02	0.98/0.02	1.00/0.03/(−)	**0.95/0.02/(+)**
area 40×40	2.000	1.03/0.02	0.98/0.02	0.97/0.03/(=)	**0.95/0.02/(+)**
area 40×40	5.000	1.03/0.02	0.91/0.06	**0.78/0.05/(+)**	**0.78/0.05/(+)**
area 40×40	7.000	1.03/0.02	0.68/0.12	**0.58/0.03/(+)**	**0.58/0.03/(+)**
area 40×40	10.000	1.03/0.02	0.42/0.06	0.43/0.03/(=)	0.43/0.03/(=)

average value / standard deviation / (statistical comparison)
(+) - the algorithm provides statistically better solutions than the PC-TSP [11,1].

distributions since the problems are random. The algorithms are considered to provide statistically different results if the P-value of the T-test is below 0.05. In this case, the algorithm with a lower average value of R is considered better (denoted by the character '+'). The computed averages, standard deviations, and results of statistical comparison are presented in Table 1.

The results in Table 1 indicate that SOM solutions of the PC-TSP provide paths with almost identical costs as the pure solution of the TSP. Even though not all sensors are visited, the cost of the final path is (on average) similar to the tour visiting all the goals. To further explore the benefit of the proposed approach, an additional set of problems was created within a larger area with dimensions 200 km × 200 km, which provides a better opportunity to avoid distant goals with small penalties. Statistical indicators for these problems are presented in Table 2, and an example of found solutions is depicted in Fig. 3.

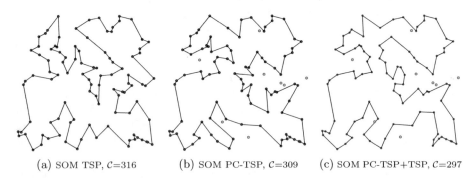

(a) SOM TSP, $\mathcal{C}$=316   (b) SOM PC-TSP, $\mathcal{C}$=309   (c) SOM PC-TSP+TSP, $\mathcal{C}$=297

Fig. 3. Found solutions by the SOM for the TSP and PC-TSP

Regarding the required computational time, the proposed SOM algorithm provides solutions of the PC-TSP (with 100 goals) in less than ten milliseconds using a single core of the iCore7 processor running at 3.4 GHz. The real computational requirements are depicted in Fig. 4. In some cases, finding the

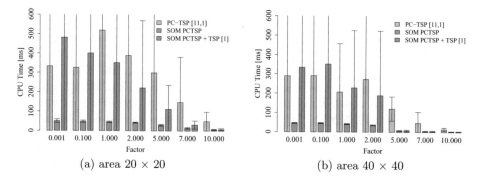

Fig. 4. Real computational time (in milliseconds) to solve the PC-TSP

optimal solution of the TSP by [1] is computationally demanding, which makes standard deviations very high. On the other hand, the proposed SOM based PC-TSP algorithm requires almost identical computational time per particular problem scenario. The computational burden decreases for lower penalties, which is caused by selection of only few goals to be visited.

Discussion – The presented results indicate the proposed SOM adaptation rules provide a feasible solution to the PC-TSP that achieves performance competitive with other approaches. Moreover, the results also indicate that for problems where the selection of the goals is more important (i.e., problems with a higher penalty factor f), the SOM approach performs better.

On the other hand, the benefit of solving the considered instances using the PC-TSP is not always evident because the solution quality is sometimes similar to the solution of the pure TSP. This is mainly caused by instances of the problem with dense goals, as is indicated by the results for goals placed within a larger area. Similar findings have been reported in [13]. All together, these results indicate that solving the motivational data collection problem as an instance of the PC-TSP provides substantial benefit when the penalties are small relative to the size of the environment.

The addressed problem provides the groundwork for studying the principles of SOM adaptation employed in routing problems. Here it is worth mentioning that an explicit consideration of the final cost of the tour represented by the ring in the selection of winners does not provide good results, and the convergence of the network becomes much slower than for the proposed approach. This also holds for different adaptation schemata (e.g., a fixed number of neurons), which do not provide stable adaptation with competitive results.

5   Conclusion

Novel adaptation rules to address the prize-collecting traveling salesman problem using a self-organizing map have been introduced in this paper. The main feature of the proposed SOM for the PC-TSP is that it provides a simultaneous selection of the goals together with a sequence of their visit. The proposed method uses SOM-based solution to the PC-TSP that provides competitive results to other methods. As such, it extends the portfolio of routing problems where SOM can be applied.

In the motivational autonomous data collection task, the PC-TSP provides substantial reduction in solution cost when the penalties are small relative to the size of the environment. Previous research indicates that considering a non-zero communication radius to read the data from the sensor station from a longer distance can improve the solution more significantly. This extension would lead to a combination of the PC-TSP with the traveling salesman problem with neighborhoods (TSPN), where SOM has also been successfully applied. Therefore, we plan to extend the proposed approach to address such combined problems. In addition, our future work is also directed towards considering dynamic evaluation of the penalty when it depends on both the current tour and the expected goals to be visited.

Acknowledgments. The presented work is supported by the Czech Science Foundation (GAČR) under research project No. 13-18316P. Support under United States National Science Foundation Grant IIS-1317815 to Geoffrey Hollinger is also gratefully acknowledged.

References

1. Applegate, D., Bixby, R., Chvátal, V., Cook, W.: CONCORDE TSP Solver (2003), http://www.tsp.gatech.edu/concorde.html (cited October 20, 2013)
2. Applegate, D., Bixby, R., Chvátal, V., Cook, W.: The Traveling Salesman Problem: A Computational Study. Princeton University Press, Princeton (2007)
3. Archer, A., Bateni, M., Hajiaghayi, M., Karloff, H.: Improved approximation algorithms for prize-collecting steiner tree and tsp. In: IEEE Symposium on Foundations of Computer Science (2009)
4. Ausiello, G., Bonifaci, V., Leonardi, S., Marchetti-Spaccamala, A.: Prize-collecting traveling salesman and related problems. In: Gonzalez, T.F. (ed.) Handbook of Approximation Algorithms and Metaheuristics. CRC Press (2007)
5. Balas, E.: The prize collecting traveling salesman problems. Networks 19, 621–636 (1989)
6. Bienstock, D., Goemans, M., Simchi-Levi, D., Williamson, D.: A note on the prize collecting traveling salesman problem. Mathematical Programming 59, 413–420 (1993)
7. Cochrane, E.M., Beasley, J.E.: The co-adaptive neural network approach to the Euclidean travelling salesman problem. Neural Networks 16(10), 1499–1525 (2003)
8. Créput, J.C., Koukam, A.: A memetic neural network for the Euclidean traveling salesman problem. Neurocomputing 72(4-6), 1250–1264 (2009)
9. Faigl, J.: Approximate Solution of the Multiple Watchman Routes Problem with Restricted Visibility Range. IEEE Transactions on Neural Networks 21(10), 1668–1679 (2010)
10. Faigl, J., Přeučil, L.: Self-Organizing Map for the Multi-Goal Path Planning with Polygonal Goals. In: Honkela, T. (ed.) ICANN 2011, Part I. LNCS, vol. 6791, pp. 85–92. Springer, Heidelberg (2011)

11. Goemans, M., Williamson, D.P.: A general approximation technique for constrained forest problems. SIAM J. Computing 24(2), 296–317 (1995)
12. Helsgaun, K.: An Effective Implementation of the Lin-Kernighan Traveling Salesman Heuristic. European Journal of Operational Research 126(1) (2000)
13. Hollinger, G., Mitra, U., Sukhatme, G.: Autonomous data collection from underwater sensor networks using acoustic communication. In: IROS, pp. 3564–3570. IEEE (2011)
14. Somhom, S., Modares, A., Enkawa, T.: A self-organising model for the travelling salesman problem. Journal of the Operational Research Society, 919–928 (1997)
15. Tucci, M., Raugi, M.: Stability analysis of self-organizing maps and vector quantization algorithms. In: IJCNN, pp. 1–5 (2010)

A Survey of *SOM*-Based Active Contour Models for Image Segmentation

Mohammed M. Abdelsamea[1], Giorgio Gnecco[1], and Mohamed Medhat Gaber[2]

[1] IMT Institute for Advanced Studies, Lucca, Italy
[2] Robert Gordon University, Aberdeen, UK
mohammed.abdelsamea@imtlucca.it,
giorgio.gnecco@imtlucca.it,
m.gaber1@rgu.ac.uk

Abstract. Self Organizing Maps (*SOM*s) have attracted the attention of many computer vision scientists, particularly when dealing with image segmentation as a contour extraction problem. The idea of utilizing the prototypes (weights) of a *SOM* to model an evolving contour has produced a new class of Active Contour Models (*ACM*s), known as *SOM*-based *ACM*s. Such models have been proposed in general with the aim of exploiting the specific ability of *SOM*s to learn the edge-map information via their topology preservation property, and overcoming some drawbacks of other *ACM*s, such as trapping into local minima of the image energy functional to be minimized in such models. In this survey paper, the main principles of *SOM*s and their application in modelling active contours are first highlighted. Then, we review existing *SOM*-based *ACM*s with a focus on their advantages and disadvantages in modelling the evolving contour via different kinds of *SOM*s. Finally, some current research directions are identified.

Keywords: Image segmentation, Self Organizing Maps, active contours, *SOM*-based *ACM*s, topology preservation, neural networks.

1 Introduction

Image segmentation is the problem of partitioning the domain Ω of an image $I(x)$, where $x \in \Omega$ is the pixel location within the image, into different subsets Ω_i, where each subset has a different characterization in terms of color, intensity, texture, and/or other features used as similarity criteria. Segmentation is a fundamental component of image processing, and plays a significant role in computer vision, object recognition, and object tracking.

Active Contour Models (*ACM*s) usually deal with the segmentation problem as an optimization problem, formulated in terms of a suitable "energy" functional, constructed in such a way that its minimum is achieved in correspondence with a contour that is a close approximation of the actual object boundary. Starting from an initial contour, the optimization is performed iteratively, evolving the current contour with the aim of approximating better and better the actual object

T. Villmann et al. (eds.), *Advances in Self-Organizing Maps and Learning Vector Quantization,* Advances in Intelligent Systems and Computing 295,
DOI: 10.1007/978-3-319-07695-9_28, © Springer International Publishing Switzerland 2014

boundary (hence the denomination "active contour" models, which is used also for models that evolve the contour but are not based on the explicit minimization of a functional [1]). In order to guide efficiently the evolution of the current contour, $ACMs$ allow to integrate various kinds of information inside the energy functional, such as: local information (e.g., features based on spatial dependencies among pixels), global information (e.g., features which are not influenced by such spatial dependencies), shape information[1], prior information, and a-posteriori information learned from examples. As a consequence, depending on the kind of information used, one can divide $ACMs$ into several categories: e.g., edge-based $ACMs$ [4–7], global region-based $ACMs$ [8,9], edge/region-based $ACMs$ [10–12], local region-based $ACMs$ [13–15], and global/local region-based $ACMs$ [16,17]. In particular, edge-based $ACMs$ make use of an edge-detector (in general, the gradient of the image intensity) to stop the evolution of the active contour on the true boundaries of the objects of interest. Instead, region-based $ACMs$ use with the same purpose statistical information about the regions to be segmented (e.g., intensity, texture, color distribution, etc.). Depending on how the active contour is represented, one can also distinguish between parametrized [18] and level set-based $ACMs$ [8].

Although $ACMs$ often provide an effective and efficient means to extract smooth and well-defined contours, trapping into local minima of the energy functional may still occur, because such a functional may be constructed on the basis of simplified assumptions on properties of the images to be segmented (e.g., the assumption of Gaussian intensity distributions for the sets Ω_i in the case of the Chan-Vese model [2,8]). Motivated by this observation and by the specific ability of $SOMs$ to learn - via their topology preservation property [19] - information about the edge map of the image (i.e., the set of points obtained by an edge-detection algorithm), a new class of $ACMs$, named SOM-based $ACMs$ [20,21], has been proposed with the aim of modelling and controlling effectively the evolution of the active contour by a Self Organizing Map (SOM), in general without relying on an explicit energy functional to be minimized. In this paper, we review some concepts of $ACMs$ with a focus on SOM-based $ACMs$, illustrating both their strengths and limitations.

The paper is organized as follows. Section 2 provides a brief discussion on parametrized and level set-based $ACMs$. In Section 3, we review the state of the art of SOM-based $ACMs$. Section 4 provides some conclusions and discuss current research directions in SOM-based $ACMs$.

2   Active Contour Models ($ACMs$)

In order to describe an active contour, there are mainly two classes of methods: parametrized methods, in which the contour is represented by a parametric curve, and variational level set methods, for which the contour is the zero level set of a suitable function.

---

[1] Due to the possible lack of a precise prior information on the shape of the objects to be segmented, in this respect most $ACMs$ make only the assumption that it is preferable to have a smooth boundary [2]. This goal is achieved by incorporating a suitable regularization term into their energy functional [3].

2.1   Parametrized *ACM*s

In parametrized *ACM*s, the contour C is represented as

$$C := \{x \in \Omega : x = (x_1(s), x_2(s)), 0 \le s \le 1\}, \tag{1}$$

where $x_1(s)$ and $x_2(s)$ are functions of the scalar parameter s. A representative parametrized *ACM* is the Snakes model, proposed by Kass et al. [3] (see also [18] for successive developments). The main drawbacks of parametrized *ACM*s are the frequent occurrence of local minima in the image energy functional to be optimized (which is mainly due to the presence of a gradient energy term inside such a functional), and the fact that topological changes of the objects (e.g., merging and splitting) cannot be handled during the evolution of the contour.

2.2   Level Set-Based *ACM*s

Level set-based *ACM*s (also called geometric *ACM*s) were first proposed by Osher and Sethian [22]. The difference between parametric and level set-based *ACM*s is that in the latter, the contour C is implemented via a variational level set method, i.e., it is implicitly represented by a function $\phi(x)$, called level set function, where x is pixel location in the image domain Ω. Then, C is defined as the zero level set of the function $\phi(x)$ by the expression

$$C := \{x \in \Omega : \phi(x) = 0\}. \tag{2}$$

A representative state-of-the-art level set-based *ACM* is the Chan-Vese model [8]. Level set-based *ACM*s have the advantage on parametrized *ACM*s of being able to model arbitrarily complex shapes, and to handle implicitly topological changes (e.g., presence/absence of internal connectedness) of the regions to be segmented. However, likewise parametrized *ACM*s, this class of models is sensitive to local minima. An effective solution to deal with this issue consists in using instead *SOM*s to model and control the evolution of the active contour, resulting in a new class of *ACM*s, called *SOM*-based *ACM*s, which are described in the following section.

3   *SOM*-Based *ACM*s

Before discussing *SOM*-based *ACM*s, we shortly review the use of *SOM*s as a tool in pattern recognition (hence, in image segmentation as a particular case).

3.1   Self Organizing Maps (*SOM*s)

The *SOM* [19], which was proposed by Kohonen, is an unsupervised neural network whose neurons update concurrently their weights in a self-organizing manner, in such a way that, during the learning process, its neurons evolve adaptively into specific detectors of different input patterns. A basic *SOM* is composed of an input layer, an output layer, and an intermediate connection layer. The input layer contains a unit for each component of the input vector. The output layer consists of neurons that are typically located either on a 1-D or a 2-D grid, and are fully connected with the units in the input layer. The intermediate connection layer is composed of weights (also called prototypes) connecting the units in the input layer and the neurons in the output layer (in practice, one has one weight vector associated with each output neuron, where the dimension of the weight vector is equal to the dimension of the input). The learning algorithm of the SOM can be summarized by the following steps:

1. initialize randomly the weights of the neurons in the output layer, and select suitable learning rate and neighborhood size around a "winner" neuron;
2. for each training input vector, find the winner neuron using a suitable rule;
3. update the weights on the selected neighborhood of the winner neuron;
4. repeat Steps 2-3 above selecting another training input vector, until learning is accomplished (i.e., a suitable stopping criterion is satisfied).

SOMs have been used extensively for image segmentation, but often not in combination with ACMs [23, 24]. In order to improve the robustness of edge-based ACMs to the blur and to ill-defined edge information, SOMs have been also used in combination with ACMs, with the explicit aim of modelling the active contour and controlling its evolution, adopting a learning scheme similar to Kohonen's learning algorithm [19], resulting in SOM-based ACMs [20, 21] (which belong, in this case, to the class of edge-based ACMs). The evolution of the active contour in a SOM-based ACM is guided by the feature space constructed by the SOM when learning the weights associated with the neurons of the map. Among other neural network models applied in combination with ACMs, we mention multilayer perceptrons [25]. One reason to prefer SOMs to other neural network models consists in the specific ability of SOMs to learn the edge-map information via their topology preservation property. A review of SOM-based ACMs is provided in the following subsections.

3.2   An Example of a SOM-Based ACM

The basic idea of existing SOM-based ACMs is to model and implement the active contour using a SOM, relying in the training phase on the edge map of the image to update the weights of the neurons of the SOM, and consequently to control the evolution of the active contour. The points of the edge map act as inputs to the network, which is trained in an unsupervised way (in the sense that no supervised samples belonging to the foreground/background, resp., are provided). As a result, during training the weights associated with the neurons in the output map move toward points belonging to the nearest salient contour. In the following, we illustrate the general ideas of using a SOM in modelling the active contour, by describing a classical example of a SOM-based ACM, which was proposed in [20] by Venkatesh and Rishikesh.

Spatial Isomorphism Self Organizing Map (*SISOM*)-Based *ACM* [20].
The *SISOM*-based *ACM* is the first *SOM*-based *ACM* which appeared in the
literature. It was proposed with the aim of localizing the salient contours in an
image using a *SOM* to model the evolving contour. The *SOM* is composed of a
fixed number of neurons (and consequently a fixed number of "knots" or control
points for the evolving curve) and has a fixed structure. The model requires
a rough approximation of the true boundary as an initial contour. Its *SOM*
network is constructed and trained in an unsupervised way, based on the initial
contour and the edge-map information. The contour evolution is controlled by
the edge information extracted from the image by an edge detector. The main
steps of the *SISOM*-based *ACM* can be summarized as follows:

1. construct the edge map of the image to be segmented;
2. initialize the contour to enclose the object of interest in the image;
3. obtain the x_1- and x_2- coordinates of the edge points to be presented as
   inputs to the network;
4. construct a *SOM* with a number of neurons equal to the number of the edge
   points of the initial contour and two weights associated with each neuron;
   the points on the initial contour are used to initialize the *SOM* weights;
5. repeat the following steps for a fixed number of iterations:
   (a) select randomly an edge point and feed its coordinates to the network;
   (b) determine the best-matching neuron;
   (c) update the weights of the neurons in the network by the classical un-
       supervised learning scheme of the *SOM* [19], which is composed of a
       competitive phase and a cooperative one;
   (d) compute a neighborhood parameter for the contour according to the
       updated weights and a threshold.

Fig. 1 illustrates the evolution procedure of the *SISOM*-based *ACM*. On the
left-side of the figure, the neurons of the map are represented by gray circles,
while the black circle represents the winner neuron associated with the current
input to the map (in this case, the red circle on the right-hand side of the figure,
which is connected by the blue segments to all the neurons of the map). On the
right-hand side, instead, the positions of the white circles represent the initial
prototypes of the neurons, whereas the positions of the black circles represent
their final values, at the end of learning. The evolution of the contour is controlled
by the learning algorithm above, which guides the evolution of the protoypes
of the neurons of the *SOM* (hence, of the active contour) using the points of
the edge map as inputs to the *SOM* learning algorithm. As a result, the final
contour is represented by a series of prototypes of neurons located near the actual
boundary of the object to be segmented.

We conclude by mentoning that, in order to produce good segmentations, the
SISOM-based *ACM* requires the initial contour (which is used to initialize the
prototypes of the neurons) to be very close to the true boundary of the object
to be extracted, and the points of the initial contour have to be assigned to
the neurons of the *SOM* in a suitable order: if such assumptions are satisfied,
the contour extraction process performed by the model is robust to the noise.

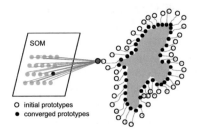

Fig. 1. The architecture of the $SISOM$-based ACM proposed in [20]

Moreover, differently from other ACMs, the model does not require a particular energy functional to be optimized.

3.3   Other SOM-Based ACMs

In this subsection, we describe other SOM-based ACMs, and highlight their advantages and disadvantages.

Time Adaptive Self Organizing Map ($TASOM$)-Based ACM [21]. The $TASOM$-based ACM was proposed by Shah-Hosseini and Safabakhsh as a development of the $SISOM$-based ACM, with the aim of inserting neurons incrementally into the SOM map or deleting them incrementally, thus determining automatically the required number of control points of the extracted contour. Moreover, each neuron is provided with its specific dynamic learning rate and neighbourhood function. As a consequence, the $TASOM$-based ACM can overcome one of the main limitations of the $SISOM$-based ACM, i.e., its sensitivity to the contour initialization, in the sense that the initial guess of the contour in the $TASOM$-based ACM can be far from the actual object boundary. Likewise the $SISOM$-based ACM, topological changes of the objects (e.g., splitting and merging) cannot be handled, since both models rely completely on the edge information (instead than on regional information) to drive the contour evolution.

Batch Self Organizing Map ($BSOM$)-Based ACM [1, 26]. This model is a modification of the $TASOM$-based ACM, and was proposed by Venkatesh et al. with the aim of dealing better with the leaking problem (i.e., the presence of a final blurred contour), which often occurs when handling images with ill-defined edges. Such a problem is due to the explicit usage by the $TASOM$-based ACM of only edge information to model and control the evolution of the contour. In the $BSOM$-based ACM, instead, the image intensity variation inside a local region is used along with the edge information to control the movement of the contour. In this way, the robustness of the model is increased in handling images with blurred edges. At the same time, the $BSOM$-based ACM is less sensitive to the initial guess of the contour, when compared to parametrized ACMs like Snakes, and to the SOM-based ACMs described above. However, likewise all such models, the $BSOM$-based ACM has not the ability to handle topological changes of the objects to be segmented. An extension of the *BSOM*-based *ACM* was proposed in [27, 28] and applied therein to the segmentation of pupil images. Such a modified version of the basic *BSOM*-based *ACM* increases the smoothness of the extracted contour, and prevents the extracted contour from being extended over the true boundaries of the object.

Fast Time Adaptive Self Organizing Map (*FTA-SOM*)-Based *ACM* [29]. This is another modification of the *TASOM*-based *ACM*, and was proposed by Izadi and Safabakhsh with the aim of decreasing its computational complexity. The *FTA-SOM*-based *ACM* is based on the observation that choosing the learning rate parameters of the prototypes of the neurons of the *SOM* in such a way that they are equal to a large fixed value when they are far from the boundary, and to a small value when they are near the boundary, can lead to a significant increase of the convergence speed of the active contour. Accordingly, in each iteration, the *FTA-SOM*-based *ACM* finds the minimum distance of each neuron from the boundary, then its sets the associated learning rate as a fraction of that distance.

Coarse to Fine Boundary Location Self Organizing Map (*CFBL-SOM*)-Based *ACM* [30]. The above *SOM*-based *ACM*s work in an unsupervised way, as the user is required only to provide an initial contour to be evolved automatically. In [30], Zeng et al. proposed the *CFBL-SOM*-based *ACM* as the first supervised *SOM*-based *ACM*, i.e., a model in which the user is allowed to provide supervised points (supervised "seeds") from the desired boundaries. Starting from this coarse information, the *SOM* neurons are then employed to evolve the contour to the desired boundaries in a "coarse-to-fine" approach. The *CFBL-SOM*-based *ACM* follows such a strategy when controlling the evolution of the contour. So, an advantage of the *CFBL-SOM*-based *ACM* over the *SOM*-based *ACM*s described above is that it allows to integrate prior knowledge on the desired boundaries of the objects to be segmented, which comes from the user interaction with the *SOM*-based *ACM* segmentation framework. When compared with such *SOM*-based *ACM*s, this property provides the *CFBL-SOM*-based *ACM* with the ability of handling objects with more complex shapes, inhomogeneous intensity distributions, and weak boundaries.

Conscience, Archiving and Mean-Movement Mechanisms Self Organizing Map (*CAM-SOM*)-Based *ACM* [31]. The *CAM-SOM*-based *ACM* was proposed by Sadeghi et al. as an extension of the *BSOM-ACM*, by introducing three mechanisms called Conscience, Archiving and Mean-Movement. The main achievement of the *CAM-SOM*-based *ACM* is to allow more complex boundaries (such as concave boundaries) to be captured, and to provide a reduction of the computational cost. By the Conscience mechanism, the neurons are not allowed to "win" too much frequently, which makes the capture of complex boundaries possible. The Archiving mechanism allows a significant reduction in the computational cost. By such mechanism, neurons whose prototypes are close to the boundary of the object to be segmented and whose values have not changed significantly in the last iterations are archived and eliminated from subsequent computations. Finally, in order to ensure a continuous movement of the active contour towards concave regions, the Mean-Movement mechanism is used in each epoch to force the winner neuron to move towards the mean of a set of feature points, instead of a single feature point. Together, the Conscience and Mean-Movement mechanisms prevent the contour from stopping the contour evolution at the entrance of object concavities.

Extracting Multiple Objects. The main limitation of various SOM-based ACMs is their inability to detect multiple contours and to recognize multiple objects. As mentioned above, a similar problem arises in parametric ACMs such as Snakes. To deal with the multiple contour extraction problem, Venkatesh et al. proposed in [26] to use a splitting criterion. However, if the initial contour is outside the objects, contours inside an object still cannot be extracted. Sadeghi et al. proposed in [31] a splitting criterion (to be checked at each epoch) such that the main contour can be divided into several sub-contours whenever the criterion is satisfied. The process is repeated until each of the sub-contours encloses one single object. However, the merging process is still not handled implicitly by the model, which reduces its scope, especially when handling images containing multiple objects in the presence of noise or ill-defined edges. Moreover, Ma et al. proposed in [32] to use a SOM to classify the edge elements in the image. This model relies first on detecting the boundaries of the objects. Then, for each edge pixel, a feature vector is extracted and normalized. Finally, a SOM is used as a clustering tool to detect the object boundaries when the feature vectors are supplied as inputs to the map. As a result, multiple contours can be recognized. However, the model shares the same limitations of other models that use a SOM as a clustering tool for image segmentation [23,33,34], resulting in disconnected boundaries and sensitivity to the presence of the noise.

4   Conclusions and Current Research Directions

In this paper, a survey has been provided about the current state of the art of SOM-based ACMs. SOM-based ACMs have been proposed with the aim of exploiting the specific ability of SOMs to learn the edge-map information via their topology preservation property, and reducing the occurrence of local minima, which is typical of parametrized ACMs such as Snakes. This is partly due to the fact that SOM-based ACMs do not rely on an explicit gradient energy term. Although SOM-based ACMs can effectively outperform other ACM models in handling complex images, existing SOM-based ACMs are still sensitive to the contour initialization compared to level set-based ACMs, especially when handling complex images with ill-defined edges. Moreover, SOM-based ACMs have not usually the ability to handle topological changes of the objects.

Among current research directions, we mention: 1) the possibility of combining the advantages of SOMs and the ones of level set-based ACMs; 2) the development of more sophisticated supervised SOM-based ACMs based, e.g., on the use of Concurrent Self Organizing Maps ($CSOM$s) [35]; 3) the possibility of constructing *SOM*-based *ACM*s relying on regional-based information (e.g., local/global statistical information about the intensity, texture, color distribution, etc.) instead of edge information, to guide the evolution of the active contour. Such issues have been recently addressed in [36], where the advantages of the *SOM*-based *ACM* proposed therein have been demonstrated experimentally through a comparison with other (*SOM*-based and non-*SOM*-based) *ACM*s.

References

1. Venkatesh, Y.V., Kumar Raja, S., Ramya, N.: A novel SOM-based approach for active contour modeling. In: Proc. of the Conf. on Intelligent Sensors, Sensor Networks and Information Processing, pp. 229–234 (2004)
2. Chen, S., Radke, R.J.: Level set segmentation with both shape and intensity priors. In: Proc. of the 12th IEEE Int. Conf. on Computer Vision, pp. 763–770 (2009)
3. Kass, M., Witkin, A., Terzopoulos, D.: Snakes: Active contour models. Int. J. of Computer Vision 1(4), 321–331 (1988)
4. Zhu, G.: Boundary-based image segmentation using binary level set method, Optical Engineering 46(5), article ID: 050501, 3 pages (2007)
5. Kim, W., Kim, C.: Active contours driven by the salient edge energy model. IEEE Trans. on Image Processing 22(4), 1667–1673 (2013)
6. Caselles, V., Kimmel, R., Sapiro, G.: Geodesic active contours. Int. J. of Computer Vision 22(1), 61–79 (1997)
7. Kichenassamy, S., Kumar, A., Olver, P.: Conformal curvature flows: from phase transitions to active vision. Archive for Rational Mechanics and Analysis 134(3), 275–301 (1996)
8. Chan, T.F., Vese, L.A.: Active contours without edges. IEEE Trans. on Image Processing 10(2), 266–277 (2001)
9. Talu, M.F.: ORACM: Online region-based active contour model. Expert Systems with Applications 40(16), 6233 – 6240 (2013)
10. Chen, L., Zhou, Y., Wang, Y., Yang, J.: GACV: Geodesic-Aided CV method. Pattern Recognition 39(7), 1391–1395 (2006)
11. Abdelsamea, M.M., Tsaftaris, S.A.: Active contour model driven by globally signed region pressure force. In: Proc. of the 18th Int. Conf. on Digital Signal Processing, pp. 1–6 (2013)
12. Tian, Y., Duan, F., Zhou, M., Wu, Z.: Active contour model combining region and edge information. In: Machine Vision and Applications, pp. 1–15 (2011)
13. Liu, S., Peng, Y.: A local region-based Chan-Vese model for image segmentation. Pattern Recognition 45(7), 2769–2779 (2012)
14. Zhang, K., Song, H., Zhang, L.: Active contours driven by local image fitting energy. Pattern Recognition 43(4), 1199–1206 (2010)
15. Tsai, A., Yezzi, A., Willsky, A.S.: Curve evolution implementation of the Mumford-Shah functional for image segmentation, denoising, interpolation, and magnification. IEEE Trans. on Image Processing 10(8), 1169–1186 (2001)
16. Wang, P., Sun, K., Chen, Z.: Local and global intensity information integrated geodesic model for image segmentation. In: Proc. of the Int. Conf. on Computer Science and Electronics Engineering, vol. 2, pp. 129–132 (2012)
17. Tran, T.-T., Pham, V.-T., Chiu, Y.-J., Shyu, K.-K.: Active contour with selective local or global segmentation for intensity inhomogeneous image. In: Proc. of the 3rd IEEE Int. Conf. on Computer Science and Information Technology, vol. 1, pp. 306–310 (2010)

18. Xu, C., Prince, J.L.: Snakes, shapes, and gradient vector flow. IEEE Trans. on Image Processing 7(3), 359–369 (1998)
19. Kohonen, T.: Essentials of the self-organizing map. Neural Networks 37, 52–65 (2013)
20. Venkatesh, Y.V., Rishikesh, N.: Self-organizing neural networks based on spatial isomorphism for active contour modeling. Pattern Recognition 33(7), 1239–1250 (2000)
21. Shah-Hosseini, H., Safabakhsh, R.: A TASOM-based algorithm for active contour modeling. Pattern Recognition Letters 24(910), 1361–1373 (2003)
22. Osher, S., Sethian, J.A.: Fronts propagating with curvature dependent speed: Algorithms based on Hamilton-Jacobi formulations. J. of Computational Physics, 12–49 (1988)
23. Vantaram, S.R., Saber, E.: Survey of contemporary trends in color image segmentation. J. of Electronic Imaging 21(4), article ID: 040901, 28 pages (2012)
24. Skakun, S.: A neural network approach to flood mapping using satellite imagery. Computing and Informatics 29(6), 1013–1024 (2010)
25. Middleton, I., Damper, R.I.: Segmentation of magnetic resonance images using a combination of neural networks and active contour models. Medical Enginering & Physics 26(1), 71–86 (2004)
26. Venkatesh, Y.V., Kumar Raja, S., Ramya, N.: Multiple contour extraction from graylevel images using an artificial neural network. IEEE Trans. on Image Processing 15(4), 892–899 (2006)
27. Vasconcelos, G.S., Bastos, C.A.C.M., Tsang, I.R., Cavalcanti, G.D.C.: BSOM network for pupil segmentation. In: Proc. of the IEEE Int. Joint Conf. on Neural Networks, pp. 2704–2709 (2011)
28. Bastos, C.A.C.M., Tsang, I.R., Vasconcelos, G.S., Cavalcanti, G.D.C.: Pupil segmentation using pulling & pushing and BSOM neural network. In: Proc. of the IEEE Int. Conf. on Systems, Man, and Cybernetics, pp. 2359–2364 (2012)
29. Izadi, M., Safabakhsh, R.: An improved time-adaptive self-organizing map for high-speed shape modeling. Pattern Recognition 42(7), 1361–1370 (2009)
30. Zeng, D., Zhou, Z., Xie, S.: Coarse-to-fine boundary location with a SOM-like method. IEEE Trans. on Neural Networks 21(3), 481–493 (2010)
31. Sadeghi, F., Izadinia, H., Safabakhsh, R.: A new active contour model based on the conscience, archiving and mean-movement mechanisms and the SOM. Pattern Recognition Letters 32(12), 1622–1634 (2011)
32. Ma, Y., Gu, X., Wang, Y.: Contour detection based on self-organizing feature clustering. In: Int. Conf. on Computing, Networking and Communications, vol. 2, pp. 221–226 (2007)
33. Teng, W.-G., Chang, P.-L.: Identifying regions of interest in medical images using self-organizing maps. J. of Medical Systems 36(5), 2761–2768 (2012)
34. Yang, Z., Bai, Z., Wu, J., Chen, Y.: Target region location based on texture analysis and active contour model. Trans. of Tianjin University 15(3), 157–161 (2009)
35. Neagoe, V.-E., Ropot, A.-D.: Concurrent self-organizing maps for pattern classification. In: Proc. of the 1st IEEE Int. Conf. on Cognitive Informatics, pp. 304–312 (2002)
36. Abdelsamea, M.M., Gnecco, G., Gaber, M.M.: A concurrent SOM-based Chan-Vese model for image segmentation. In: submitted to the 10th Workshop on Self-Organizing Maps (2014) (provisionally accepted)

A Biologically Plausible SOM Representation of the Orthographic Form of 50,000 French Words

Claude Touzet[1], Christopher Kermorvant[2], and Hervé Glotin[3]

[1] Aix-Marseille University (AMU),
Lab. de Neurosciences Intégratives et Adaptatives,
LNIA UMR-CNRS 7260, Pôle Cerveau-Comportement-Cognition, Marseille, France
[2] A2iA SA (Analyse d'Image & Intelligence Artificielle), Paris, France
[3] Institut Univ. de France (IUF) & Univ. Aix-Marseille (AMU), Univ. Toulon
(UTLN), ENSAM, Lab. des Sciences de l'Information et des Systèmes (LSIS), UMR
CNRS 7296, Toulon, France
claude.touzet@univ-amu.fr,
christopher.kermorvant@a2ia.com,
glotin@univ-tln.fr

Abstract. Recently, an important aspect of human visual word recognition has been characterized. The letter position is encoded in our brain using an explicit representation of order based on letter pairs: the open-bigram coding [15]. We hypothesize that spelling has evolved in order to minimize reading errors. Therefore, word recognition using bigrams — instead of letters — should be more efficient. First, we study the influence of the size of the neighborhood, which defines the number of bigrams per word, on the performance of the matching between bigrams and word. Our tests are conducted against one of the best recognition solutions used today by the industry, which matches letters to words. Secondly, we build a cortical map representation of the words in the bigram space — which implies numerous experiments in order to achieve a satisfactory projection. Third, we develop an ultra-fast version of the self-organizing map in order to achieve learning in minutes instead of months.

Keywords: Handwriting recognition, word recognition, open-bigram coding, orthographic representation, cortical representation.

1 Introduction

Visual handwritten word recognition is an active field, attracting hundreds of researchers [1], starting as early as 1929. A huge amount of ideas have been implemented and tested including algorithms (such as dynamic programming [2]) or holistic approaches (such as considering only the global characteristics of the word [3]), statistical methods (such as hidden Markov models (HMM) [4]), contextual approaches (such as contextual character geometry [5]), and artificial neural networks (such as multiple layer perceptron (MLP) [6] and error-backpropagation training [7] or self-organizing maps [8]).

Since 2009, connectionist models such as multi-dimensional LSTM (Long Short-Term Memory) recurrent neural networks [9-10], deep feed-forward neural

T. Villmann et al. (eds.), *Advances in Self-Organizing Maps and Learning*
Vector Quantization, Advances in Intelligent Systems and Computing 295,
DOI: 10.1007/978-3-319-07695-9_29, © Springer International Publishing Switzerland 2014

networks [11] and various mixtures of these have won several international connected handwriting competitions (such as the International Conference on Document Analysis and Recognition) without any prior knowledge about the various languages (French [17], Arabic [24]) to be learned. GPU-based deep learning methods for feed-forward networks were the first artificial pattern recognizers to achieve human-competitive performance [12] on the famous MNIST handwritten digits problem [13].

Such results support the claim that we are currently experiencing a second Neural Network ReNNaissance (the first one happened between 1985 and 1993). In many applications, deep NNs are now outperforming all other methods, including support vector machines (SVM).

Deep and recurrent neural networks refer explicitly to the brain architectures, and mimic some of the principles that are known about the way that human brain implement word reading. Dehaene *et al.* have proposed a biologically plausible model of the cortical organization of reading [14] that assumes seven successive steps of increasing complexity — from the retinal ganglion cells to a cortical map of the orthographic word forms.

Cognitive psychology has done a tremendous amount of work relatively to reading, one among the most important cognitive abilities. However, these discoveries have not been considered by pattern recognition researchers, most certainly because of field boundaries between soft and hard science. One of the most recent successes of experimental psychology was the demonstration that human visual word recognition uses an explicit representation of letter position order based on letter pairs: the open-bigram coding [15].

In its simplest form, an open-bigram (OB) coding assumes a limit of 2 intervening letters (see Bigrams 2 in Fig. 1). For example, TABLE bigrams amount to 9: TA, TB, TL, (not TE), AB, AL, AE, BL, BE, LE. The weighting of each bigram is 1 if present (0 otherwise) in binary OB models. In graded OB models, weights decrease with the distance between letter positions.

1.1   Why Bigrams Are Better

Various measures of distance can be used to ascertain the orthographic proximity of two words.

1. For example, the orthographic distance (D1) between two words (X, the number of shared letters):

$$D1 \ (word1, word2) = (2 \ X) \ / \ (word\_length1 + word\_length2)$$

Distance D1 is an increasing arithmetic function of X. This distance is a logical choice when using a letters coding model.

2. Another possibility is the distance (D2):

$$D2 \ (word1, word2) = (X * (X+1)) \ /$$
$$(word\_length1 * (word\_length1 + 1) + word\_length2 * (word\_length2 + 1) \ )$$

Distance D2 is an increasing geometric function of X. This distance is a logical choice for OB coding since the number of bigrams in common between two words is given by:

$$(X * (X+1)) / 2$$

A geometric increase in distance between words is interesting because it allows to take into account the respective length of the words. For example, in the case of two words of respective length 5 and 8 letters, sharing 3 letters, D1 = 6/13 and D2 = 12/102. In the case of two words of respective length 3 and 10 letters sharing 3 letters, D1 remains unchanged (6/13), where D2 = 12/122.

Using D2, when the number of shared letters is equivalent, lower ratios (word_length1 / word_length2) are privileged. In a representation that takes into account the distance between neighbors, D2 privileges neighbor words with the same length. To resume, the bigram representation (resp. to a 'letter' representation) allows for a greater continuity of the representation when the length of the words is also taken into account.

The Levenshtein distance (Edit-distance) takes into account the position of the letters in the word. Therefore, it is less biologically plausible.

1.2   How Many Bigrams per Word?

Using the RIMES data-set [16] (7400 words) and the letters extracted by A2iA [17] (first proposal) we test the influence of the size of the bigram set over the word recognition. It is important to note that the 'poor' quality of the letter extraction only allows a Word Recognition Rate of 28%.

When we use a nearest neighbor convergence with distance D1 (because we know the whole vocabulary), a performance of 44% is achieved.

Fig2. shows that the performance using bigrams are better, depending on the size of the bigram set. We vary this size from a bigram set with no intervening letter (bigrams 0: TABLE = TA, AB, BL, LE) to the whole letters of the word (TA, TB, TL, TE, AB, AL, AE, BL, BE, LE).

Letters	Bigrams 0	Bigrams 1	Bigrams 2	Bigrams 3	Bigrams (whole)
44%	45%	48%	49%	50%	51%

Fig. 1. A bigram representation of the word — in the case of the RIMES data-set — allows a much better performance in recognition (improvement from 44% to 51%)

Because the bigram representation is an over-coding, missing or wrongly labeled letters have less impact on the recognition procedure. Bigrams increase the size of the representation (compared to a letter representation), which allows to resist to failures. This seems to imply that existing words in the language (French) have evolved in order for this bigram over-coding to be pertinent (at least more than a pure letter representation is).

2   Cortical Map Model

A Kohonen map (also known as a Self-Organized Map - SOM [20]) is a model of the cortical map. We will use it to implement a biologically plausible representation of the orthographic form of words.

2.1   Not Uniform Representation Despite Uniform Frequency

The following figure illustrates the performances of the SOM learning of 25 (French) words (uniform frequency distribution for all words). α and β (learning coefficients for the winner and its four neighbors) are initially set to 0.6 and 0.15, and decrease with the number of iterations (by 1/total_nb_of_iterations to 0.1 and 0.05 resp.). Each node has four neighbors (North, South, East and West), nodes on the border of the map have only three neighbors, nodes at the corners have only two neighbors. The size of the map is 25 nodes (5 x 5). The number of iterations is set to 50. Learning samples selection is random. Number of input dimensions: 193 (binary OB). Non-null inputs average only a few dozens per sample. Figure 2 displays the nodes associated to each word.

The words (Fig. 2) represented by the same node are similar, but nevertheless quite different. In particular, the length of the words may be very different, and it seems that the short words (e.g., "action") are somewhat pulled by the long ones. This comes from the fact that only a fraction of the inputs are non null (e.g., 15 out of 193 in the case of "action"), and the impact of the (null) input weights are important.

	amélio-		ALBESTROFF	
allée		adapté adaptée		acheter
	aboie aboiements aboit		agents	affectant annexe
Abonné	Alors	allemand animal		accidenté accidentée action
abonnés	Ainsi aisé		ACTUELLEMENT	accompagnée amenée

Fig. 2. SOM learning of 25 words using their bigram representations. Several nodes have no matching correspondence with any words of the learning base, when at the same time several nodes are the prototypes for several words (such as: "accidenté accidentée action").

2.2   Long Words Pull Shorter Ones

We introduce a difference among the non-null and null inputs by using different values of α and β when the weight update relates to null inputs. They are fixed during all the learning and set to 0.05 and 0.01 respectively. If these coefficients were set to 0 then the weights associated to these null inputs could not be updated, which ends-up with a bias (favoring long words) since these connections are nevertheless updated from time to time by non-null inputs. As shown in Fig. 3, the lengths of the various words belonging to the same node are closer. However, as in the previous case (Fig.2), there are numerous non-used nodes, and an exaggeration of the distance between nodes.

amenée			allée allemand animal	
				ACTUELLEMENT
amélio-	Alors		accompagnée adapté adaptée	
	aboie Ainsi aisé	aboiements		ALBESTROFF
annexe	Abonné abonnés	aboit action agents	accidenté accidentée	acheter affectant

Fig. 3. α and β associated to null inputs are fixed set to 0.05 and 0.01 respectively. The lengths of the various words belonging to the same node are more similar (e.g., "aboit action agents").

2.3 Equi-selection of the Winners

To alleviate the defect shown on the previous Fig. 3, we modify the learning algorithm in order to impose that the each node wins as often as any other, only once per iteration (Fig. 4).

affectant	Alors	action			accidenté accidentée
aisé	Ainsi	acheter			adapté adaptée
allée	ACTUELLEMENT	ALBESTROFF			aboie
animal	amélio-			aboit agents	aboiements
accompagnée		allemand amenée annexe			Abonné abonnés

Fig. 4. Forcing the learning on each node has improved the occupancy of the map. The number of unused nodes is reduced by a factor of 2 (compared to Fig. 3). However, there are still errors in the sense that a node may represent several words (e.g., "allemand amenée annexe").

2.4 Increasing Map Size to Add Flexibility

One possibility that would explain this overuse of several nodes — and non-use of several others — may be related to the fact that the distribution of the words (and their bigrams) is highly constrained by the size of the map (25 nodes for 25 words). A larger map helps to spread the words without losing the neighborhood property (Fig. 5).

	amenée		aboit	aboie	
ACTUELLEMENT		acheter annexe		aboiements	
	ALBESTROFF		adapté adaptée		amélio-
Ainsi aisé		allemand		action	Alors
	abonnés	allée		affectant	
Abonné	animal	agents	accompagnée		accidenté accidentée

Fig. 5. A 36 nodes map (6 x 6) representing the 25 learning samples. The number of learning iterations has increased to 100 (instead of 50), in order to allow the same amount of modifications per weight. The larger map allows a better separation between words that are not true neighbors. Only 2 nodes representing more than one word ask for explanations: "Ainsi aisé" and "acheter annexe".

2.5   A Correct Cortical Map Representation

Continuing with the idea of extending the map in order to separate what is different, it is of tremendous importance to clearly see the frontiers between the words (in order to implement an efficient word recognition system: one node/one word). Fig. 6 displays the word associated to each node (not just the winning node associated to a given input). A given word may now be represented by several nodes.

Nodes 1 to 10	11 – 20	21-30	31-40	41-50
aboit	aboie	allée	allée	amélio-
aboit	aboit	Abonné	amélio-	amélio-
aboit	Abonné	Abonné	Abonné	amélio-
abonnés	abonnés	Abonné	abonnés	agents
abonnés	aisé	aisé	adapté	agents
abonnés	aisé	adapté	adapté	adaptée
abonnés	aisé	aisé	adaptée	adaptée
abonnés	aisé	aisé	aisé	adaptée
abonnés	abonnés	aisé	accidenté	**accidentée**
abonnés	allée	Abonné	accidenté	accidenté

51-60	61-70	71-80	81-90	91-100
allée	allemand	allemand	accompagnée	accompagnée
amélio-	animal	aboiements	aboiements	annexe
Ainsi	Ainsi	aboiements	aboiements	annexe
Ainsi	Alors	acheter	amenée	amenée
agents	acheter	acheter	ALBESTROFF	amenée
agents	adaptée	acheter	ALBESTROFF	ACTUELLEMENT
adaptée	**accidentée**	affectant	ACTUELLEMENT	ACTUELLEMENT
accidentée	**accidentée**	affectant	affectant	affectant
accidentée	**accidentée**	**accidentée**	affectant	amenée
accidentée	action	action	action	action

Fig. 6. A 10x10 map (100 nodes) representing the 25 samples of the learning base. Due to space constraints, the map has been cut in two equal parts. In fact, there is only one map of 10 columns. Again, due to the increase of the map size, the number of iterations has been set to 200. As we can see, the frontiers between the various words have a clear semantics. The respective occupancy size (measured using the number of nodes associated to a given word) contains also some information. Similar words (e.g., **"accidentée"** (in bold) and accidenté") occupy larger regions than "isolated" words (such as "animal" or "Alors").

3   Ultra-fast Building of the SOM

We try to build a SOM (with 4 neighbors per neuron) using a D2 (bigram) distance for the 50 000 words of the French (using the eManulex database [18]). Computing requirements are huge, since a matrix of the D2 measures (50 000 x 50 000 — about 20 Go of RAM) must be computed and kept into memory [19]. An on-the-fly computing does not solve the problem because each iteration requires about 2.5 GFLOPS, and several thousands of iterations are required. It would take about 6 months on a standard PC using a Python written software to generate the SOM representing the 50 000 French words. Obviously, acceleration procedures must be found.

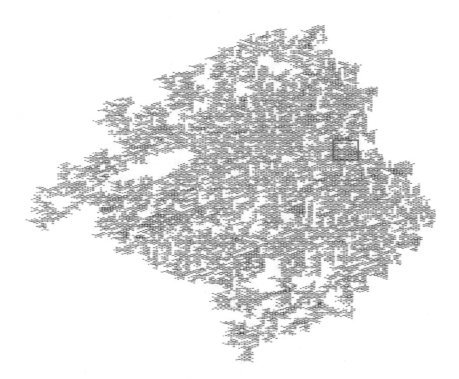

Fig. 7. A cortical map coding 50 000 French words. The size of the map (if every word was readable — font size "6") is about three meters long. The complete map is accessible at: http://www.touzet.org/Claude/Cognilego/FSOM44025-3.pdf ; an animated version is accessible at: http://www.touzet.org/Claude/Cognilego/FSOM44025-3.avi . Zones in white are regions without words. They constitute frontiers among regions of similar words. Following our initial hypothesis that the orthographic form of words optimizes bigrammic recognition, it is tempting to make the hypothesis that these zones are available for future creations of words which will be easily recognized — except that we must be remembered that 50 000 is just a fraction of all already existing French words (total number supposed to be around 200 000 words). Each experiment generates a different map, but the textures of the maps look similar (black and white patterns).

310   C. Touzet, C. Kermorvant, and H. Glotin

Therefore, we have proposed and developed various optimizations / simplifications such as to keep only the best 100 scores (D2) for each word (instead of 50 000 score values), and to compute the self-organizing map using a stochastic crystal growing algorithm, instead of the classical but costly Kohonen algorithm:

1. A first word is selected and associated to the node at the center of the map.
2. One of its neighbor nodes is randomly selected. The best matching word is found: its distance to the already placed neighbor words is minimum (the summation of all D2 distances). Its weights are adjusted (alpha = 1.0, beta = 0.0).
3. Repeat from 2 until last word.

The final result is not the result of a global optimization process, but the duration of the (self-) learning is reduced to 40 minutes for the 50 000 words.

Using this ultra-fast learning map, we build a bigram representation of the 50 000 words (Fig. 7 & 8). Note that we also changed the number of neighbors, from 4 to 6 (hexagonal lattice), following the original formulation of the SOM [20]. This allows for a more compact map, with less frontiers and discontinuities. Also the hexagonal lattice appears to be more biologically plausible, and more efficient. Our ultra-fast SOM shares a number of similarities with the SOM of symbol strings [25], a much earlier work. However, among other differences, where the SOM of symbol strings involves successive training and growing phases, our proposal integrates learning and growing in one step.

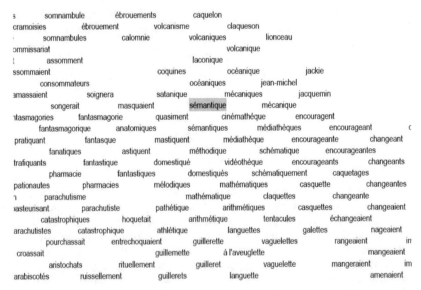

Fig. 8. Enhancement of the (fig. 7) cortical map. The neighborhood size is 6, e.g., the word "sémantique" has 6 neighbors: "satanique, océaniques, mécaniques, cinémathèque, sémantiques, quasiment".

4   Conclusion

Open bigrams (OB) allow an over-coding of the orthographic form of words that facilitates recognition. OB coding favors same length words (i.e., neighbors of similar lengths). Using OB description, a cortical map has been built in order to visualize (the most frequent) 50 000 French words. This visualization of the cortical representation of (OB) words is highly pedagogic, allowing to really appreciate the fact that neighbor words are somewhat different from what we would naively think (i.e., letter-based distance). In future work, we may consider weigthed metrics in the bigramic space, taking into consideration their uncertainty. The incertainty of a bigram may simply be defined by a bayesian approach based on the counts of the bigram and the letter frequency. The most informative bigrams(x,y) are the ones with small probability that y follows x. Then, if our assumption is correct, languages may have evolved to separate words in the bigrammic space according to distance based on the most informative bigrams. A weighted cosine metrics shall still be fast enought to compute such soft bigrammig map.

A realistic developmental database that takes into account the order of presentation of the words to the children would certainly generates a different kind of maps [19], less optimal (because neighbor words may be seen at different ages and end up in very different locations on the map), but closer to biological cortical map.

It is important to remember that the ultra-fast learning allows only for a local optimization and does not take into account the sampling frequency of the learning samples (each sample is represented on the map). Last, but not least, this ultra-fast learning is very important since it allows to consider the implementation of the Theory of neuronal Cognition [21-23]. In this case, the difficulty is no more in the learning duration, but in the availability of the learning data for each of the 500 cortical maps.

Acknowledgements. Work supported by the French Research Agency ANR 2010-CORD-013 "Cognilego — From pixels to semantics: a cognitive approach".

References

1. Impedovo, S.: More than twenty years of advancements on Frontiers in Handwriting Recognition. Pattern Recognition (June 12, 2013) (in press)
2. Chen, W., Gader, P., Shi, H.: Lexicon-driven handwritten word recognition using optimal linear combinations of order statistics. IEEE Trans. Pattern Anal. Mach. Intell. 21(1), 77–82 (1999)
3. Salome, J., Leroux, M., Badard, J.: Recognition of cursive script words in a small lexicon. In: Proc. of ICDAR 2011, pp. 774–782 (1991)
4. Cho, W., Lee, S., Kim, J.H.: Modeling and recognition of cursive words with hidden Markov models. Pattern Recognition 28(12), 1941–1953 (1995)
5. Xue, H., Govindaraju, V.: Incorporating Contextual Character Geometry in Word Recognition. In: Proceedings of the Eighth International Workshop on Frontiers in Handwriting Recognition, pp. 123–127 (2002)

6. Oh, I.-S., Suen, C.Y.: A class-modular feedforward neural network for handwriting recognition. Pattern Recognition 35(1), 229–244 (2002)
7. Senior, A.W., Fallside, F.: An off-line cursive script recognition system using recurrent error propagation networks. In: Proc. Third Intl. W. F. Hand-Writing Recog., pp. 132–141 (1993)
8. Laaksonen, J.: Subspace classifiers in recognition of handwritten digits, PhD thesis, Helsinki University of Technology (1997)
9. Graves, A., Schmidhuber, J.: Offline Handwriting Recognition with Multidimensional Recurrent Neural Networks. In: Bengio, Y., Schuurmans, D., Lafferty, J., Williams, C.K.I., Culotta, A. (eds.) Advances in Neural Information Processing Systems 22 (NIPS 22), Vancouver, BC, pp. 545–552 (2009)
10. Graves, A., Liwicki, M., Fernandez, S., Bertolami, R., Bunke, H., Schmidhuber, J.: A Novel Connectionist System for Improved Unconstrained Handwriting Recognition. IEEE Trans. Pattern Analysis and Machine Intelligence 31(5), 855-868 (2009)
11. Ciresan, D.C., Meier, U., Gambardella, L.M., Schmidhuber, J.: Convolutional Neural Network Committees For Handwritten Character Classification. In: Proc. of ICDAR 2011, Beijing, China, pp. 1135–1139 (2011)
12. Ciresan, D.C., Meier, U., Schmidhuber, J.: Multi-column Deep Neural Networks for Image Classification. In: IEEE CVPR, pp. 3642–3649 (2012)
13. LeCun, Y., Bottou, L., Bengio, Y., Haner, P.: Gradient-based learning applied to document recognition. Proc. IEEE 86, 2278–2324 (1998)
14. Dehaene, S., Cohen, L., Sigman, M., Vinckier, F.: The neural code for written words: A proposal. Trends in Cognitive Sciences 9, 335–341 (2005)
15. Whitney, C., Bertrand, D., Grainger, J.: On coding the position of letters in words: A test of two models. Experimental Psychology 59(2), 109–114 (2012)
16. RIMES: Reconnaissance et Indexation de données Manuscrites et de fac-similÉS / Recognition and Indexing of handwritten documents and faxes, http://www.rimes-database.fr
17. Menasri, F., Louradour, J., Bianne-Bernard, A.-L., Kermorvant, C.: The A2iA French handwriting recognition system at the Rimes-ICDAR2011 competition. In: Chien, L.-C., Lee, S.-D., Wu, M.H. (eds.) Document Recognition and Retrieval Conference, Proceedings of the SPIE, vol. 8297, p. 8 (2012)
18. Ortga, É., Lété, B.: eManulex: Electronic version of Manulex and Manulex-infra databases (2010), http://www.manulex.org
19. Dufau, S., Lété, B., Touzet, C., Glotin, H., Ziegler, J., Grainger, J.: Developmental Perspective on Visual Word Recognition: New Evidence and a Self-Organizing Model. European Journal of Cognitive Psychology 22(5), 669–694 (2010)
20. Kohonen, T.: Self-organizing maps. 3rd Extended edn. Springer, Heidelberg (2001)
21. Touzet, C.: Why Neurons are Not the Right Level of Abstraction for Implementing Cognition. In: BICA 2012: Annual Int. Conf. on Biologically Inspired Cognitive Architectures, Palermo, Italy, pp. 317–318 (2012)
22. Touzet, C.: The Illusion of Internal Joy. In: Schmidhuber, J., Thórisson, K.R., Looks, M. (eds.) AGI 2011. LNCS, vol. 6830, pp. 357–362. Springer, Heidelberg (2011)
23. Touzet, C.: Consciousness, Intelligence, Free-Will? The answers from the Theory of neuronal Cognition. La Machotte Ed., Auriol, France (2010) (in French)
24. Bluche, T., Louradour, J., Knibbe, M., Moysset, B., Benzeghiba, F., Kermorvant, C.: The A2iA Arabic Handwritten Text Recognition System at the OpenHaRT2013 Evaluation (submitted, 2014)
25. Kohonen, T., Somervuo, P.: Self-Organizing Maps of Symbol Strings with Application to Speech Recognition. In: Proc. of WSOM 1997, Espoo, FI, pp. 2–7 (1997)

Author Index

Abdelsamea, Mohammed M.   199, 293

Backhaus, Andreas   167
Bahrmann, Frank   133
Bellas, Anastasios   145
Bendhaiba, Laura   219
Biehl, Michael   121
Boelaert, Julien   45, 219
Böhme, Hans-Joachim   133, 157
Bouveyron, Charles   145
Buldain, David   35

Choe, Yoonsuck   187
Choi, Jinho   187
Cottrell, Marie   145

Dozono, Hiroshi   89

Estévez, Pablo A.   229

Faigl, Jan   281
Fischer, Lydia   109
Fix, Jérémy   25

Gaber, Mohamed Medhat   199, 293
Geweniger, Tina   99
Glotin, Hervé   303
Gnecco, Giorgio   199, 293

Hammer, Barbara   109, 123
Hellbach, Sven   133, 157
Hermann, W.   77
Himstedt, Marian   133
Hollinger, Geoffrey A.   281

Honkela, Timo   209
Huijse, Pablo   229

Kaden, Marika   77, 157
Kermorvant, Christopher   303
Klingner, Mathias   157
Knauer, Uwe   167
Korhonen, Jaakko   209

Lacaille, Jerome   145
Lagus, Krista   209
Lange, Mandy   259
Lee, John A.   65
Lötsch, Jörn   249

Mariette, Jérôme   45
Matsuda, Nobuo   271
Merényi, Erzsébet   181

Nebel, David   109, 123, 259
Nova, David   229

Ohkita, Masaaki   239
Olier, Iván   55
Olteanu, Madalina   45, 219
Oyabu, Matashige   239

Pelayo, Enrique   35
Peluffo-Ordóñez, Diego H.   65

Riedel, Martin   123, 133, 157
Rossi, Fabrice   3

Saarinen, Esa   209
Sato, Hedeaki   271

Schleif, Frank-Michael   99
Seiffert, Udo   167

Tajima, Fumiaki   271
Tokutaka, Heizo   239
Tosi, Alessandra   55
Touzet, Claude   303

Ultsch, Alfred   249

Vellido, Alfredo   55
Verleysen, Michel   65
Villa-Vialaneix, Nathalie   45, 219
Villmann, Thomas   77, 99, 109, 123, 133,
    157, 259

Wersing, Heiko   109

Yoo, Jaewook   187
Yoshihara, Kazuhiro   239

Printed in the United States
by Baker & Taylor Publisher Services